Basic ENVIRONMENTAL AND ENGINEERING GEOLOGY

F. G. Bell

Research Associate, British Geological Survey, UK;
formerly Professor and Head of Department of Geology and Applied Geology,
University of Natal, South Africa

Whittles Publishing

Published by
Whittles Publishing Limited,
Dunbeath,
Caithness, KW6 6EY,
Scotland, UK
www.whittlespublishing.com

Distributed in North America by
CRC Press LLC,
Taylor and Francis Group,
6000 Broken Sound Parkway NW, Suite 300,
Boca Raton, FL 33487, USA

ISBN-10   1-904445-02-0
ISBN-13   978-1-904445-02-9
USA ISBN   978-1-4200-4470-6

Printed by Athenæum Press Ltd., Gateshead, UK

Contents

CONTENTS

Preface

As society becomes more aware of the significance of the environment, so environmental issues assume more importance. Environmental geology plays a vital role in the physical environment of humans and so should be appreciated by all those who are involved in some way with this aspect of the environment from environmental geologists, through environmental scientists and managers, and planners to civil engineers, mining engineers and architects. In addition, engineering geology is intimately associated with the human environment, in particular, with its infrastructure from routeways to reservoirs. Again many of those individuals referred to should find the book of value.

With this in mind, the book examines the influence of geohazards on the environment; the significance of soil and water resources; and the impact of mining, waste disposal and pollution and contamination on the environment. It also considers various aspects of construction that are involved in the development of our infrastructure that allow society to function as it does. Land evaluation and geological construction materials are therefore taken into account. Hence, the book provides not only an integration of the basic essentials of environmental and engineering geology that should be of interest to the well informed reader but it also should provide a text for undergraduates who are studying these two aspects of geology. In addition, it should provide material for those students following courses in those subjects mentioned previously. Last but not least, numerous suggestions for further reading are provided for each chapter to allow the interested and conscientious reader to pursue the subject matter further.

F. G. Bell

1

Basic Geology

1.1 Igneous rocks

Rocks are divided according to their origin, into three groups, namely, igneous, metamorphic and sedimentary rocks – the igneous rocks being the most common in the Earth's crust (Table 1.1). Igneous rocks are formed when hot molten rock material called magma solidifies. Magmas are developed either within the Earth's crust or the uppermost part of the mantle. They comprise hot solutions of several liquid phases, the most conspicuous of which is a complex silicate phase. Accordingly, the igneous rocks are composed principally of silicate minerals. Furthermore, of the silicate minerals, six families, the olivines $(Mg,Fe)_2SiO_4$, the pyroxenes (e.g. augite $(Ca,Mg,Fe^{+2},Fe^{+3},Ti,Al)_2(Si,Al)_2O_6$), the amphiboles (e.g. hornblende $(Ca,Na,K)_{2-3}(Mg,Fe^{+2},Fe^{+3}Al)_5Si_6(Si,Al)_2O_{22}(OH,F)_2$), the micas (e.g. biotite $K_2(Mg,Fe^{+2})_{6-4}(Fe^{+3},Al,Ti)_{0-2}(Si_{6-5}Al_{2-3}O_{20})O_{0-2}(OH,F)_{4-2}$ and muscovite $K_2Al_4(Si_6Al_2O_{20})(OH,F)_4$), the feldspars (e.g. orthoclase $KAlSi_3O_8$, albite $NaAlSi_3O_8$ and anorthite $CaAl_2Si_2O_8$) and the silica minerals (e.g. quartz SiO_2), are quantitatively by far the most important constituents. Fig. 1.1 shows the approximate distribution of these minerals in the commoner igneous rocks.

Igneous rocks may be divided into intrusive and extrusive types according to their mode of occurrence. In the former type, the magma crystallizes within the Earth's crust, whereas in the latter it solidifies at the surface, having been erupted as lavas and/or pyroclasts from a volcano. The intrusions may be further subdivided on a basis of their size, into major and minor categories; the former are developed in a plutonic (i.e. deep-seated) environment, the latter in a hypabyssal environment. Most of the plutonic intrusions have a granite–granodiorite composition and basaltic rocks account for the majority of the extrusives.

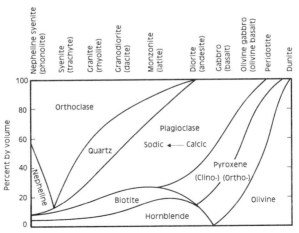

Fig. 1.1 Approximate mineral composition of the commoner igneous rock types (plutonic rocks without brackets; volcanic equivalents in brackets).

Table 1.1 Rock type classification (after International Association of Engineering Geology).

Genetic/group	Detrital sedimentary		At least 50% of grains are of carbonate*		Pyroclastic (At least 50% of grains are of fine-grained igneous rock)	Chemical/organic
Usual structure	Bedded					
Composition	Grains of rock, quartz, feldspar and clay minerals					
very coarse-grained (Rudaceous) — Boulders, Cobbles	Grains are of rock fragments					
60						
coarse-grained (Rudaceous) — Gravel	Rounded grains: conglomerate	Angular grains: breccia	Carbonate gravel	Calci-rudite	Rounded grains: agglomerate; Angular grains: Volcanic breccia; Lapilli tuff	Saline rocks; Halite; Anhydrite; Gypsum
2						
medium-grained (Arenaceous) — Sand	Grains are mainly mineral fragments. Sandstone: grains are mainly mineral fragments. Quartz arenite: 95% quartz, voids empty or cemented. Arkose: 75% quartz, up to 25% feldspar; voids empty or cemented. Greywacke: 75% quartz, 15% fine detrital material: rock and feldspar fragments		Carbonate sand	Calc-arenite	Ash; Tuff	Limestone; Dolomite; Chert; Flint
fine-grained (Argillaceous) — Silt	Siltstones: 50% fine-grained particles — Mudstone		Carbonate silt	Calci-siltite chalk	Fine-grained ash or tuff	
0.06						
very fine-grained (Argillaceous) — Clay	Claystones: 50% very fine-grained particles — Shale		Carbonate mud	Calci-lutite	Very fine-grained ash or tuff	Peat; Lignite; Coal
0.002						

Grain size (mm)

* limestone and dolomite undifferentiated

Genetic/group	Metamorphic		Igneous			
Usual structure	Foliated		Massive			
Composition	Quartz, feldspars, micas, acicular dark minerals		Light coloured minerals are quartz, feldspar, mica		Dark and Light minerals	Dark minerals
Grain size (mm)			Acid rocks >65% silica	Intermediate 55–65% silica	Basic rocks 45–54% silica	Ultrabasic <45% silica
60 — very coarse-grained			Pegmatite			Pyroxenite and Peridotite
2 — coarse-grained	Gneiss (ortho-, para-. Alternate layers of granular and flaky minerals)	Marble / Granulite	Granite	Diorite	Gabbro	Serpentinite
medium-grained	Migmatite / Schist	Quartzite / Hornfels / Amphibolite	Microgranite	Microdiorite	Dolerite	
0.06 — fine-grained	Phyllite		Rhyolite	Andesite	Basalt	
0.002 — very fine-grained	Slate / Mylonite					
glassy amorphous			Obsidian and pitchstone (volcanic glasses)		Tachylyte	

3

Fig. 1.2 A dyke on the south coast of the Isle of Skye, Scotland.

Fig. 1.3 A sill resting above rocks of Old Red Sandstone (Devonian) age – Salisbury Crags, Edinburgh, Scotland

1.1.1 Igneous intrusions

Dykes and sills are minor intrusions. Dykes are discordant igneous intrusions, that is, they traverse their host rocks at an angle and are steeply dipping (Fig. 1.2). As a consequence, their surface outcrop is little affected by topography and, in fact, they usually strike in a straight course. Dykes range in width up to several tens of metres but their average width is a few metres. Their length also varies but they can range up to a few hundred kilometres long. Dykes often occur along faults, which provide a natural path of escape for the injected magma. Most dykes are of basaltic composition. Multiple dykes are formed by two or more injections of the same material that occurred at different times whereas composite dykes involve two or more injections of magma of different composition.

Generally, sills are comparatively thin (although they can be up to several hundred metres thick). They are more or less parallel-sided igneous intrusions which have been injected in an approximately horizontal direction, although their attitude may subsequently be altered by folding. They occur over relatively extensive areas. When sills occur in a sequence of sedimentary rocks the magma was injected along the bedding planes (Fig. 1.3). Nevertheless, an individual sill may transgress upwards from one horizon to another. Because sills are intruded along bedding planes they are said to be concordant and their outcrop is similar to that of the country rocks. Most sills are composed of basic igneous material but may be multiple or composite in character.

The major intrusions include batholiths, stocks and bosses. Batholiths are very large in size, and are generally of granitic or granodioritic composition. Indeed, many batholiths have an immense surface exposure. Batholiths frequently appear to have no visible base, and their contacts are well defined and dip steeply outwards. However, some granitic batholiths appear to be made up of composite irregular sheets. They are more or less stratified, are not bottomless, and have been termed stratiform granitic massifs. Bosses are distinguished from stocks in that they have a circular outcrop. Both their surface exposures are of limited size, frequently less than 100 km². They probably represent upward extensions from deep-seated batholiths.

Certain structures are associated with granite massifs, tending to be best developed at the margins. Most joints and minor faults in batholiths possess a relationship with the shape of the intrusion. Cross joints or Q joints lie at right angles to the flow lines (Fig. 1.4). Joints that strike parallel to the flow lines and are steeply dipping are known as longitudinal or S joints. Diagonal joints are orientated at 45° to the direction of the flow lines. Flat-lying joints may be developed during or after formation of an intrusion and they may be distinguished as primary and secondary respectively. Normal faults and thrusts occur in the marginal zones of large intrusions and the adjacent country rocks. Flat-lying normal faults form as a result of tension developed parallel to the flow lines. They generally are restricted to the upper parts of a massif.

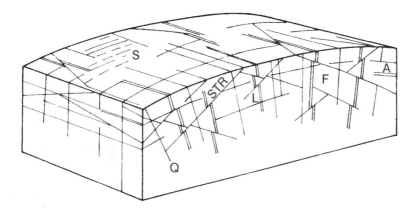

Fig. 1.4 Types of structures in a batholith. Q = cross joints; S = longitudinal joints; L = flat-lying joints; STR = planes of stretching; F = linear flow structures; A = aplite dykes.

1.1.2 Volcanic activity and extrusive rocks

Volcanic zones are associated with the boundaries of the crustal plates (Fig. 1.5). Plates can be largely continental, oceanic, or both oceanic and continental. Oceanic crust is composed of basaltic material whereas continental crust varies from granitic to basaltic in composition. At destructive plate margins oceanic plates are overridden by continental plates. The descent of the oceanic plate, together with any associated sediments, into zones of higher temperature leads to melting and the formation of magmas. Such magmas vary in composition but some may be richer in silica, that is, they may be andesitic or even rhyolitic. The latter magmas often are responsible for violent eruptions. By contrast, at constructive plate margins, where plates are diverging, the associated volcanic activity is a consequence of magma formation in the upper mantle. This magma is of basaltic composition and, as it is less viscous than andesitic or rhyolitic magma, there is relatively little explosive activity and associated lava flows are more mobile. Certain volcanoes, for example, those of the Hawaiian Islands, are located in the centres of plates and owe their origins to hot spots in the Earth's internal structure that have 'burned' holes through the overlying plates.

Volcanic activity occurs when magma travels to the Earth's surface and is erupted either from a fissure or a central vent. In some cases instead of flowing from the volcano as a lava, the magma is exploded into the air by the rapid escape of the gases from within it. The fragments produced by such explosive activity are collectively known as pyroclasts. Eruptions from volcanoes are spasmodic rather than continuous. Between eruptions, activity may take the form of steam and vapours issuing from small vents named fumaroles or solfataras but in some volcanoes even this activity does not take place and such a dormant state may continue for centuries.

When a magma is erupted it separates at low pressures into incandescent lava and a gaseous phase. Steam may account for over 90% of the gases emitted during a volcanic eruption. Other gases present include carbon dioxide, carbon monoxide, sulphur dioxide, sulphur trioxide, hydrogen sulphide, hydrogen chloride and hydrogen fluoride. Small quantities of methane, ammonia, nitrogen, hydrogen thiocyanate, carbonyl sulphide, silicon tetrafluoride, ferric chloride, aluminium chloride, ammonium chloride and argon also have been noted in volcanic gases.

At high pressures gas is held in solution, but as pressure falls gas is released by the magma. The rate at which gas escapes determines the explosiveness of an eruption. An explosive eruption occurs when, because of its high viscosity, magma cannot readily allow gas to escape. It is related only secondarily to the amount of gas a magma holds.

5

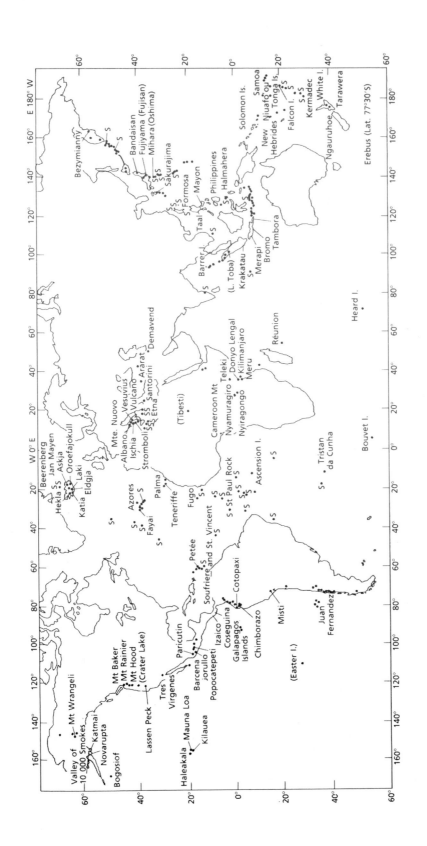

Fig. 1.5 Distribution of active volcanoes of the world (s = submarine eruptions).

Fig. 1.6 A cinder covered landscape, Lassen Volcanic National Park, California. The cinders were erupted from a nearby cinder cone.

As mentioned above, pyroclasts are material that has been fragmented by explosive action. They may consist of fragments of lava, fragments of pre-existing solidified lava or pyroclasts, or fragments of country rock that, in both latter instances, have been blown from the neck of the volcano. The size of pyroclasts varies enormously. The largest blocks thrown into the air are referred to as 'bombs' and may weigh over 100 tonnes whereas the smallest consist of very fine ash. Bombs consist of clots of lava or of fragments of wall rock. The term lapilli is applied to pyroclastic material that has a diameter in the range from approximately 10 to 50 mm. Cinder or scoria is irregular shaped material of lapilli size (Fig. 1.6). It usually is glassy and fairly to highly vesicular, and represents the ejected froth of a magma. The finest pyroclastic material is called ash. Much more ash is produced on eruption of acidic than basaltic magma. This is because acidic material is more viscous and so gas cannot escape as readily from it as it can from basaltic lava. Beds of ash commonly show lateral as well as vertical variation. In other words, with increasing distance from a vent the ash becomes finer and, in the second case, because the heavier material falls first, ashes frequently exhibit graded bedding (coarser material occurring at the base of a bed whilst it becomes finer towards the top). The spatial distribution of ash is influenced by wind direction and deposits on the leeside of a volcano may be much more extensive than on the windward. Rocks that consist of fragments of volcanic ejectamenta set in a fine-grained groundmass are referred to as agglomerate or volcanic breccia, depending on whether the fragments are rounded or angular respectively. After pyroclastic material has fallen back to the surface it eventually becomes indurated. It is then described as tuff. Tuffs are usually well bedded and the deposits of individual eruptions may be separated by thin bands of fossil soil or old erosion surfaces. Pyroclast deposits that accumulate beneath the sea are often mixed with a varying amount of sediment and are referred to as tuffites.

When clouds or showers of intensely heated, incandescent lava spray fall to the ground, they weld together. Because the particles become intimately fused with each other they attain a largely pseudo-viscous state, especially in the deeper parts of the deposit. The term ignimbrite is used to describe the rocks that are formed. If ignimbrites are deposited on a steep slope, they begin to flow. Hence, they resemble lava flows. Ignimbrites are associated with nuées ardentes (see Chapter 3).

Lavas are emitted from volcanoes at temperatures only slightly above their 'freezing' points. During the course of their flow the temperature falls from within outwards until solidification takes place somewhere between 600 and 900°C, depending upon their chemical composition and gas content. Basic lavas solidify at a higher temperature than do acidic lavas. The rate of flow of a lava is determined by the

Fig. 1.7 Pahoehoe (ropy) lava (foreground) and
aa lava (background), Hawaii.

Fig. 1.8 Pipes and vesicles in basaltic lava,
Transfer Tunnel, Lesotho Highlands Water
Project, Lesotho.

gradient of the slope down which it moves and by its viscosity – that, in turn, is governed by its composition (notably silica content), temperature and volatile content. Thus, basic lavas flow faster and further than acid lavas. Indeed, basic lavas have been known to travel at speeds of up to 80 km h^{-1}.

The upper surface of a recently solidified lava flow develops various structures which are referred to as: hummocky or ropy (pahoehoe); rough, fragmental, clinkery or spiny (aa); or blocky (Fig. 1.7). Obviously, the surface of lava solidifies before the main body of the flow beneath. Pipes, vesicle trains or spiracles may be developed in a lava depending on the amount of gas given off, the resistance offered by the lava and the speed at which the flow is travelling. Pipes are tubes that project upwards from the base and are usually several centimetres in length and 1 centimetre or less in diameter (Fig. 1.8). Vesicle trains form when gas action has not been strong enough to produce pipes.

Thin lava flows may be broken by joints that run either at right angles or parallel to the direction of flow. Joints do occur with other orientations but they are much less common. Those joints that are normal to the surface usually display a polygonal arrangement but only rarely do they give rise to columnar jointing. The joints develop as the lava cools. Typical columnar jointing is developed in thick flows of basalt (Fig. 1.9). The columns in columnar jointing are interrupted by cross joints that may be either flat or saucer-shaped. The latter may be convex up or down. These are not to be confused with platy joints that are developed in lavas as they become more viscous on cooling, so that slight shearing occurs along flow planes.

Fig. 1.9 Columnar basalt, Giant's
Causeway, Northern Ireland.

1.1.3 Texture of igneous rocks

The degree of crystallinity is one of the most important aspects of texture. An igneous rock may be composed of an aggregate of crystals, of natural glass, or of crystals and glass in varying proportions. This depends on the rate of cooling and composition of the magma/lava. If a rock is completely composed of crystalline mineral material, then it is described as holocrystalline. Most rocks are holocrystalline. Conversely, rocks that consist entirely of glassy material are referred to as holohyaline. The terms hypo-, hemi- or merocrystalline are given to rocks that are made up of intermediate proportions of crystalline and glassy material. The size of individual crystals are described as cryptocrystalline if they can just be seen under the highest resolution of the microscope or as microcrystalline if they can be seen at a lower magnification. These two types, together with glassy rocks, are collectively described as aphanitic, which means that the individual minerals cannot be distinguished with the naked eye. When the minerals of which a rock is composed are mega- or macroscopic, that is, they can be recognized with the unaided eye, the rock is described as phanerocrystalline. Three grades of megascopic texture usually are distinguished, that is, fine-grained, medium-grained and coarse-grained, the limits being under 1 mm diameter, 1–5 mm diameter, and over 5 mm diameter respectively.

A granular texture is one in which there is no glassy material and the individual crystals have a grain-like appearance. If the minerals are approximately the same size, then the texture is described as equigranular whereas if this is not the case it is referred to as inequigranular. Equigranular textures are more typical of plutonic igneous rocks. Many volcanic and hypabyssal rocks display inequigranular textures, the two most important types being the porphyritic and poikilitic textures. In the former case large crystals or phenocrysts are set in a fine-grained groundmass. A porphyritic texture may be distinguished as macro- or microporphyritic according to whether or not it may be observed with the unaided eye respectively. The poikilitic texture is characterized by the presence of small crystals enclosed within larger ones.

The most important rock-forming minerals are often referred to as felsic or mafic, depending upon whether they are light or dark coloured respectively. Felsic minerals include quartz, muscovite, feldspars and feldspathoids, whilst olivines, pyroxenes, amphiboles and biotite are mafic minerals. The colour index of a rock is an expression of the percentage of mafic minerals that it contains. Leucocratic rocks contain less than 30% dark minerals, mesocratic rocks contain between 30 and 60% dark minerals, melanocratic rocks contain between 60 and 90% dark minerals, and hypermelanic rocks contain over 90% dark minerals. Usually, acidic rocks are leucocratic whilst basic and ultrabasic rocks are melanocratic and hypermelanic respectively.

1.1.4 Igneous rock types

Granites and granodiorites are by far the commonest plutonic igneous rocks. They are characterized by a coarse grained, granular texture (Fig. 1.10). Although the term granite lacks precision, a normal granite has been defined as a rock in which quartz forms more than 5% and less than 50% of the quarfeloids (quartz, feldspar, feldspathoid content), potash feldspar constitutes 50 to 95% of the total feldspar content, the plagioclase is sodi-calcic, and the mafites form more than 5% and less than 50% of the total constituents. In granodiorite the plagioclase is oligoclase or andesine and is at least double the amount of alkali feldspar present, the latter forming 8 to 20% of the rock. The plagioclases are set in a quartz-potash-feldspar matrix. The term pegmatite refers to coarse- or very coarse-grained rocks commonly associated with granitic rocks. Pegmatites occur as dykes, sills, veins, lenses or irregular pockets in the host rocks. Aplites occur as veins, usually several tens of millimetres thick, typically in granites. They possess a fine-grained, equigranular texture. There is no important difference in composition between aplite and pegmatite, both having crystallized from residual magmatic fluids.

Rhyolites are acidic extrusive rocks that are commonly associated with andesites. They generally are regarded as the volcanic equivalents of granite, are usually leucocratic and sometimes exhibit flow banding. They may be holocrystalline but very often they contain an appreciable amount of glass. Rhyolites are frequently porphyritic, the phenocrysts varying in size and abundance, and may occur in a glassy,

Fig. 1.10 A thin section of granite containing feldspar, quartz and biotite (x 20).

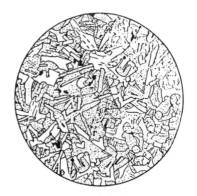

Fig. 1.11a A thin section of basalt containing augite (pyroxene) and laths of calcic placioclase, with some magnetite and partially altered glass (x 24). **b** (below) Olivine basalt showing olivine (centre) surrounded by laths of plagioclae feldspar and augite.

cryptocrystalline or microcrystalline groundmass. Vesicles frequently are present in these rocks. Acidic rocks of hypabyssal occurrence are often porphyritic, quartz porphyry being the commonest example. Quartz porphyry is similar in composition to rhyolite but it occurs in sills and dykes.

Syenites are plutonic rocks that have a granular texture and consist of potash feldspar, a subordinate amount of acid plagioclase and some mafic minerals, usually biotite or hornblende. Diorite has been defined as an intermediate plutonic, granular rock composed of plagioclase and hornblende, although at times the latter may be partially or completely replaced by biotite and/or pyroxene. Plagioclase, in the form of oligoclase and andesine, is the dominant feldspar. If orthoclase

is present it acts only as an accessory mineral. Trachytes and andesites are the fine-grained equivalents of syenites and diorites respectively. Andesite is the commoner of the two types. Trachytes are often porphyritic with most phenocrysts being composed of alkali feldspar. The groundmass is usually a holocrystalline aggregate of sanidine (a high temperature potash feldspar) laths. Andesites are commonly porphyritic with a holocrystalline groundmass. Plagioclase (oligoclase-andesine), which is the dominant feldspar, forms most of the phenocrysts. The plagioclases of the groundmass are more sodic than those of the phenocrysts. Sanidine occurs in the groundmass and may encircle some of the plagioclase phenocrysts. Hornblende is the commonest of the ferro-magnesian minerals and may occur as phenocrysts or in the groundmass, as may biotite and pyroxene.

Gabbros and norites are the commonest basic plutonic igneous rocks with granular textures. They are dark in colour. Plagioclase, commonly labradorite (but bytownite also occurs) is usually the dominant mineral in gabbros and norites. Augite is the most common type of pyroxene found in gabbros and usually is subhedral or anhedral. Norites, unlike gabbros, contain orthopyroxenes instead of clinopyroxenes (e.g. augite), hypersthene being the principle pyroxene. Basalts are the extrusive equivalents of gabbros and norites, and are composed principally of basic plagioclase and pyroxene in roughly equal amounts, or there may be an excess of plagioclase (Fig. 1.11). It is by far the most important type of extrusive rock. Basalts also occur in dykes, sills and volcanic plugs. Basalts exhibit a great variety of textures and may be holocrystalline or merocrystalline, equigranular or macro- or microporphyritic. Dolerites are of similar mineral composition to basalt and occur as minor intrusions. They are fine- to medium-grained and usually are equigranular, but as they grade towards basalts, they tend to become porphyritic. Nevertheless, the phenocrysts generally constitute less than 10% of the rock.

1.2 Metamorphism and metamorphic rocks

Metamorphic rocks are derived from pre-existing rock types and have undergone mineralogical, textural and structural changes that have taken place in the physical and chemical environments in which the rocks existed. The processes responsible for change give rise to progressive transformation that takes place in the solid state. Changing temperature and/or pressure conditions are the primary agents causing metamorphic reactions in rocks. Individual minerals are stable over limited temperature–pressure conditions. This means that when these limits are exceeded mineralogical adjustment has to be made to establish equilibrium with the new environment. Two major types of metamorphism may be distinguished on the basis of geological setting. One type is of local extent whereas the other extends over a large area. The first type includes thermal or contact metamorphism and the latter refers to regional metamorphism.

1.2.1 Thermal or contact metamorphism

Thermal metamorphism occurs around igneous intrusions so that the principal factor controlling the reactions involved is temperature, the encircling zone of metamorphic rocks being referred to as a contact aureole (Fig. 1.12). The size of an aureole depends upon the temperature and size of the intrusion, the quantity of hot gases and hydro-thermal solutions that emanated from it, and the type of country rocks involved. Aureoles developed in argillaceous sediments are more extensive than those found in arenaceous or calcareous rocks. This is because clay minerals, which account for a large proportion of the composition

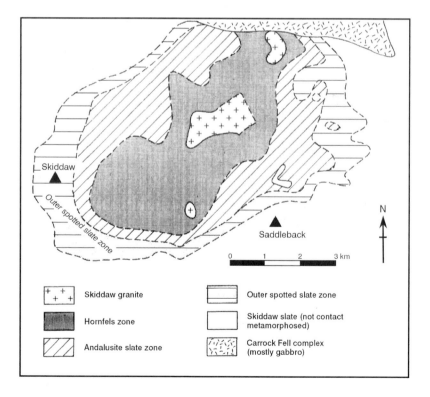

Fig. 1.12 Geological map of Skiddaw granite and its contact aureole, Cumbria, England.

11

of argillaceous rocks, are more susceptible to temperature changes than quartz (common in most sandstones) or calcite (in limestones). Aureoles formed in igneous or previously metamorphosed terrains are also less significant than those developed in argillaceous sediments. Within a contact aureole there is usually a sequence of mineralogical changes from the country rocks to the intrusion, which have been brought about by the effects of an increasing thermal gradient whose source was in the hot magma. A frequently developed sequence varies inward from spotted slates to schists then hornfelses. Hornfelses are characteristic products of high grade thermal metamorphism. They are dark coloured rocks with a fine-grained interlocking texture, containing andalusite [Al_2OSiO_4], cordierite [$(Mg,Fe^{+2})_2Al_3Si_5AlO_{18}$], quartz, biotite, muscovite, microcline or orthoclase, and sodic plagioclase (Fig. 1.13).

The metamorphism of a quartzose sandstone leads to the recrystallization of quartz to form quartzite and the higher the metamorphic grade, the coarser the fabric. At high grades foliation tends to develop and a gneissose rock is produced. The acid and intermediate igneous rocks are resistant to thermal

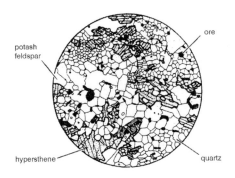

Fig. 1.13   A thin section of hornfels (x 35).

metamorphism, indeed they usually are only affected at very high grades when gneissose rock may be formed. On the other hand, basic igneous rocks give rise to hornblende and pyroxene hornfelses when subjected to thermal metamorphism.

1.2.2   Dynamic metamorphism

Stress is the most important factor in dynamic metamorphism. Dynamic metamorphism is produced on a comparatively small scale and usually is highly localized, for example, its effects may be found in association with large faults or thrusts. On a larger scale it is associated with folding, however, in the latter case it is difficult to distinguish between the processes and effects of dynamic metamorphism and those of low grade regional metamorphism. Recrystallization is at a minimum at low temperatures and the texture of a rock is governed primarily by mechanical processes. Brecciation is the process by which a rock is fractured, the angular fragments produced being of varying size. It commonly is associated with faulting and thrusting. The fragments of a crush breccia may themselves be fractured and the mineral components may exhibit permanent strain phenomena. If during the process of fragmentation pieces are rotated, then they are eventually rounded and embedded in the worn-down powdered material. The resultant rock is referred to as a crush conglomerate. Mylonites are produced by the pulverization of rocks, which not only involves extreme shearing stress but also considerable confining pressure. Mylonitization therefore is associated with major faults. Mylonites are composed of strained porphyroclasts set in an abundant matrix of fine-grained or cryptocrystalline material. In the most extreme cases of dynamic metamorphism the resultant crushed material may be fused to produce a vitrified rock referred to as a pseudotachylite.

1.2.3 Regional metamorphism

Metamorphic rocks extending over hundreds or even thousand of square kilometres are found exposed in the Precambrian shields and the eroded roots of fold mountains. As a consequence, the term regional has been applied to this type of metamorphism. Regional metamorphism involves both the processes of changing temperature and stress. The principal factor is temperature of which the maximum figure concerned in regional metamorphism is probably in the neighbourhood of 800°C. Igneous intrusions are found within areas of regional metamorphism, but their influence is restricted. Regional metamorphism may be regarded as taking place when the confining pressures are in excess of three kilobars, whilst below that figure, certainly below two kilobars, falls within the field of contact metamorphism.

Regional metamorphism is a progressive process, that is, in any given terrain formed initially of rocks of the same composition, zones of increasing grade may be defined by different mineral assemblages. Each zone is defined by a significant mineral and their mineralogical variation can be correlated with changing temperature–pressure conditions. The boundaries of each zone therefore can be regarded as isograds, lines of equal metamorphic conditions.

Slates are the products of low grade regional metamorphism of argillaceous sediments. As the metamorphic grade increases slate gives way to phyllite in which somewhat larger crystals of chlorite $[(Mg,Al,Fe)_{12}(Si,Al)_8O_{20}(OH)_{16}]$ and mica occur. This, in turn, gives way to mica schists. A variety of minerals such as garnet $[(Mg,Fe^{+2},Mn,Ca)_3Al_2(SiO_4)_3]$, kyanite $[Al_2SiO_5]$, staurolite $[2Al_2SiO_5.Fe^{+2}(OH)_2]$ and sillimanite $[Al_2SiO_5]$ may be present in these schists indicating formation at increasing temperatures. At high temperatures schists give way to gneisses. When sandstones are subjected to regional metamorphism a quartzite develops that has a granoblastic (granular) texture. A micaceous sandstone, or one in which there was an appreciable amount of argillaceous material, on metamorphism yields a quartz-mica schist. Metamorphism of arkoses and feldspathic sandstones leads to the recrystallization of feldspar and quartz so that granulites, with a granoblastic texture, are produced. Pure carbonate rocks when subjected to regional metamorphism simply recrystallize to form either a calcite or dolomite marble with a granoblastic texture. Any silica present in carbonate rocks usually recrystallizes as quartz. The presence of micas in these rocks tends to give them a schistose appearance, schistose marbles or calc-schists being developed. Where mica is abundant it forms lenses or continuous layers giving the rock a foliated structure.

Quartz and muscovite mica are important components in regionally metamorphosed rocks derived from acid igneous parents, muscovite-quartz schist being a typical product of the lower grades. At higher grades muscovite is converted to potash feldspar so that quartzo-feldspathic gneisses and granulites are common. Some gneisses may be strongly foliated (Fig. 1.14). Basic rocks are converted into greenschists by low grade regional metamorphism, to amphibolites at medium grade and pyroxene granulites and eclogites at high grades.

1.2.4 Metamorphic textures and structures

Most deformed metamorphic rocks possess some kind of preferred orientation. Preferred orientations commonly are exhibited as mesoscopic linear or planar structures that allow the rocks to split more easily in one direction than others. Perhaps the most familiar example is cleavage in slate, a similar type of structure in metamorphic rocks of higher grade is schistosity. Foliation comprises segregation of particular minerals into inconstant bands or contiguous lenticles that exhibit a common parallel orientation.

Slaty cleavage occurs in rocks of low metamorphic grade (Fig. 1.15a) and is characteristic of slates and phyllites. It is independent of bedding, which it commonly intersects at high angles, and it reflects a highly developed preferred orientation of minerals, particularly of those belonging to the mica family. Fracture cleavage is a parting defined by closely spaced parallel fractures that are usually independent of any planar preferred orientation of minerals that may be present in a rock mass (Fig. 1.15b). Unlike slaty cleavage, fracture cleavage is not restricted to one type of rock.

Fig. 1.14 Folded foliated gneiss, north of Mosjoen, Norway.

(a)

Fig. 1.15 (a) An exposure of slate near Barmouth, Wales, showing cleavage dipping out of face. (b) Fracture cleavage developed in folded Horton Flags, Silurian, near Stainforth, North Yorkshire. The inclination of the fracture cleavage is indicated by the vertical hammer, the other hammer indicates the dip of the bedding.

(b)

Schistosity develops in a rock when it is subjected to increasing temperatures and stress that involve its reconstitution, which is brought about by localized solution of mineral material and recrystallization. When recrystallization occurs under conditions that include shearing stress, then a directional element is imparted to the newly formed rock. Minerals are arranged in parallel layers along the direction normal to the plane of shearing stress giving the rock its schistose character (Fig. 1.16). The most important minerals responsible for the development of schistosity are those that possess an acicular, flaky or tabular habit, the micas being the principal family involved. The more abundant flaky and tabular minerals are in such rocks, then the more pronounced is the schistosity.

14

Fig. 1.16 Folded schist with quartz rod, Isle of Arran, Scotland.

Foliation in a metamorphic rock is a most conspicuous feature consisting of parallel bands or tabular lenticles formed of contrasting mineral assemblages such as quartz-feldspar and mica-chlorite-amphibole (Fig. 1.14). This parallel orientation agrees with the direction of schistosity, if any is present in nearby rocks. Foliation therefore would seem to be related to the same system of stress and strain responsible for the development of schistosity. However, at higher temperatures the influence of stress becomes less and so schistosity tends to disappear in rocks of high-grade metamorphism. By contrast, foliation becomes a more significant feature. What is more, minerals of flaky habit are replaced in the higher grades of metamorphism by minerals like garnet, kyanite, sillimanite, diopside (a pyroxene $Ca(Mg,Fe)(SiO_3)_2$) and orthoclase.

1.2.5 Metasomatism
Metasomatic activity involves the introduction of material into, as well as removal from, a rock mass by a gaseous or aqueous medium, the resultant chemical reactions leading to mineral replacement. Thus, two types of metasomatic action can be distinguished, namely, pneumatolytic that involves hot gases, and hydrothermal that involves hot solutions. Replacement is brought about by atomic or molecular substitution so that there is usually little change in rock texture. The composition of the transporting medium is continuously changing because of the exchange of material between it and the rocks that are affected. The gases and hot solutions involved usually emanate from an igneous source and the effects of metasomatism often are particularly notable about an intrusion of granitic character since there is a greater concentration of volatiles in acid than in basic magmas.

Both gases and solutions make use of any structural weaknesses such as faults or joint planes, in the rocks they invade. Because these provide easier paths of escape, metasomatic activity is concentrated along them. They also travel through the pore spaces in rocks, the rate of infiltration being affected by the porosity, the shape of the pores and the temperature–pressure gradients. Metasomatic action, especially when it is concentrated along fissures and veins, may bring about severe alteration of certain minerals. For instance, feldspars in granite or gneiss may be highly kaolinized as a result of metasomatism, and limestone may be reduced to a weakly bonded granular aggregate.

1.3 Sedimentary rocks
Sedimentary rocks form an outer skin on the Earth's crust, covering three-quarters of the continental areas and most of the sea floor. They vary in thickness up to ten kilometres. Most sedimentary rocks are of secondary origin in that they consist of detrital material derived from the breakdown of pre-existing rocks. The agents responsible for breakdown of rocks include weathering, rivers, the sea, wind and ice;

and water, wind and ice, together with gravity, transport the broken down material elsewhere before it is deposited. It has been variously estimated that shales and sandstones, both of mechanical derivation, account for between 75 and 95% of all sedimentary rocks. Certain sedimentary rocks are the products of chemical or biochemical precipitation whilst others are of organic origin. Hence, the sedimentary rocks can be divided into two principal groups, namely, the clastic or exogenetic and the non-clastic or endogenetic types. However, one factor that all sedimentary rocks have in common is that they are deposited and this gives rise to their most noteworthy characteristic, that is, they are bedded or stratified.

The particles of which most sedimentary rocks are composed have undergone varying amounts of transportation. The amount of transport, together with the agent responsible, play an important role in determining the character of a sediment. For instance, transport over short distances usually means that the sediment is unsorted (the exception being beach sands), as does transportation by ice. With lengthier transport by water or wind not only does the material become sorted but it is further reduced in size. The character of a sedimentary rock is also influenced by the type of environment in which it has been deposited. For example, ripple marks and cross bedding are present in sediments that have been deposited in shallow water.

The composition of a sedimentary rock depends partly on the composition of the parent material and the stability of its component minerals, and partly on the type of action to which the parent rock was subjected and the length of time it had to undergo such action. Because of its stability, quartz is the commonest constituent of igneous and metamorphic rocks that is found in abundance in sediments. Most of the other silicate minerals ultimately give rise to clay minerals.

In order to turn an unconsolidated sediment into a solid rock it must be lithified. Lithification involves two processes, consolidation and cementation. The amount of consolidation that takes place within a sediment depends, firstly, on its composition and texture and, secondly, on the pressures acting on it, notably that due to the weight of overburden. Consolidation of sediments deposited in water also involves dewatering. The porosity of a sediment is reduced as consolidation takes place and as the individual particles become more closely packed they may even be deformed. Pressures developed during consolidation may lead to the differential solution of minerals and the authigenic growth of new ones. Fine-grained sediments possess a higher porosity than do coarser types and therefore undergo a greater amount of consolidation. For instance, muds and clays may have original porosities ranging up to 80% compared with 45 to 50% in sands and silts.

Cementation involves the bonding together of sedimentary particles by the precipitation of material in the pore spaces. This reduces the porosity. The cementing material may be derived by partial solution of grains or may be introduced into the pore spaces from an extraneous source by circulating waters. Conversely, cement may be removed from a sedimentary rock by leaching. The type of cement and, more importantly the amount, affect the strength of a sedimentary rock. The type also influences its colour. For example, sandstones with a siliceous or calcium carbonate cement are usually whitish grey, those with a sideritic (iron carbonate) cement are buff coloured, whilst a red colour is indicative of a hematitic (iron oxide) cement and brown of limonite (hydrated iron oxide). However, sedimentary rocks frequently are cemented by more than one material. The matrix of a sedimentary rock refers to the fine material trapped within the pore spaces between the particles. It helps to bind the latter together.

The texture of a sedimentary rock refers to the size, shape and arrangement of its constituent particles. The size of the particles of a clastic sedimentary rock allows it to be placed in one of three groups that are termed rudaceous (e.g. gravels, conglomerates), arenaceous (e.g. sands and sandstones) and argillaceous (i.e. mudrocks). However, grain size is a property that is not easy to assess accurately, for the grains and pebbles of which clastic sediments are composed are irregular, three-dimensional objects. Because of their smallness, the size of sand and silt grains is measured indirectly by sieving and sedimentation techniques respectively. If individual particles of clay have to be measured, then this is done with the aid of an electron microscope. If a rock is strongly indurated, then its disaggregation is impossible without fracturing many of the grains. In such a case, a thin section of the rock is made and size analysis is carried out with the aid of a petrological microscope and micrometer. The results of a size

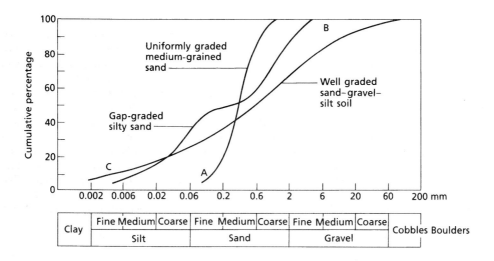

Fig. 1.17 Cumulative curves of well graded, uniformly graded and gap-graded sediments.

analysis may be represented graphically by a frequency curve or histogram. More frequently, however, the results are used to draw a cumulative curve (Fig. 1.17). The degree of sorting can be estimated from the curve in that the steeper the curve, the better the sediment is sorted. Shape is probably the most fundamental property of any particle but unfortunately it is one of the most difficult to quantify. Shape frequently is assessed in terms of roundness and sphericity that may be estimated visually by comparison with standard images. A sedimentary rock is an aggregate of particles, some of which may exhibit a degree of orientation, as for example, platy minerals in shales and laminated sandstones.

1.3.1 Bedding and sedimentary structures

Sedimentary rocks are characterized by their stratification, and bedding planes are frequently the dominant discontinuity in sedimentary rock masses (Fig. 1.18). As such their spacing and character (irregular, waved or straight, tight or open, rough or smooth) are of particular importance in terms of engineering behaviour. An individual bed may be regarded as a thickness of sediment of the same composition that was deposited under the same conditions. However, lamination can result from minor fluctuations in the velocity of the transporting medium or the supply of material, both of which produce alternating thin layers of slightly different grain size. Generally, lamination is associated with the presence of thin layers of platy minerals, notably micas. These have a marked preferred orientation, usually parallel to the bedding planes, and are responsible for the fissility of the rock. The surfaces of these laminae are usually smooth and straight. Although lamination is most characteristic of shales, it also may be present in siltstones and sandstones, and occasionally in some limestones.

Cross or current bedding is a depositional feature that occurs in sediments of fluvial, littoral, marine and aeolian origin, and most notably is found in sandstones (Fig. 1.19). In windblown sediments it generally is referred to as dune bedding. Cross bedding is confined within an individual sedimentation unit and consists of cross laminae inclined frequently between 20 and 30° to the true bedding planes. The size of the sedimentation unit in which they occur varies enormously. For example, in micro-cross bedding it measures only a few millimetres, whilst in dune bedding the unit may exceed 100 m.

Although graded bedding occurs in several different types of sedimentary rock, it is characteristic of greywackes. The sedimentation unit is graded from coarser grain size at the bottom to finer at the top.

Fig. 1.18 Well developed bedding in the Caithness Flagstones, Latheronwheel, Scotland.

Individual graded beds range in thickness from a few millimetres to several metres. Usually, the thicker the bed, the coarser is its grain size.

1.3.2   Sedimentary rock types

A gravel is an unconsolidated accumulation of rounded fragments, the lower size limit of which is 2 mm. The term rubble has been used to describe those deposits with angular fragments. The composition of a gravel deposit reflects not only the source rocks of the area from which it was derived but also is influenced by the agent(s) responsible for its formation and the climatic regime in which it was, or is being deposited. The latter two factors have a varying tendency to reduce the proportion of unstable material. Relief also influences the nature of a gravel deposit. For example, gravel production under low relief is small and the pebbles tend to be inert residues such as vein quartz, quartzite, chert [SiO_2] and flint [SiO_2]. Conversely,

Fig. 1.19 Cross-bedding (i.e. dune beding) in the Coconino Sandstone, Nevada.

high relief and accompanying rapid erosion yield coarse, immature gravels. When a gravel is indurated it forms a conglomerate, when a rubble is indurated it is termed a breccia.

Sands consist of loose mixtures of mineral grains and rock fragments. Generally, they tend to be dominated by a few minerals, the chief of which is quartz. When sand is indurated it forms sandstone. Silica is a common cementing agent in sandstones, as are carbonate cements, especially calcite [$CaCO_3$]. Ferruginous and gypsiferous cements also are found in sandstones. Cement, notably the carbonate and gypsiferous types, may be removed in solution by percolating pore fluids. Quartz, feldspar and rock fragments are the principal detrital components of which sandstones are composed, and consequently they have been used, along with the amount of matrix material present, to define the major classes of sandstone. Those sandstones with less than 15% matrix can be classed into three families. The orthoquartzites or quartz arenites contain 95% or more of quartz; 25% or more of the detrital material in arkoses consists of feldspar; and in lithic sandstones 25% or more of the detrital material consists of rock fragments. A greywacke is a sandstone with more than 15% matrix material.

Silts are clastic sediments derived from pre-existing rocks, chiefly by mechanical breakdown processes. They commonly are found in alluvial, lacustrine, and marine deposits, where they tend to interdigitate with deposits of sand and clay. Silts also are present with sands and clays in estuarine and deltaic sediments. Silts are mainly composed of fine quartz material. Loess is a wind-blown deposit that is mainly of silt size and consists mostly of quartz particles, with lesser amounts of feldspar and clay minerals. Siltstones are lithified silts and may be massive or laminated. Micro-cross bedding is developed in some siltstones and the laminae may be convoluted. Siltstones have a high quartz content with a predominantly siliceous cement. Frequently, siltstones are interbedded with shales or fine-grained sandstones, the siltstones occurring as thin ribs.

Deposits of clay are composed principally of clay minerals and fine quartz. The former represent the commonest breakdown products of most of the chief rock forming silicate minerals. Clay minerals are hydrated aluminium silicates with a platy structure, that is, they are phyllosilicates. The three major types of clay minerals are kaolinite, illite and montmorillonite. Kaolinite [$Al_4Si_4O_{10}(OH)_8$] is formed by the alteration of feldspars, feldspathoids and other aluminium silicates due to hydrothermal action. Weathering under acidic conditions also is responsible for kaolinization. Kaolinite is the chief clay mineral in most residual and transported clays, is important in shales, and is found in variable amounts in fire-clays, laterites and soils. It is the most important clay mineral in china clays or kaolin and ball clays. Deposits of kaolin are associated with granites, granodiorites, gneisses and granulites. Illite [$K_{2-3}Al_8(Al_{2-3}, Si_{13-14})O_{40}(OH)_8$] commonly occurs in clays and shales, and is found in variable amounts in tills and loess, but is less common in soils. It develops as an alteration product of feldspars, micas or ferromagnesian silicates upon weathering or may form from other clay minerals during diagenesis. Illite also may be of hydrothermal origin. The development of illites, both under weathering and by hydrothermal processes is favoured by an alkaline environment. Montmorillonite [$Al_4Si_8O_{20}(OH)_4.nH_2O$] develops when basic igneous rocks, in badly drained areas, are subjected to weathering. The presence of magnesium is necessary for this mineral to form, if the rocks were well drained, then it would be carried away and kaolinite would develop. An alkaline environment also favours the formation of montmorillonite. Montmorillonites occur in soils and argillaceous sediments such as shales formed from basic igneous rocks. It is the principal constituent of bentonitic clays, which are formed by the weathering of basic volcanic ash. Hydrothermal action also may lead to the development of montmorillonite. Residual clay deposits develop in place and are the products of weathering. In humid regions residual clays tend to become enriched in hydroxides of ferric iron and aluminium, and impoverished in lime, magnesia and alkalis. Even silica is removed in hot humid regions, resulting in the formation of hydrated alumina or iron oxide, as in laterite. The composition of transported clay deposits varies because these materials consist mainly of abrasion products (usually silty particles) and transported residual clay material.

Shale is the commonest sedimentary rock and is characterized by its lamination. Sedimentary rock of similar size range and composition, but which is not laminated usually is referred to as mudstone. In fact, there is no sharp distinction between shale and mudstone, one grading into the other. An increasing content of siliceous or calcareous material decreases the fissility of a shale whereas shales that have a high organic content are finely laminated. Laminae range from 0.05 to 1.0 mm in thickness, with most in the range of 0.1 to 0.4 mm. Clay minerals and quartz are the principal constituents of mudstones and shales. Feldspars often occur in the siltier shales. Shales also may contain appreciable quantities of carbonate, particularly calcite, and gypsum. Indeed, calcareous shales frequently grade into shaly limestones. Carbonaceous black shales generally are rich in organic matter, contain a varying amount of pyrite [FeS_2], and are finely laminated.

The term limestone is applied to those rocks in which the carbonate fraction exceeds 50%, over half of which is calcite or aragonite [$CaCO_3$]. If the carbonate material is made up chiefly of dolomite [$(Ca,Mg)CO_3$], then the rock is named dolostone (this rock is generally referred to as dolomite but this term can be confused with that of the mineral of the same name). Limestones and dolostones constitute about 20 to 25% of the sedimentary rocks. They are polygenetic. Some are of mechanical origin representing carbonate detritus that has been transported and deposited or that has accumulated *in situ*. Others represent chemical or biochemical precipitates that have formed in place. Allochthonous or transported limestone has a fabric similar to that of sandstone and also may display current structures such as cross bedding and ripple marks. By contrast, carbonate rocks that have formed *in situ*, that is, autochthonous types, show no evidence of sorting or current action and at best possess poorly developed stratification. Exceptionally, some autochthonous limestones show growth bedding, the most striking of which is stromatolitic bedding as seen in algal limestones. Carbonate rocks may contain various amounts of impurities, notably quartz and clay minerals. The latter give rise to fine-grained argillaceous limestones, which frequently may be wavy bedded or nodular bedded, the beds being separated by thin shaly partings.

Evaporitic deposits are quantitatively unimportant as sediments. They are formed by precipitation from saline waters, the high salt content being brought about by evaporation from inland seas or lakes in arid regions. Salts also can be deposited from subsurface brines, brought to the surface of a playa or sabkha (see Chapter 9) flat by capillary action. Experimental work has shown that when the original volume of sea water is reduced by evaporation to about half, then a little iron oxide and some calcium carbonate are precipitated. Gypsum [$CaSO_4.nH_2O$] begins to form when the volume is reduced to about one-fifth of the original, rock salt (halite, NaCl) begins to precipitate when about one-tenth of the volume remains, and finally when only 1.5% of the sea water is left potash and magnesium salts start to crystallize. This order agrees in a general way with the sequence found in some evaporitic deposits. However, many complex replacement sequences occur amongst evaporitic rocks, for example, carbonate rocks may be replaced by anhydrite [$CaSO_4$] and sulphate rocks by halite.

Organic residues that accumulate as sediments are of two major types, namely, peaty material that when buried gives rise to coal, and sapropelic residues. Sapropel is either rich in silt or composed wholly of organic compounds that collect at the bottom of still bodies of water. Such deposits may give rise to cannel or boghead coals. Sapropelic coals usually contain a significant amount of inorganic matter as opposed to humic coals in which the inorganic content is low. The former generally are not extensive and are not underlain by seat earths (i.e. fossil soils). Peat deposits accumulate in poorly drained environments where the formation of humic acid gives rise to deoxygenated conditions. These inhibit the bacterial decay of organic matter. Peat accumulates wherever the deposition of plant debris exceeds the rate of its decomposition. A massive deposit of peat is required to produce a thick seam of coal, for example, a seam 1 m thick probably represents 15 m of peat.

Chert and flint are the two most common siliceous sediments of chemical origin. Chert is a dense rock composed of one or more forms of silica such as opal, chalcedony or microcrystalline quartz. It may occur as thin beds or as nodules in carbonate host rocks.

Some sediments may have a high content of iron. The iron carbonate siderite [$FeCO_3$] often occurs interbedded with chert or mixed in varying proportions with clay, as in clay ironstones. Some iron-bearing formations are formed mainly of iron oxide, hematite [Fe_2O_3] being the most common mineral. Bog iron ore is chiefly an earthy mixture of ferric hydroxides.

1.4 Stratigraphy and stratification

Stratigraphy is that branch of geology that deals with the study and interpretation of stratified rocks, and with the identification, description, sequence (both vertical and horizontal), mapping and correlation of stratigraphic rock units. As such it begins with the discrimination and description of stratigraphical units such as formations. This is necessary so that the complexities present in every stratigraphical section may be simplified and organized.

Deposition of sediments involves the build-up of material on a given surface and so this surface influences the attitude of the beds that form. At the start of deposition the layers of sediment more or less conform to the surface on which accumulation is occurring, provided this is not too irregular. With continued deposition any irregularities in the original surface are filled and the strata that then form tend to lie in a horizontal plane. However, once a layer is formed, and before lithification occurs, it may be disturbed by subsequent deposition. Furthermore, differential consolidation of different materials, for example, sand and mud, or differential consolidation over buried hills may give rise to inclined bedding.

The changes that occur during deposition are responsible for stratification, that is, the layering that characterizes sedimentary rocks. The simple cessation of deposition ordinarily does not produce stratification. The most obvious change that gives rise to stratification is that in the composition of the material being deposited. Even minor changes in the type of material may lead to distinct stratification, especially if they affect the colour of the rocks concerned. Changes in grain size also may cause notable layering and changes in other textural characteristics may help distinguish one bed from another, as may variations in the degree of consolidation or cementation.

The extent and regularity of beds of sedimentary rocks vary within wide limits. This is because lateral persistence and regularity of stratification reflect the persistence and regularity of the agent responsible for deposition. For instance, sands may have been deposited in one area whilst in a neighbouring area muds were being deposited. Hence, lateral changes in lithology reflect differences in the environments in which deposition took place. On the other hand, a formation with a particular lithology, which is mappable as a stratigraphic unit, may not have been laid down at the same time wherever it occurs. The base of such a formation is described as diachronous. Diachronism is brought about when a basin of deposition is advancing or retreating, as for example, in a marine transgression or regression. In an expanding basin the lowest sediments to accumulate are not as extensive as those succeeding. The latter are said to overlap the lowermost deposits. Conversely, if the basin of deposition is shrinking the opposite situation arises in that succeeding beds are less extensive. This phenomenon is termed offlap. Agents that are confined to channels or deposited over small areas produce irregular strata that are not persistent. By contrast, strata that are very persistent are produced by agents that operate over wide areas. In addition, folding and faulting of strata, along with subsequent erosion, give rise to discontinuous outcrops.

Since sediments are deposited it follows that the topmost layer in any succession of strata is the youngest. Also, any particular stratum in a sequence can be dated by its position in the sequence relative to other strata. This is the Law of Superposition. This principle applies to all sedimentary rocks except, of course, those that have been overturned by folding or where strata have been thrust over younger rocks. When fossils are present in the beds concerned their correct way up can be discerned. However, if fossil evidence is lacking the correct way up of the succession can be determined from evidence provided by 'way-up' structures such as ripple marks, cross bedding, graded bedding, mud-cracks, scour and fill channels, etc.

1.4.1 Unconformities

An unconformity represents a time break in the stratigraphical record and occurs when no deposition took place or rocks were removed by erosion over a period of time that can vary from short to very long. The beds above and below the surface of unconformity are described as unconformable. The structural relationship between unconformable units allows four types of unconformity to be distinguished. In Fig.

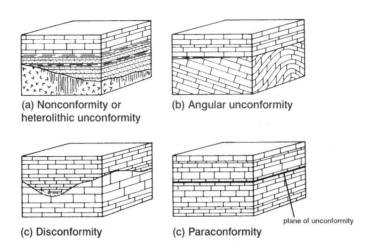

(a) Nonconformity or heterolithic unconformity

(b) Angular unconformity

(c) Disconformity

(c) Paraconformity

plane of unconformity

Fig. 1.20   Types of unconformity.

1.20(a), stratified rocks rest upon igneous or metamorphic rocks. This type of feature has been referred to as a nonconformity (it has also been called a heterolithic unconformity). An angular unconformity is shown in Fig. 1.20(b), where an angular discordance separates the two units of stratified rocks. In an angular unconformity the lowest bed in the upper sequence of strata often rests on beds of differing ages. This is referred to as overstep. In a disconformity, as illustrated in Fig. 1.20(c), the beds lie parallel both above and below the unconformable surface but the contact between the two units concerned is an uneven surface of erosion. When deposition is interrupted for a significant period but there is no apparent erosion of sediments or tilting or folding, then subsequently formed beds are deposited parallel to those already existing. In such a case, the interruption in sedimentation may be demonstrable only by the incompleteness of the fossil sequence. This type of unconformity has been termed a paraconformity (Fig. 1.20(d)).

Evidence of the presence of an unconformity may take the form of pronounced irregularities in its surface caused by erosion. Also, the existence of weathered strata beneath an unconformity provides evidence of its presence, fossil soils providing a good example. The abrupt truncation of bedding planes, folds, faults, dykes, joints, etc. in the beds below an unconformity offers yet other evidence of its presence, although large-scale thrusts give rise to a similar structural arrangement. Post-unconformity sediments often commence with a conglomeratic deposit and its pebbles may be derived from the older rocks below the unconformity.

1.4.2   Geological time

Stratigraphy distinguishes rock units and time units. A rock unit, such as a stratum or a formation, possesses a variety of physical characteristics that enable it to be recognized as such, and so measured, described, mapped and analysed. A rock unit is sometimes termed a lithostratigraphical unit. A particular rock unit required a certain interval of time to form. Hence, stratigraphy not only deals with strata but also deals with age, and the relationship between strata and age. Accordingly, time units and time-rock units have been recognized. Time units are intervals of time, the largest of which are eons, of which there are two, namely, Pre-Cambrian time and Phanerozoic time. Eons are divided into eras, and eras into periods (Table 1.2). Periods are in turn divided into epochs and epochs into ages. Time units and time-rock units are directly comparable, that is, for each time unit there is a corresponding time-rock unit. For example, the time-rock unit corresponding to a period is a system. Indeed, the time allotted to a time

Table 1.2 The broad divisions of the geological column and timescale.

	Era	Period	Age range (Ma)
PHANEROZOIC	CENOZOIC	Quaternary	2 – 0
		Tertiary	66 – 2
	MESOZOIC	Cretaceous	144 – 66
		Jurassic	208 – 144
		Triassic	245 – 208
	LATE PALAEOZOIC	Permian	286 – 245
		Carboniferous	360 – 286
		Devonian	408 – 360
	EARLY PALAEOZOIC	Silurian	438 – 408
		Ordovician	495 – 438
		Cambrian	545 – 495
	PROTEROZOIC	Precambrian	2500 – 545
	ARCHAEAN		3800 – 2500
	PRE-ARCHAEAN		3800 – 4600

unit is determined from the rocks of the corresponding time-rock unit. A time-rock unit has been defined as a succession of strata bounded by theoretically uniform time planes, regardless of the local lithology of the unit. Fossil evidence usually provides the basis for the establishment of time planes. Ideal time-rock units would be bounded by completely independent time planes, however, practical time-rock units depend on whatever evidence is available.

The geological systems are time-rock units that are based on stratigraphical successions present in certain historically important areas. In other words, in their type localities the major time-rock units are also rock units. The boundaries of major time-rock units generally are important structural or faunal breaks or are placed at highly conspicuous changes in lithology. Major unconformities frequently are chosen as boundaries. Away from their type areas major time-rock units may not be so distinctive or easily separated. In fact, although systems are regarded as of global application there are large regions where the recognition of some of the systems has not proved satisfactory. The formation of the geological systems evolved gradually as geology developed, being finally established towards the end of the nineteenth century. Systems are divided into series.

1.4.3 Correlation

The process by which the time relationships between strata in different areas is established is referred to as correlation. Correlation is therefore the demonstration of equivalency of stratigraphical units. Palaeontological and lithological evidence are the two principal criteria used in correlation. Physical continuity may be of some use in local correlation, that is, it can be assumed that a given bed, or bedding plane, is roughly contemporaneous throughout an outcrop of bedded rocks. Tracing of bedding planes laterally, however, is severely limited since individual beds or bedding planes die out, are interrupted by faults, are missing in places due to removal by erosion, are concealed by overburden, or merge with others laterally. Consequently, outcrops are rarely good enough to permit an individual bed to be traced laterally over an appreciable distance. A more practicable procedure is to trace a member of a formation. However, this also can prove misleading if beds are diachronous.

Where outcrops are discontinuous, physical correlation depends on lithological similarity, that is, on matching rock types across the breaks in the hope of identifying the beds involved. The lithological characters used to make comparison in such situations include gross lithology, subtle distinctions within one rock type such as a distinctive mineral suite, notable microscopic features or distinctive key beds. The greater the number of different, and especially unusual, characters that can be linked, the better are the chances of reliable correlation. Even so, such factors must be applied with caution and wherever possible such correlation should be verified by the use of fossils.

As a stratal formation may be identified by a distinctive suite of fossils, this allowed the Law of Faunal Succession to be developed. It states that strata of different ages are characterized by different fossils or suites of fossils. In this way, particular or index fossils can be used as a method of recognizing strata of equivalent age. As far as correlation is concerned, index fossils should have a wide geographical distribution and a limited stratigraphical range. Generally, organisms that possessed complicated structures provide the best guides for correlation. The presence of characteristic species in strata of a particular age allows the distinction of zones. A zone may be defined as that strata deposited during a particular interval of time when a given fauna or flora existed. In some cases zones have been based on the complete faunal assemblage present whilst in other instances they have been based on the members of a particular phylum or class. Although a faunal or floral zone is defined by reference to an assemblage of fossils, it usually is named after some characteristic species and this fossil is known as the zone fossil. It is assumed that zonal species have time ranges that are similar in different areas.

1.5 Geological structures

The two most important features that are produced when strata are deformed by earth movements are folds and faults, that is, the rocks are buckled or fractured respectively. A fold is produced when a more or less planar surface is deformed to give a waved surface whereas a fault represents a surface of discontinuity along which the strata on either side have been displaced relative to each other.

1.5.1 Folds

There are two important directions associated with folded strata, namely, dip and strike. True dip gives the maximum angle at which a bed of rock is inclined and should always be distinguished from apparent dip (Fig. 1.21). The latter is a dip of lesser magnitude whose direction can run anywhere between that of true dip and strike. Strike is the trend of a fold and is orientated at right angles to the true dip, it has no inclination (Fig. 1.21).

Fig. 1.21 A coastal exposure near Cambrai, California, illustrating dip and strike in sedimentary rocks.

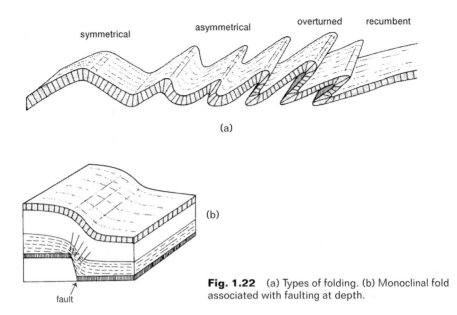

(a)

(b)

Fig. 1.22 (a) Types of folding. (b) Monoclinal fold associated with faulting at depth.

Folds are wave-like in shape and vary enormously in size. Simple folds are divided into two types, anticlines and synclines. In the former, the beds are convex upwards, whereas in the latter they are concave upwards. Anticlines and synclines are symmetrical if both limbs are arranged equally about the axial plane so that the dips on opposing flanks are the same, otherwise they are asymmetrical (Fig. 1.22). In symmetrical folds the axis is vertical whilst in asymmetrical folds it is inclined. The crestal line of an anticline is the line that joins the highest parts of a fold whilst the trough line runs through the lowest parts of a syncline (Fig. 1.23). The amplitude of a fold is defined as the vertical difference between the crest and the trough, whilst the length of a fold is the horizontal distance from crest to crest or trough to

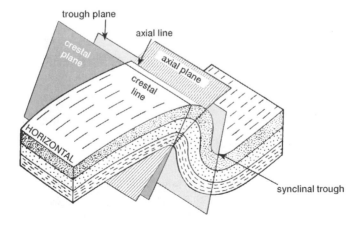

Fig. 1.23 Block diagram of a non-plunging overturned anticline and syncline, illustrating various fold elements.

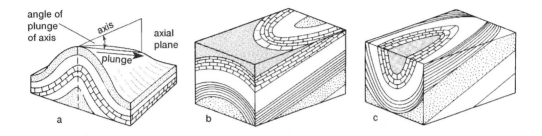

Fig. 1.24 (a) Block diagram of an anticlinal fold illustrating plunge. (b) Plunging anticline. (c) Plunging syncline.

trough. The hinge of a fold is the line along which the greatest curvature exists and it can be either straight or curved. However, the axial line is another term that has been used to describe the hinge line. The limb of a fold occurs between the hinges, all folds having two limbs. The axial plane of a fold is commonly regarded as the plane that bisects the fold and passes through the hinge line. Folds are of limited extent and as one fades out the attitude of its axial line changes, that is, it dips away from the horizontal. This is referred to as the plunge or pitch of the fold (Fig. 1.24). The amount of plunge can change along the strike of a fold and a reversal of plunge direction can occur. The axial line then is waved, concave upwards areas being termed depressions whilst convex upwards areas are known as culminations. As folding movements become intensified, overfolds are formed in which both limbs are inclined, together with the axis, in the same direction but at different angles (Fig. 1.22a). In a recumbent fold the beds have been completely overturned so that one limb is inverted, and the limbs, together with the axial plane, dip at a low angle (Fig. 1.22a). If beds that are horizontal or nearly so, suddenly dip at a high angle, then the feature they form is termed a monocline (Fig. 1.22b). When traced along their strike, monoclines may eventually flatten out or pass into a normal fault, indeed they often are formed as a result of faulting at depth.

Fig. 1.25 The Howick fault occurs in strata of Lower Carboniferous age and is exposed along the coast at Cullernose Point, Northumberland, England. This fault has an estimated throw of 200 m, the downthrown side being on the left. The fault zone is about a metre wide and the beds on either side have been up- or down-turned by drag caused by the relative movements along the fault.

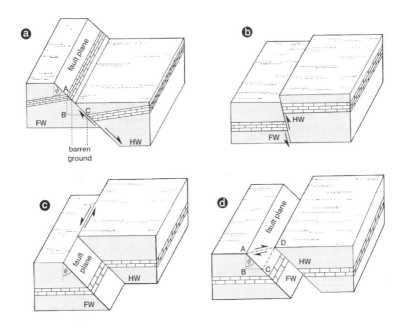

Fig. 1.26 Types of fault: (a) normal fault (b) reverse fault (c) wrench or strike-slip fault (d) oblique-slip fault. FW = footwall; HW = hanging wall; AB = throw; BC = heave; ϕ = angle of hade.

1.5.2 Faults

Faults are fractures in strata along which the adjacent rock has been displaced (Fig. 1.25). The amount of displacement may vary from only a few tens of millimetres to several hundred kilometres. In many faults the fracture is a clean break but in others the displacement is not restricted to a simple fracture but is developed throughout a fault zone.

The dip and strike of a fault plane can be described in the same way as those of a bedding plane. The angle of hade is the angle enclosed between the fault plane and the vertical. The hanging-wall of a fault refers to the upper rock surface along which displacement has occurred whilst the foot-wall is the term given to that below. The vertical shift along a fault plane is called the throw whilst the term heave refers to the horizontal displacement. Where the displacement along a fault has been vertical, then the terms downthrow and upthrow refer to the relative movement of strata on opposite sides of the fault plane.

Faults can be classified according to the direction in which movement has taken place along the fault plane, to the relative movement of the hanging and foot-walls, and to the attitude of the fault in relation to the strata involved. If the direction of slippage along the fault plane is used to distinguish between faults, then three types may be recognized, namely, dip-slip faults, strike-slip faults, and oblique-slip faults. In the dip-slip fault the slippage occurred along the dip of the fault, in the strike-slip fault it took place along the strike and in the oblique-slip fault movement occurred diagonally across the fault plane (Fig. 1.26). When the relative movement of the hanging and foot-walls is used as a basis of classification, then normal, reverse and wrench faults can be recognized. The normal fault is characterized by the occurrence of the hanging-wall on the downthrown side whilst in the reverse fault the foot-wall occupies the downthrown side. Reverse faulting involves a vertical duplication of strata, unlike normal faults where the displacement gives rise to a region of barren ground (Fig. 1.26). In the wrench fault neither the foot-wall nor the hanging-wall have moved up or down in relation to one another (Fig. 1.26).

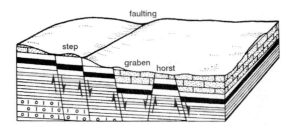

Fig. 1.27 Block diagram illustrating step faulting, and horst and graben structures.

Considering the attitude of the fault to the strata involved, strike faults, dip (or cross) faults, and oblique faults can be distinguished. A strike fault is one that trends parallel to the beds it displaces, a dip or cross fault is one that follows the inclination of the strata and an oblique fault runs at angle with the strike of the rocks it intersects. In areas that have not undergone intense tectonic deformation reverse and normal faults generally dip at angles in excess of 45°. Their low-angled equivalents, termed thrusts and lags respectively, are inclined at less than that figure and tend to be associated with strongly folded terrain. When a series of normal faults run parallel to one another with their downthrows all on the same side, the area involved is described as being step faulted (Fig. 1.27). Horsts and rift valleys (graben) are also illustrated in Fig. 1.27.

Although a fault may not be observable its effects may be reflected in the topography. For example, if blocks are tilted by faulting, then a series of scarps are formed. If the rocks on either side of a fault are of different hardness, then a scarp may form along the fault as a result of differential erosion. Triangular facets occur along a fault scarp associated with an upland region. They represent the remnants left behind after swift flowing rivers have cut deep valleys into the scarp. Scarplets are indicative of active faults and are found near the foot of mountains where they run parallel to the base of the range. On the other hand, natural escarpments may be offset by cross faults. Stream profiles may be interrupted by faults or, in a region of recent uplift, their courses may in fact be relatively straight due to them following faults. Springs often occur along faults. Faults may be responsible for the formation of waterfalls in the path of a stream. However, it must be emphasized that the physiographical features noted above may be developed without the aid of faulting and consequently they do not provide a fool-proof indication of stratal displacement. Faults provide a path of escape and they are therefore frequently associated with mineralization, silicification and dyke intrusion.

1.5.3 Discontinuities

A discontinuity represents a plane of weakness within a rock mass across which the rock material is structurally discontinuous. Although discontinuities are not necessarily planes of separation, most are and they possess little or no tensile strength. Discontinuities vary in size from small fissures on the one hand to huge faults on the other. The most common discontinuities are joints and bedding planes (Fig. 1.28). Other important discontinuities are planes of cleavage and schistosity.

Joints are fractures along which little or no displacement has occurred and are present within all types of rock. At the ground surface, joints may open as a consequence of denudation, especially weathering, or the dissipation of residual stress.

A group of joints that run parallel to each other are termed a joint set whilst two or more joints sets that intersect at a more or less constant angle are referred to as a joint system. If one set of joints is dominant, then these joints are known as primary joints, the other set or sets of joints being termed secondary. If joints are planar and parallel or sub-parallel they are described as systematic, conversely

Fig. 1.28 Joint systems in argillaceous limestone, south of Cullernose point, Northumberland, England.

when they are irregular they are termed non-systematic. Joints can be divided, on the basis of size, into master joints that penetrate several rock horizons and persist for hundreds of metres; major joints are smaller joints but are still well defined structures and minor joints do not transcend bedding planes.

Joints are formed through failure in tension, in shear, or through some combination of both. Most joints are post-compressional structures, formed as a result of the dissipation of residual stress after folding occurred. Some spatially restricted small joints associated with folds, such as radial tension joints, are probably initiated during folding. Joints also are formed within igneous rocks when they initially cool down, and in wet sediments when they dry out. The most familiar of these are the columnar joints in lava flows, sills and some dykes. Cross joints, longitudinal joints, diagonal joints and flat-lying joints are associated with large granitic intrusions have been described above. Sheet or mural joints have a similar orientation to flat-lying joints. When they are closely spaced and well developed they impart a pseudostratification to the host rock. It has been noted that the frequency of sheet jointing is related to the depth of overburden, in other words, the thinner the rock cover the more pronounced the sheeting. This suggests a connection between removal of overburden by denudation, dissipation of residual stress and the development of sheeting.

The shear strength of a rock mass and its deformability are very much influenced by the discontinuity pattern, its geometry and how well it is developed. A record of discontinuity spacing, whether in a field

Table 1.3 Description of bedding plane and joint spacing.

Description of bedding-plane spacing	Description of joint spacing	Limits of spacing
Very thickly bedded	Extremely wide	over 2 m
Thickly bedded	Very wide	0.6–2 m
Medium bedded	Wide	0.2–0.6 m
Thinly bedded	Moderately wide	60 mm–0.2 m
Very thinly bedded	Moderately narrow	20–60 mm
Laminated	Narrow	6–20 mm
Thinly laminated	Very narrow	under 6 mm

exposure or in a core stick, aids appraisal of rock mass structure. Bedding planes are usually the dominant discontinuity in sedimentary rocks and their spacing can be described as shown in Table 1.3. The same boundaries can be used to describe the spacing of joints. As joints represent surfaces of weakness, the larger and more closely spaced they are, the more influential they become in reducing the effective strength of a rock mass. The persistence of a joint plane refers to its continuity. This is one of the most difficult properties to quantify since joints frequently continue beyond the rock exposure and consequently in such instances it is impossible to estimate their continuity. Be that as it may, persistence can be described as follows:

Very low persistence	less than 1 m
Low persistence	1–3 m
Medium persistence	3–10 m
High persistence	10–20 m
Very high persistence	greater than 20 m

Block size provides an indication of how a rock mass is likely to behave, since block size and inter-block shear strength determine the mechanical performance of a rock mass under given conditions of stress. The following descriptive terms of block size are used:
(1) Massive – few joints or very wide spacing
(2) Blocky – approximately equidimensional
(3) Tabular – one dimension considerably shorter than the other two
(4) Columnar – one dimension considerably larger than the other two
(5) Irregular – wide variations of block size and shape
(6) Crushed – heavily jointed to 'sugar cube'.
The block size may be described quantitatively as shown in Table 1.4.

Table 1.4 Quantitative description of block sizes.

Term	Block size	Equivalent discontinuity spacing in blocky rock	Volumetric joint count (J_v) (joints/m³)
Very large	over 8 m³	Extremely wide	less than 1
Large	0.2–8 m³	Very wide	1–3
Medium	0.008–0.2 m³	Wide	3–10
Small	0.0002–0.008 m³	Moderately wide	10–30
Very small	less than 0.0002 m³	Less than moderately wide	over 30

Discontinuities, especially joints, may be open or closed. How open they are is of importance in relation to the overall strength and permeability of a rock mass, and this often depends largely on the amount of weathering that the rocks have suffered (Table 1.5). Some joints may be partially or completely filled. The type and amount of filling not only influence the effectiveness with which the opposing joint surfaces are bound together, thereby affecting the strength of the rock mass, but also influence permeability. If the infilling is sufficiently thick, then the walls of the joint are not in contact and hence the strength of the joint plane depends on that of the infill material.

The nature of opposing joint surfaces also influences rock mass behaviour as the smoother they are, the more easily can movement take place along them. However, joint surfaces are usually rough. Hence, the nature of a joint surface may be considered in relation to its waviness, roughness and the condition of the walls. Waviness and roughness differ in terms of scale and their effect on the shear strength of the joint. Waviness refers to first order undulations of a joint surface that are not likely to shear off during movement. Therefore, the effects of waviness do not change with displacements along a joint surface.

Table 1.5 Description of the aperture of discontinuity surfaces.

Description	Width of aperture
Tight	zero
Extremely narrow	less than 2 mm
Very narrow	2–6 mm
Narrow	6–20 mm
Moderately narrow	20–60 mm
Moderately wide	60–200 mm
Wide	over 200 mm

Waviness does not affect the frictional properties of the discontinuity. On the other hand, roughness refers to second order irregularities that are sufficiently small to be sheared off during movement. Increasing roughness of the discontinuity walls increases the effective friction angle along the joint surface. These effects diminish or disappear when infill is present.

The compressive strength of the rock comprising the walls of a discontinuity is a very important component of shear strength and deformability, especially if the walls are in direct rock to rock contact. Weathering (and alteration) frequently is concentrated along the walls of discontinuities, thereby reducing their strength. The weathered material can be assessed in terms of its grade and index tests (see Section 1.6).

Seepage of water through rock masses usually takes place via the discontinuities, although in some sedimentary rocks seepage through the pores also may play an important role. The prediction of groundwater level, probable seepage paths and approximate groundwater pressures frequently indicates construction problems. Seepage from open or filled discontinuities can be assessed qualitatively as shown in Table 1.6.

Joints in a rock mass reduce its effective shear strength at least in a direction parallel with the discontinuities. Hence, the strength of jointed rock masses is highly anisotropic. Joints offer no resistance to tension whereas they offer high resistance to compression. Nevertheless, they may deform under compression if there are crushable irregularities, compressible filling or apertures along the joint or if the wall rock is altered.

Several attempts have been made to relate the numerical intensity of fractures to the quality of unweathered rock masses and to quantify their effect on deformability. For example, the rock quality designation (RQD) is based on the percentage core recovery when drilling rock with NX (57.2 mm) or larger diameter diamond core drills. Assuming that a consistent standard of drilling can be maintained, the percentage of solid core obtained depends on the strength and degree of discontinuities in the rock mass concerned. The RQD is the sum of the core sticks in excess of 100 mm expressed as a percentage of the total length of core drilled. However, the RQD does not take account of the joint opening and condition, a further disadvantage being that with fracture spacings greater than 100 mm the quality is excellent irrespective of the actual spacing (Table 1.7). This particular difficulty can be overcome by using the fracture spacing index, which refers to the frequency per metre that natural fractures occur within a rock mass (Table 1.7). The effect of discontinuities in a rock mass can be estimated by comparing the *in situ* compressional wave velocity with the laboratory sonic velocity of an intact core sample obtained from the rock mass. Any difference in these two velocities is due to the presence of the discontinuities in the field. The velocity ratio (V_{cf}/V_{cl}), where V_{cf} and V_{cl} are the compressional wave velocities of the rock mass *in situ* and of the intact specimen respectively, approaches unity for a high quality massive rock mass with only a few tight joints but as the degree of jointing and fracturing becomes more severe, the velocity ratio is reduced (Table 1.7). An estimate of the numerical value of the deformation modulus of a jointed rock mass can be obtained from various *in situ* tests (see Chapter 8). The values derived from

Table 1.6 Seepage via discontinuities.

Seepage rating	Open discontinuities Description	Filled discontinuities Description
1	The discontinuity is very tight and dry, water flow along it does not appear possible.	The filling material is heavily consolidated and dry, significant flow appears unlikely due to very low permeability.
2	The discontinuity is dry with no evidence of water flow.	The filling materials are damp but no free water is present.
3	The discontinuity is dry but shows evidence of water flow, i.e. rust staining, etc.	The filling materials are wet, occasional drops of water.
4	The discontinuity is damp but no free water is present.	The filling materials show signs of outwash, continuous flow of water (estimate l min^{-1}).
5	The discontinuity shows seepage, occasional drops of water but no continuous flow.	The filling materials are washed out locally, considerable water flow along outwash channels (estimate l min^{-1} and describe pressure, i.e. low, medium, high).
6	The discontinuity shows a continuous flow of water (estimate l min^{-1} and describe pressure, i.e. low, medium, high).	The filling materials are washed out completely, very high water pressures are experienced, especially on first exposure (estimate l min^{-1} and describe pressure).

Table 1.7 Classification of rock quality in relation to the incidence of discontinuities.

Quality classification	RQD (%)	Fracture frequency per metre	Velocity ratio(V_{cf}/V_{cl})	Mass factor(j)
Very poor	0–25	over 15	0.0–0.2	–
Poor	25–50	15–8	0.2–0.4	less than 0.2
Fair	50–75	8–5	0.4–0.6	0.2–0.5
Good	75–90	5–1	0.6–0.8	0.5–0.8
Excellent	90–100	less than 1	0.8–0.10	0.8–0.10

such tests are normally smaller than those determined in the laboratory from intact core specimens and the more heavily the rock mass is jointed, the larger the discrepancy between the two values. Thus, if the ratio between these two values of deformation modulus is obtained from a number of locations on a site, the rock mass quality can be evaluated by the rock mass factor (j), that is, the ratio of the deformability of a rock mass to that of the deformability of the intact rock (Table 1.7).

One of the most widely used methods of collecting discontinuity data is simply by direct measurement on the ground. A direct survey can be carried out subjectively in that only those structures that appear to be important are measured and recorded. In a subjective survey the effort can be concentrated on the apparently significant joint sets. Conversely, in an objective survey all discontinuities intersecting a fixed line (a line scan) or area (fracture set mapping) of the rock face are measured and recorded. In line scanning the distance along a tape at which each discontinuity intersects is noted, as is the direction of

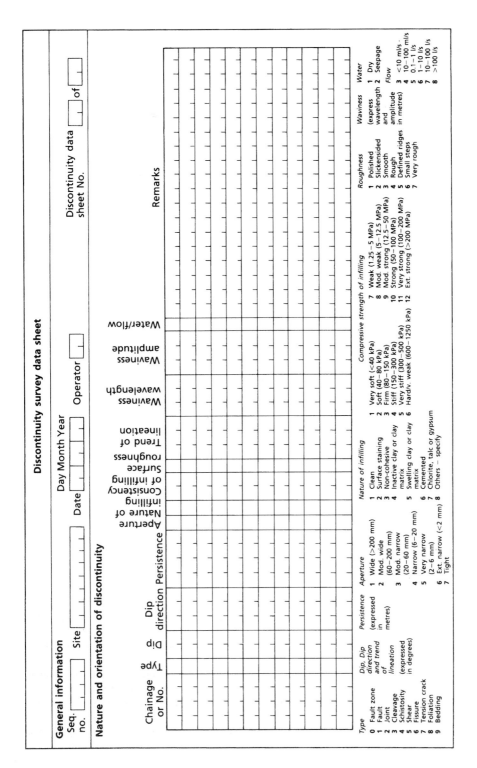

Fig. 1.29 A discontinuity survey chart recommended by the Geological Society of London (courtesuy Geologial Society).

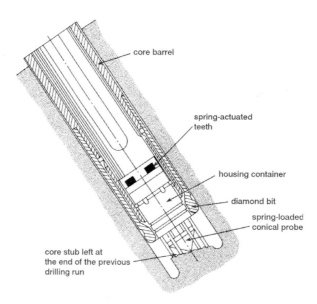

Fig. 1.30 A core orientator.

the pole to each discontinuity (this provides an indication of the dip direction). The dip of the pole from the vertical is recorded as this is equivalent to the dip of the plane from the horizontal. The strike and dip directions of discontinuities in the field can be measured with a compass and the amount of dip with a clinometer. Measurement of the length of a discontinuity provides information on its continuity. Measurements should be taken continuously over distances of about 30 m to ensure that the survey is representative. A minimum of at least 200 readings per locality is required to ensure statistical reliability. A summary of the other details that should be recorded regarding discontinuities is given in Fig. 1.29.

The value of discontinuity data gathered from orientated cores from drill holes depends in part on the quality of the rock concerned, in that poor quality rock is likely to be lost during drilling. However, it is impossible to assess the persistence, degree of separation or the nature of the joint surfaces. What is more, infill material, especially if it is soft, is not recovered by the drilling operations. Core orientation can be achieved by using a core orientator (Fig. 1.30). Integral sampling provides a method of obtaining information on the spacing, opening and infilling of discontinuities (Fig. 1.31). The method has been used with success in all types of rock masses, from massive to highly weathered varieties.

Drill hole inspection techniques can be used to obtain discontinuity data. Such techniques include the use of drill hole periscopes, drill hole cameras or closed-circuit television. The drill hole periscope affords direct inspection and can be orientated from outside the hole. However, its effective use is limited to about 30 m depth below the surface. The drill hole camera can be orientated prior to photographing a section of the wall of a drill hole. The television camera provides a direct view of the drill hole and a recording can be made on videotape. These three systems are limited in that they require relatively clear conditions and so may be of little use below the water table, particularly if the water in a drill hole is murky. The televiewer produces an acoustic picture of the drill hole wall. One of its advantages is that drill holes need not be flushed prior to its use.

Much data relating to discontinuities can be obtained from photographs of exposures. Photographs may be taken looking horizontally at a rock mass from the ground or they may be taken from the air

34

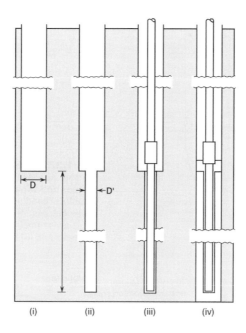

Fig. 1.31 Stages of integral sampling: (i) start position (ii) drilling hole for connecting rod (iii) fixing connecting rod (iv) overcoring for integral sample.

looking vertically, or occasionally obliquely, down at the outcrop. These photographs may or may not have survey control. Uncontrolled photographs are taken using hand-held cameras. Stereo-pairs are obtained by taking two photographs of the same face from positions about 5% of the distance from the face apart, along a line parallel to the face. Delineation of major discontinuity patterns and preliminary subdivision of the face into structural zones can be made from these photographs. Unfortunately, data cannot be transferred with accuracy from them onto maps and plans. On the other hand, discontinuity data can be accurately located on maps and plans by using controlled photographs. Controlled photographs are obtained by aerial photography with complementary ground control or by ground-based phototheodolite surveys. Aerial photographs, with a suitable scale, have proved useful in the investigation of discontinuities. Photographs taken with a phototheodolite can be used with a stereo-comparator that produces a stereoscopic model. Measurements of the locations or points in the model can be made with an accuracy of approximately 1 in 5000 of the mean object distance. As a consequence, a point on a face photographed from 50 m can be located to an accuracy of 10 mm. In this way the frequency, orientation and continuity of discontinuities can be assessed. Such techniques prove particularly useful when faces that are inaccessible or unsafe have to be investigated.

Data from a discontinuity survey usually are plotted on a stereographic projection. The use of spherical projections, commonly the Schmidt or Wulf net, means that traces of the planes on the surface of the 'reference sphere' can be used to define the dips and dip directions of discontinuities. In other words, the inclination and orientation of a particular plane is represented by a great circle or a pole, normal to the plane, which are traced on an overlay placed over the stereonet (Fig. 1.32a). When recording field observations of the direction and amount of dip of discontinuities it is convenient to plot the poles rather than the great circles. The poles then can be contoured in order to provide an expression of orientation concentration. This affords a qualitative appraisal of the influence of the discontinuities on the engineering behaviour of the rock mass concerned (Fig. 1.32b).

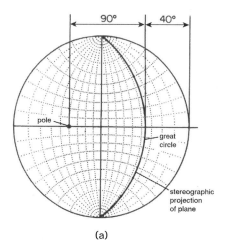

(a)

Fig. 1.32 (a) Stereonet with plane and pole plotted. (b) Main types of slope failure in rock masses and appearance of stereoplots of structural conditions likely to give rise to these failures. (From Hoek and Bray, 1981, courtesy of the Institute of Materials, Minerals and Mining.)

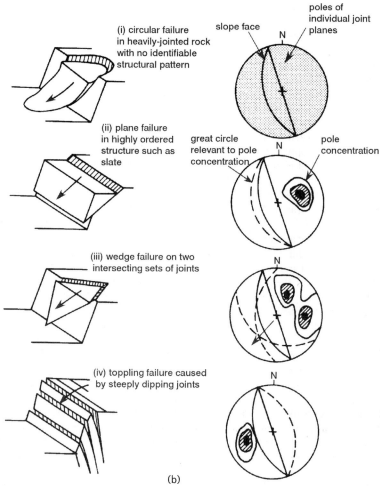

(i) circular failure in heavily-jointed rock with no identifiable structural pattern

(ii) plane failure in highly ordered structure such as slate

(iii) wedge failure on two intersecting sets of joints

(iv) toppling failure caused by steeply dipping joints

(b)

1.6 Weathering

Weathering of rocks is brought about by physical disintegration, chemical decomposition and biological activity. The agents of weathering, unlike those of erosion, do not provide transportation of debris from a rock surface. Therefore, unless this rock waste is otherwise removed it eventually acts as a protective cover, preventing further weathering from taking place. If weathering is to be continuous, fresh rock exposures must be constantly revealed, which means that the weathered debris must be removed by the action of gravity, running water, wind or moving ice.

1.6.1 Types of weathering

Mechanical or physical weathering is particularly effective in climatic regions that experience significant diurnal changes of temperature. This does not necessarily imply a large range of temperature, as freeze-thaw action can proceed where the range is limited.

As far as frost susceptibility is concerned the porosity, pore size and degree of saturation all play an important role. When water turns to ice it increases in volume by up to 9% thus giving rise to an increase in pressure within the pores, especially of saturated rock. This action is further enhanced by the displacement of pore water away from the developing ice front. Once ice has formed the ice pressures rapidly increase with decreasing temperature, so that at approximately $-22°C$ ice can exert a pressure of up to 200 MPa. Usually, coarse-grained rocks withstand freezing better than fine-grained types. The critical pore size for freeze-thaw durability appears to be about 0.005 mm. In other words, rocks with larger mean pore diameters allow outward drainage and escape of fluid from the frontal advance of the ice line and are therefore less frost susceptible. Alternate freeze-thaw action causes cracks, fissures, joints and some pore spaces to be widened. As the process advances, angular rock debris gradually is broken from the parent body.

The mechanical effects of weathering are well displayed in hot deserts, where wide diurnal ranges of temperature cause rocks to expand and contract. Because rocks are poor conductors of heat these effects are mainly localized in their outer layers where alternate expansion and contraction creates stresses that eventually rupture the rock. In this way flakes of rock break away from the parent material, the process being termed exfoliation. The effects of exfoliation are concentrated at the corners and edges of rock masses so that their outcrops gradually become rounded.

Chemical weathering leads to mineral alteration and the solution of rocks. Alteration is principally effected by oxidation, hydration, hydrolysis and carbonation whilst solution is brought about by acidified or alkalized waters. Chemical weathering also aids rock mass disintegration by weakening the rock fabric and by emphasizing any structural weaknesses, however slight, that it possesses (Fig. 1.33). When decomposition occurs within a rock, the altered material frequently occupies a greater volume than that from which it was derived and in the process internal stresses are generated. If this swelling occurs in the outer layers of a rock mass, then it eventually causes them to peel off from the parent body. The increase in volume that occurs in a rock mass on weathering leads to a reduction in its density with corresponding increase in porosity and loss of strength.

In dry air rocks decay very slowly. The presence of moisture hastens the rate tremendously, firstly, because water is itself an effective agent of weathering and, secondly, because water holds in solution substances that react with the component minerals of the rock. The most important of these substances are free oxygen, carbon dioxide, organic acids and nitrogen acids. Free oxygen is an important agent in the decay of all rocks that contain oxidizable substances, iron and sulphur being especially suspect. The rate of oxidation is quickened by the presence of water; indeed it may enter into the reaction itself, as for example, in the formation of hydrates. However, its role is chiefly that of a catalyst. Carbonic acid is produced when carbon dioxide is dissolved in water and it may possess a pH value of about 5.7. The principal source of carbon dioxide is not the atmosphere but the air contained in the pore spaces in the soil where its proportion may be one hundred or so times greater than it is in the atmosphere. An abnormal concentration of carbon dioxide is released when organic material decays. Furthermore, humic acids are formed by the decay of humus in soil

Fig. 1.33 Spheroidal weathering in basalt, Tideswell, Derbyshire, England.

waters; they ordinarily have pH values between 4.5 and 5.0 but occasionally they may be less than 4.0.

The simplest reactions that take place on weathering are the solution of soluble minerals and the addition of water to substances to form hydrates. Solution commonly involves ionization, for example, this takes place when gypsum and carbonate rocks are weathered. Hydration and dehydration take place amongst some substances, a common example being gypsum and anhydrite:

$$CaSO_4 + 2H_2O = CaSO_4 2H_2O$$

$$\text{(anhydrite)} \qquad \text{(gypsum)}$$

The above reaction produces an increase in volume of approximately 6% and accordingly causes the enclosing rocks to be wedged further apart. Iron oxides and hydrates are conspicuous products of weathering, usually the oxides are a shade of red and the hydrates yellow to dark brown.

Sulphur compounds are readily oxidized by weathering. Because of the hydrolysis of the dissolved metal ion, solutions forming from the oxidation of sulphides are acidic. For instance, when pyrite is oxidized initially, ferrous sulphate and sulphuric acid are formed. Further oxidation leads to the formation of ferric sulphate. Very insoluble ferric oxide or hydrated oxide is formed if highly acidic conditions are produced.

Limestones are composed chiefly of calcium carbonate and they are suspect to acid attack because CO_3 readily combines with H to form the stable bicarbonate, HCO_3:

$$CaCO_3 + H_2CO_3 = Ca(HCO_3)_2$$

In water with a temperature of 25°C the solubility of calcium carbonate ranges from 0.01 to 0.05 g l^{-1}, depending upon the degree of saturation with carbon dioxide. Dolostone is somewhat less soluble than limestone.

Weathering of the silicate minerals is primarily a process of hydrolysis. Much of the silica that is released by weathering forms silicic acid but where it is liberated in large quantities some of it may form colloidal or amorphous silica. Mafic silicates usually decay more rapidly than felsic silicates and in the process they release magnesium, iron and lesser amounts of calcium and alkalis. Olivine is particularly unstable, decomposing to form serpentine, which on further weathering forms talc and carbonates. Chlorite is the commonest alteration product of augite (the principal pyroxene) and of hornblende (the principal amphibole).

When subjected to chemical weathering feldspars decompose to form clay minerals, the latter are consequently the most abundant residual products. The process is effected by the hydrolyzing action of weakly carbonated water that leaches the bases out of the feldspars and produces clays in colloidal form. The alkalis are removed in solution as carbonates from orthoclase (K_2CO_3) and albite (Na_2CO_3), and as bicarbonate from anorthite ($Ca(HCO_3)_2$). Some silica is hydrolyzed to form silicic acid. Although the exact process is not fully understood the equation given below is an approximation towards the truth:

$$2KAlSi_3O_6 + 6H_2O + CO_2 = Al_2Si_2O_5(OH)_4 + 4H_2SiO_4 + K_2CO_3$$

(orthoclase)   (kaolinite)

The colloidal clay eventually crystallizes as an aggregate of minute clay minerals.

Clays are hydrated aluminium silicates and when they are subjected to severe chemical weathering in humid tropical regimes they break down to form laterite or bauxite. The process involves the removal of siliceous material and this again is brought about by the action of carbonated waters. Intensive leaching of soluble mineral matter from surface rocks takes place during the wet season. During the subsequent dry season groundwater is drawn to the surface by capillary action and minerals are precipitated there as the water evaporates. The minerals generally consist of hydrated peroxides of iron, and sometimes of aluminium, and very occasionally of manganese. The precipitation of these insoluble hydroxides gives rise to an impermeable lateritic soil. When this point is reached the formation of laterite ceases as no further leaching can occur. As a consequence lateritic deposits are usually less than 7 m thick.

Plants and animals play an important role in the breakdown and decay of rocks, indeed their part in soil formation is of major significance. Tree roots penetrate cracks in rocks and gradually wedge the sides apart whilst the adventitious root system of grasses breaks down small rock fragments to particles of soil size. Burrowing rodents also bring about mechanical disintegration of rocks. The action of bacteria and fungi is largely responsible for the decay of dead organic matter. Other bacteria are responsible, for example, for the reduction of iron or sulphur compounds.

1.6.2 Tests of durability and weathering grade

A number of tests have been devised in an attempt to assess the durability of rocks. One that often is used is the slake-durability test, which estimates the resistance to wetting and drying, particularly of mudstones and rocks that have undergone a certain degree of alteration. The sample is placed in a test drum, oven dried and then weighed, after which drum and sample are half immersed in a tank of water, attached to a rotor arm and rotated for 10 min (Fig. 1.34). The cylindrical periphery of the drum is formed of 2 mm sieve mesh so that broken down material is lost whilst the test is in progress. After slaking, the drum and the material retained are dried and weighed. The slake-durability index is obtained by dividing the weight of the sample retained by its original weight and expressing the answer as a percentage. The following scale of slake-durability is used:

Very low	under 25%
Low	25–50%
Medium	50–75%
High	75–90%
Very high	90–95%
Extremely high	over 95%

Failure of consolidated and poorly cemented rocks occurs during saturation when the swelling pressure (or internal saturation swelling stress σ_s), developed by capillary suction pressures exceeds their tensile strength. An estimate of σ_s can be obtained from the modulus of deformation (E).

$$E = \sigma_s/\varepsilon_D \tag{1.1}$$

where ε_D is the free-swelling coefficient. The latter is determined by a sensitive dial gauge recording the amount of swelling of an oven-dried core specimen along the vertical axis during saturation in water for 12 hours, ε_D being obtained as follows:

Fig. 1.34 Slake-durability apparatus.

$$\varepsilon_D = \frac{\text{Change in length after swelling}}{\text{Initial length}} \tag{1.2}$$

The geodurability classification is based on the free-swelling coefficient and uniaxial compressive strength (Fig. 1.35).

Several attempts have been made to devise engineering classifications of weathered rock masses. This has been done by attempting to assess the grade of weathering by reference to some simple index test. When coupled with a grading system, this means that the disadvantages inherent in these simple index tests are largely overcome. For instance, a coefficient of weathering (K) has been based upon the ultrasonic velocities of the rock material according to the expression:

$$K = (V_u - V_w)/V_u \tag{1.3}$$

where V_u and V_w are the ultrasonic velocities of the fresh and weathered rock respectively. The grade of weathering as determined from the ultrasonic velocity is as follows:

Grade of weathering	Ultrasonic velocity (ms^{-1})	Coefficient of weathering
Fresh	over 5000	0
Slightly weathered	4000 – 5000	0 – 0.2
Moderately weathered	3000 – 4000	0.2 – 0.4
Strongly weathered	2000 – 3000	0.4 – 0.6
Very strongly weathered	under 2000	0.6 – 1.0

The quick absorption, Schmidt hammer, and point load strength tests also have been used to assess the grade of weathering (Table 1.8).

Petrographic techniques have been used to evaluate successive stages in mineralogical and textural changes brought about by weathering. The factors used in such an assessment can include the amount of discoloration, decomposition and disintegration shown by the rock mass. In a microscopic analysis, the mineral composition, degree of alteration and frequency of microfractures can be used.

An alternative means of assessing the grade of weathering is based on a simple description of the geological character of the rock mass concerned as seen in the field, the description embodying different grades of weathering that are related to engineering performance (Fig. 1.36). Most such classifications of weathered rock have been based on the degree of chemical decomposition exhibited by a rock mass and primarily directed towards weathering in granitic rocks. Nonetheless, classifications of other weathered rock types have been developed. Usually, the grades lie one above the other in a weathered profile developed from a single rock type, the highest grade being at the surface. This, however, is not necessarily the case in complex geological conditions. Such classifications can be used to produce maps showing the distribution of the grade of weathering at a particular engineering site.

Table 1.8 Weathering indices for granite.

Type of weathering	Quick absorption(%)	Bulk density (Mg m⁻³)	Point-load strength(MPa)	Unconfined compressive strength(MPa)
Fresh	Less than	Over 2.61	Over 10	Over 250
Partially stained*	0.2	2.56–2.61	6–10	150–250
Completely stained*	0.2–1.0	2.51–2.56	4–6	100–150
Moderately weathered	1.0–2.0	2.05–2.51	0.1–4	2.5–100
Highly/completely weathered	2.0–10.0 over 10	Less than 2.05	Less than 0.1	Less than 2.5

*Slightly weathered

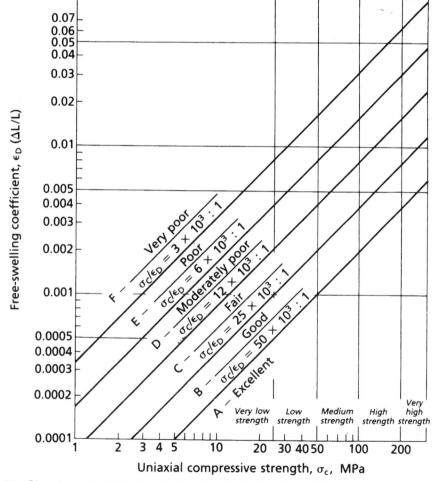

Fig. 1.35 Olivier's geodurability classification of mudrocks (courtesy of Elsevier).

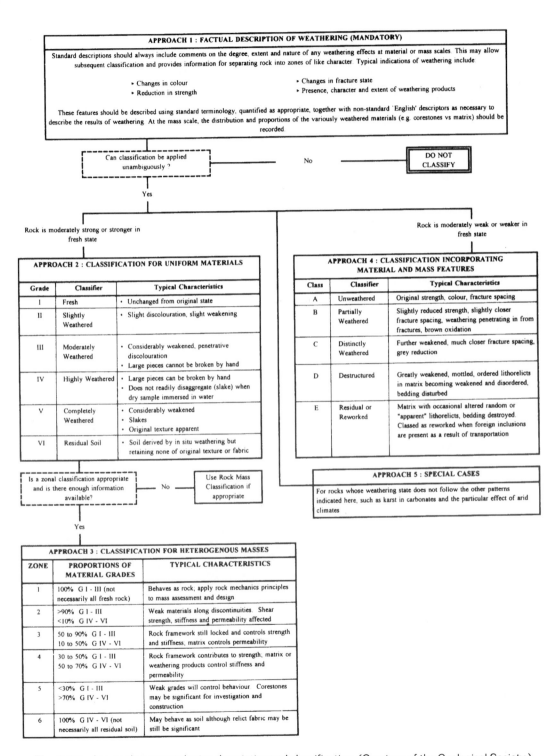

Fig. 1.36 Approaches to weathering description and classification. (Courtesy of the Geological Society.)

2
Geology and Planning

2.1 Introduction

Human activities and land uses are spatially distributed, with some locations being very specifically determined. For instance, the location of mineral resources determines where mining may occur, if a deposit is economically viable or strategically necessary. Conversely, many land uses may have no specific connection with locality and so may be situated in different places with similar suitability. Therefore, land-use planning must seek to establish acceptable criteria for the location of each activity. So the problem of location of each activity may involve competing claims from different areas in order to bring about an overall optimization. Of course, the same area of land may be suitable for different uses and so its use has to be resolved by the planning process.

Land-use planning represents an attempt to reduce the number of conflicts and adverse environmental impacts both in relation to society and nature. In the first instance, land-use planning involves the collection and evaluation of relevant data from which plans can be formulated. The policies that result depend on economic, sociological and political influences in addition to the physical factors involved in planning. In the geological context, sufficient geological data should be provided to planners and engineers so that, ideally, they can develop the environment in harmony with nature. Indeed, geological information is required at all levels of planning and development from the initial identification of a social need to the construction stage. Even after construction, further involvement may be necessary in the form of advice on hazard monitoring, maintenance, or remedial works.

The importance of geology in planning physical facilities and individual structures cannot be over-stressed since land is the surface expression of underlying geology. Consequently, land-use planning can only be effective if there is a proper understanding of the geology concerned. In addition, the development of land must be planned with the full realization of the natural forces that have brought it to its present state and take into account the dynamic character of nature so that development does not upset the delicate balance any more than is essential. Geology therefore should be the starting point of all planning. Accordingly, it is important that geological data should be readily appreciated by the planner, developer and civil engineer.

Maps are one of the principal ways of representing geological data. They are a means of storing and transmitting information, in particular, of conveying specific information about the spatial distribution of given factors or conditions. Unfortunately, the conventional geological map often is inadequate for the needs of planners, developers and civil engineers. Recently, however, various types of maps have been developed for planning purposes that incorporate geological data. Such maps include morphological maps, engineering geomorphological maps, environmental geological maps and engineering geological maps (see Section 2.2). They are simple maps that provide an indication of areas where there are (or are not) geological constraints on development. They also can indicate resources with mineral, groundwater or agricultural potential that might be used in development or that should not be sterilized by building over.

Many older industrial and urban areas are undergoing redevelopment and therefore there is a need to plan the new environment so that land with difficult ground conditions is avoided by building development – or the cost of building on such land is fully appreciated. Again, planners require relevant information. Data from site investigations frequently are available in many urban areas and can be used to prepare thematic geological maps for planners. Other methods of using, evaluating and portraying geological information for planning purposes include geographical information systems (see Section 2.3).

Since land use inevitably involves the different development of particular areas, some type of land classification(s) constitutes the basis on which land-use planning is carried out. However, land should also be graded according to its potential uses and capabilities (see land capability analysis, Chapter 5). In other words, indices are required to assess the environmental status of natural resources and their potential. Such indices should establish limits, trends and thresholds, as well as providing insight that offers some measure of success in relation to national and local programmes that are addressing environmental problems.

Laws are now in force in many countries that require the investigation and evaluation of potential consequences to the environment as a result of large building projects. An environmental impact assessment therefore usually involves a description of the proposed scheme, its impact on the environment, particularly noting any adverse effects, and alternatives to the proposed scheme. The aim of the environmental impact process is to improve the effectiveness of existing planning policy. It should bring together the requirements of the development and the constraints of the environment so that conflicts can be minimized. As such, an environmental impact assessment must provide objective information on which decisions can be made. The information it contains must be comprehensive so that, after analysis and interpretation, it not only informs but also helps direct development planning.

Checklists are frequently used in environmental assessment. These comprehensive lists relate to specific environmental parameters and specific human actions. They facilitate data gathering and presentation, help to order thinking and alert against omission of possible environmental impacts. The matrix approach greatly expands the scope of checklists. Such an approach may have one data set related to environmental parameters and the other to human actions with the two sets forming a cross-tabulation. A matrix system of analysis can act as a super checklist for those involved in the preparation of environmental impact assessments and also tends to make assessment more objective. The approach involves quantification of data, establishment of cause and effect relationships, and weighting of impacts. The matrix approach can highlight areas of particular concern or those where further investigation is necessary. The method is adaptable. All cells in the matrix that represent a possible impact are evaluated individually, being scored in terms of the magnitude of the impact and its social significance. This helps in the assessment of the risk or uncertainty related to a particular impact.

One of the aspects of planning that intimately involves geology is the control or prevention of geological hazards that militate against the interests of humans (see Section 2.4). The development of planning policies for dealing with hazards in particular, requires an assessment of the severity, extent and frequency of the hazard in order to evaluate the degree of risk. Both the probable intensity and frequency of the hazard(s), and the susceptibility (or probability) of damage to human activities in the face of such hazards are integral components of risk assessment. Vulnerability analyses comprising risk identification and evaluation should be carried out in order to make rational decisions on how best to reduce or overcome the effects of potentially disastrous events through systems of permanent controls on land development. The geological data needed in planning and decision making are given in Table 2.1.

Once the risk has been assessed methods whereby the risk can be reduced have to be investigated and evaluated in terms of public costs and benefits (see Section 2.5). The risks associated with geological hazards may be reduced, for instance, by control measures carried out against the hazard-producing agent; by pre-disaster community preparedness, monitoring, surveillance and warning systems that allow evacuation; by restrictions on development of land; and by the use of appropriate building codes together with structural reinforcement of property. In addition, the character of the ground conditions can affect both the viability and implementation of planning proposals. The incorporation of geological information into planning processes should mean that proposals can be formulated that do not conflict with the ground conditions present.

Table 2.1 Data required to reduce losses from geological hazards*.

Reduction decisions	Technical information needed about the hazards from earthquakes, floods, ground failures, and volcanic eruptions
Avoidance	Where has the hazard occurred in the past? Where is it occurring now? Where is it predicted to occur in the future? What is the frequency of occurrence?
Land-use zoning	Where has the hazard occurred in the past? Where is it occurring now? Where is it predicted to occur in the future? What is the frequency of occurrence? What is the physical cause? What are the physical effects of the hazard? How do the physical effects vary within an area? What zoning within the area will lead to reduced losses to certain types of construction?
Engineering design	Where has the hazard occurred in the past? Where is it occurring now? Where is it predicted to occur in the future? What is the frequency of occurrence? What is the physical cause? What are the physical effects of the hazard? How do the physical effects vary within an area? What engineering design methods and techniques will improve the capability of the site and the structure to withstand the physical effects of a hazard in accordance with the level of acceptable risk?
Distribution of losses	Where has the hazard occurred in the past? Where is it occurring now? Where is it predicted to occur in the future? What is the frequency of occurrence? What is the physical cause? What are the physical effects of the hazard? How do the physical effects vary within an area? What zoning has been implemented in the area? What engineering design methods and techniques have been adopted in the area to improve the capability of the structure to withstand the physical effects of a hazard in accordance with the level of acceptable risk? What annual loss is expected in the area? What is the maximum probable annual loss?

*Hays, W.W. and Shearer, C.F. 1981. Suggestions for improving decision making to face geologic and hydrologic hazards. In *Facing Geologic and Hydrologic Hazards – Earth Science Considerations*, Hays, W.W. (Ed.), United States Geological Survey, Professional Paper 1240-B, Washington, D.C

2.2 Applied geological maps

Both topographical and geological maps (together with accompanying memoirs) contain much basic data that can be used for a feasibility project, or during the planning stage of a construction operation when a decision regarding the choice of sites has to be made. However, from the point of view of the planner or engineer the shortcomings of conventional geological maps are that the boundaries are stratigraphical and more than one type of rock may be included in a single mappable unit. What is more, a geological map is lacking in quantitative information concerning data like the physical properties of the rocks and soils represented, the nature of the discontinuities, the amount and degree of weathering, the hydrogeological conditions, etc. However, special maps can be prepared from a geological map. For example, maps that can interpret the geology in terms of potential supplies of construction materials. They can show the spatial distribution of each kind of material, its topographical position and accessibility to existing means of transportation. Cross-sections, symbols and abbreviated logs can show the approximate thickness of each unit and whether it is at or near the surface. A geological map also can be used to indicate those rocks that should be investigated as potential producers of groundwater. Some idea of the thickness of potential aquifers and the depth at which they occur can be obtained from sections. Perhaps some idea of the quality of the water may be gleaned from the type of rock forming an aquifer, as for example, hard water from a limestone aquifer.

Various types of geological maps are now produced by some national geological surveys with a view to aiding planning and development. These include various types of hazard maps, engineering geomorphological maps, environmental geological maps and engineering geological maps.

2.2.1 Engineering geomorphological maps

Engineering geomorphological mapping involves the recognition of landforms along with their delimitation in terms of size and shape. The maps produced therefore enable a rapid appreciation of the nature of the ground and so help the design of more detailed investigations, and are of value as far as land-use planning and construction projects are concerned. Identification of the characteristics of the terrain of an area affords a basis for evaluation of potential locations for development and for avoidance of the worst hazard areas. The recognition of both the interrelationships between landforms on site and their individual or combined relationships to landforms beyond the site is fundamental. This is necessary in order to appreciate not only how the site conditions will affect the project but, just as importantly, how the project will affect the site and the surrounding environment. In particular, knowledge of ground hazards is required for good design of a project. What is more, an understanding of the past and present geomorphological development of an area is likely to aid prediction of its behaviour during and after any construction operations. Accordingly, engineering geomorphological maps should show how surface expression will influence a project and should provide an indication of the general environmental relationship of the area concerned.

The amount of information that can be obtained from a literature survey, which includes the various types of maps and aerial photographs available, varies with location. In some developing countries little or nothing may be available, even worthwhile topographical maps, which are normally a prerequisite for a geomorphological mapping programme, may not exist. Base maps in such instances, can be made from aerial photographs, which can be specially commissioned. Much preliminary data can be obtained from aerial photographs, for instance, they allow many of the significant landforms and their boundaries to be defined prior to the commencement of fieldwork. The scale of the photographs is usually 1:10 000. Field mapping permits the correct identification of landforms, recognized on aerial photographs, as well as geomorphological processes, and indicates how they will affect the project. Mapping a site provides data on the nature of the surface materials. Further precision can be afforded geomorphological interpretations by obtaining details from climatic, hydrological or other records and by analysis of the stability of landforms.

The scale of an engineering geomorphological map is influenced by the project requirement and the map should focus attention on the information relevant to the particular project. Maps produced for

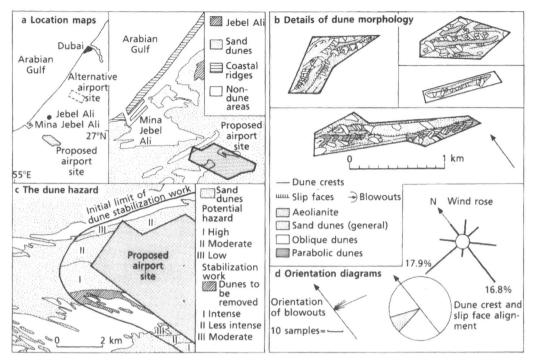

Fig. 2.1 Geomorphological analysis of a proposed airport site in Dubai with respect to the threat from mobile sand dunes (Courtesy of the Geological Society of London.)

extended areas, such as needed for route selection, are drawn at a small scale. Small-scale maps can be used for planning purposes, land use evaluation, land reclamation, flood plain management, coastal conservation etc. General engineering geomorphological maps concentrate on portraying the form, origin, age and distribution of landforms, along with their formative processes, rock type and surface materials. In addition, if information is available, details of the actual frequency and magnitude of the processes can be shown by symbols, annotation, accompanying notes or successive maps of temporal change. On the other hand, large scale maps and plans of local surveys provide an accurate portrayal of surface form, drainage characteristics and the properties of surface materials, as well as an evaluation of currently active processes. Derivative maps can be compiled from the geomorphological sheets. Such derivative maps generally are concerned with some aspect of ground conditions, for example, landslip areas, areas prone to flooding or land over which sand dunes migrate (Fig. 2.1).

2.2.2 Environmental geological maps

Environmental geological maps have been produced to meet the needs of planners and incorporate geological data that can be of value in terms of planning and land-use purposes. Such maps are essentially simple and provide some indication of those areas where there are least geological constraints on development and so they can be used by the planner or engineer at the feasibility stage of a project (Fig. 2.2). The location of exploitable mineral resources also is of interest to planners. The obvious reason for presenting geological data in a way that can be understood by planners, administrators and engineers is that they then can seek appropriate professional advice, and in this way bring about safer and more cost effective development and design of land, especially in relation to urban growth and redevelopment. In fact, two types of environmental geological maps may be produced, one for the specialist and the other for the non-specialist.

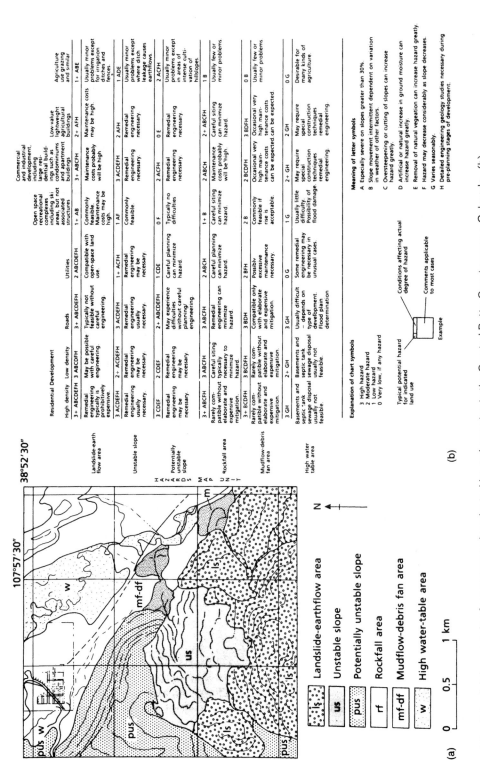

Fig. 2.2 (a) An example of environmental geological hazard mapping in the Crested Butte-Gunnison area, Colorado. (b) The matrix indicates that several geological and geology related factors should be considered when considering the land use in the area concerned. (Courtesy of the IAEG.)

Topics that are included on environmental geology maps vary but may include solid geology, unconsolidated deposits, geotechnical properties of soils and rocks, landslides, depth to rockhead, hydrogeology, mineral resources, shallow undermining and opencast workings, floodplain hazards, etc. Indeed, each aspect of geology can be presented as a separate theme on a basic or element map. Derivative maps display two or more elements combined to show, for example, foundation conditions, ease of excavation, aggregate potential, landslide susceptibility, subsidence potential, groundwater resources or capability for solid waste disposal. Environmental-potential maps are compiled from basic data maps and derived maps. They present, in general terms, the constraints on development such as areas where land is susceptible to land-slipping or subsidence, or land likely to be subjected to flooding. They can also indicate resources that might be used in development or even contamination from landfill sites. Again, these maps should be capable of being readily understood by non-geologists. Reports containing comprehensive suites of environmental geological maps are now produced that can be used by local and central planners. They represent useful sources of information for civil engineers, developers, and mineral extraction companies. Such maps should help planners avoid making poor decisions at an early stage and allow areas to be developed sequentially (e.g. mineral resources can be developed before an area is built over). In addition, they can facilitate the development of a multi-use plan of a particular area and avoid one factor adversely affecting another.

Environmental geological maps have their limitations, which stem from a number of factors such as the scale of presentation, and the availability and reliability of data. Like all geological maps, environmental geological maps are transitory and inevitably need amendment, as new or previously unavailable data become accessible. Unfortunately, erroneous conclusions can be drawn by those who are not aware of these shortcomings. This is particularly the case with regard to ground stability in urban areas. In such areas, properties that have been constructed adequately to take account of ground conditions when they were built can be blighted if it is later revealed that they are located on ground that may be suspect – for example, it may have been undermined by workings long since abandoned. Hence, it may become more difficult to obtain insurance cover or mortgages for such properties, and their values may decline. In the United States there can be legal implications involved in the production of such maps so that they have to carry disclaimers. Consequently, every attempt must be made in the preparation of thematic maps to ensure that the finished product cannot be misconstrued. The limitations on their accuracy, and the interpretation and use of data in their production must be made clear.

2.2.3 Engineering geological maps

An engineering geological map is produced from information collected from various sources (literature survey, aerial photographs and imagery, and fieldwork). Information from site investigation reports, records of past and present mining activity, successive editions of topographical maps, etc. may all prove extremely useful. Once the data have been gathered, there then follow the questions of how the data should be represented on the map and at which scale should the map be drawn. The latter is influenced very much by the requirement. The major differences between maps of different scales are, first, the amount of data they show (the more detailed a map needs to be, the larger its scale) and, second, the manner in which it is presented. Engineering geological maps usually are produced at a scale of 1:10 000 or smaller whereas engineering geological plans, being produced for particular engineering purposes, have larger scales. As far as presentation is concerned, this may involve not only the choice of colours and symbols, but also the use of overprinting. Overprinting frequently takes the form of striped or stippled shading, both of which can be varied, for instance, according to frequency, pattern, dimension or colour.

A map represents a simplified model of the facts and the complexity of various geological factors can never be portrayed in their entirety. The amount of simplification required is governed principally by the purpose and scale of the map; the relative importance of particular engineering geological factors or relationships; the accuracy of the data; and the techniques of representation employed. Engineering geological maps should be accompanied by cross-sections, and an explanatory text and legend. More than one map of an area may be required to record all the information that has been collected during a survey. In such instances, a series of overlays or an atlas of maps can be produced. Preparation of a series of engineering geological maps can reduce the amount of effort involved in the preliminary stages of a site investigation, and indeed may allow site investigations to be designed for the most economical confirmation of the ground conditions.

Engineering geological maps may be geological maps to which engineering geological data have been added. If the engineering information is extensive, then it can be represented in tabular form, perhaps accompanying the map on the reverse side. The superficial deposits and bedrock should be described in detail. Where possible rocks and soils should be classified according to their engineering behaviour. Details of geological structures should be recorded, especially faults and shear zones, as should the nature of the discontinuities and grade of weathering where appropriate.

Engineering geological maps may serve a special purpose or be multipurpose. Special purpose maps provide information on one specific aspect of engineering geology such as grade of weathering, jointing patterns, mass permeability or foundation conditions. Alternatively, special purpose maps may serve one particular purpose, as for instance, the engineering geological conditions at a dam site or along a routeway, or for zoning for land use in urban development. In addition, engineering geological maps may be analytical or comprehensive. Analytical maps provide details, or evaluate individual components, of the geological environment. Examples of such maps include those indicating degree of weathering or seismic hazard. Comprehensive maps either depict all the principal components of the engineering geological environment or are maps of engineering geological zoning, delineating individual units on a basis of uniformity of the most significant attributes of their engineering geological character. Multipurpose maps cover various aspects of engineering geology and provide general information for planning or engineering purposes, providing an impression of the geological environment, surveying the range and type of engineering geological conditions, their individual components and their interrelationships (Fig. 2.3). These maps provide planners and engineers with information that will assist them in land-use planning and the location, construction and maintenance of engineering structures of all types.

Alternatively, the rocks and soils in the area concerned may be presented as mapped units defined in terms of engineering properties. This type of map often is referred to as a geotechnical map. In such cases, the unit boundaries are drawn for changes in a particular property. Frequently, the boundaries of such units coincide with stratigraphical boundaries. In other instances, as for example, where rocks are deeply weathered, they may bear no relation to geological boundaries. Unfortunately, one of the fundamental difficulties in preparing such maps arises from the fact that changes in the physical properties of rocks and soils frequently are gradational. As a consequence, regular checking of visual observations by *in situ* testing or sampling is essential to produce a map based on engineering properties.

There are two basic types of geotechnical plan, namely, the site investigation plan, and the construction or foundation stage plan. In the case of the former type the scale varies from 1:5000 to as large as 1:500 or even 1:100, depending on the size and nature of the site and the engineering requirement. The foundation plan records the ground conditions exposed during construction operations. It may be drawn to the same scale as the site investigation plan or the construction drawings. Plans may be based on large-scale topographic maps or large-scale base maps produced by surveying or photogrammetric methods.

2.3 Geographical information systems

One means by which the power, potential and flexibility of mapping may be increased is by developing a geographical information system. Geographical information systems (GIS) represent a form of technology that is capable of capturing, storing, retrieving, editing, analyzing, comparing and displaying spatial environmental information. In other words, a geographical information system consists of four fundamental components, namely, data acquisition and verification, data storage and manipulation, data transformation and analysis, and data output and presentation. The GIS software is designed to manipulate spatial data in order to produce maps, tabular reports or data files for interfacing with numerical models. An important feature of a GIS is the ability to generate new information by the

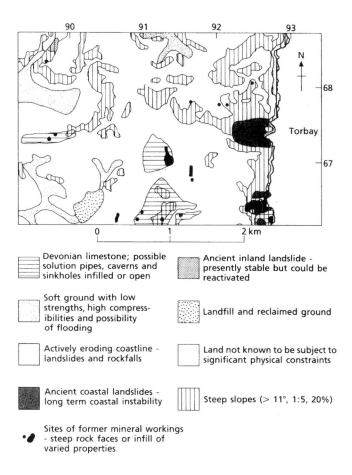

Fig. 2.3 Ground characteristics for planning and development for part of the Torbay area of South West England.

integration of existing diverse data sets sharing a compatible referencing system. Data can be obtained from remote sensing imagery, aerial photographs, aero-magnetometry, gravimetry, various types of maps, surveys and investigations. These data are recorded in a systematic manner in a computer database. Each type of data input refers to the characteristics of recognizable point, linear or areal geographical features. Details of the features usually are stored in either vector (points, lines and polygons) or raster (grid cell) formats. The manipulation and analysis of data allows it to be combined in various ways to evaluate what will happen in certain situations. The advantages of using GIS compared with conventional spatial analysis techniques are summarized in Table 2.2.

An ideal GIS for many environmental/engineering geological situations combines conventional GIS procedures with image processing capabilities and a relational database. Because frequent map overlaying, modelling, and integration with scanned aerial photographs and satellite images are required, a raster system is preferred. The system should be able to perform spatial analysis on multiple-input maps and connected attribute data tables. Necessary GIS functions include map overlay, reclassification, and a variety of other spatial functions incorporating logical, arithmetic, conditional, and neighbourhood

Table 2.2 Advantages and disadvantages of GIS.

Advantages	Disadvantages
1. A much larger variety of analytical techniques is available. Because of the speed of calculation, complex techniques requiring a large number of map overlays and table calculations become feasible.	1. A large amount of time is needed for data entry. Digitizing is especially time consuming.
2. It is possible to improve models by evaluating their results and adjusting the input variables. Users can achieve the optimum results by a process of trial and error, running the models several times, whereas it is difficult to use these models even once in the conventional manner. Therefore, more accurate results can be expected.	2. There is a danger in placing too much emphasis on data analysis as such at the expense of data collection and manipulation based on professional experience. A large number of different techniques of analysis are theoretically possible, but often the necessary data are missing. In other words, the tools are available but cannot be used because of the lack of certainty of input data.
3. In the course of a hazard assessment project, the input maps derived from field observations can be updated rapidly when new data are collected. Also, after completion of the project, the data can be used by others in an effective manner.	

operations. In many cases modelling requires the iterative application of similar analyses using different parameters. Consequently, the GIS should allow for the use of batch files and macros to assist in performing these iterations.

Geographical information systems have been used for geological hazard and vulnerability assessment, so producing hazard-susceptibility maps (Fig. 2.4). In this way decision support systems for planning purposes can be developed which evaluate a number of variables by use of GIS. Such systems can delineate areas of high risk from those where future development can take place safely. An appreciation of the physical (e.g. flood, landslip, seismic, subsidence) and chemical hazards (e.g. heavy metals, toxic components, gases, organic compounds, groundwater contamination) affecting an area can help identify those areas requiring further investigation or possible remedial action. These systems can be manipulated and updated to produce customized thematic maps, for example, for ranking geohazards in relation to foundation conditions or for epidemiological evaluation of possible public health risks.

2.4 Geological hazards and planning

Geological hazards can be responsible for devastating large areas of the land surface and so can pose serious constraints on development. However, geological processes such as volcanic eruptions, earthquakes, landslides and floods cause disasters only when they impinge upon people or their activities. Nevertheless, both the number of recorded disaster events and the number of people killed are increasing each year. In terms of financial implications, it has been estimated that natural hazards cost the global economy over $50 000 million per year. Two-thirds of this sum is accounted for by damage and the remainder represents the cost of predicting, preventing and mitigating against disasters. Man-made hazards such as groundwater pollution, soil erosion and subsidence, together with damage caused by hazards like problematic soils add to this figure. It is estimated that some 250 000 lives may be lost annually.

A natural hazard has been defined by UNESCO as the probability of the occurrence within a specified period of time, and within a given area, of a potentially damaging phenomenon. However, not all natural

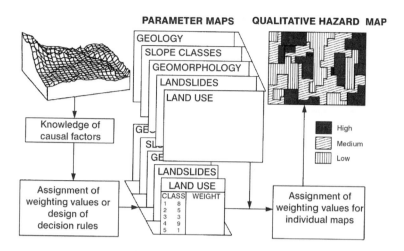

PARAMETER MAPS   QUALITATIVE HAZARD MAP

Fig. 2.4 Use of GIS for qualitative map combination.

hazards are geological and not all geological hazards are natural, some are influenced by or brought about by human activity. As hazards pose a threat to humans, their property and the environment, they are of more significance when they occur in highly populated areas. Moreover, as the global population increases so the significance of hazard is likely to rise. For instance, the number of people affected by natural disasters increases by some 6% annually. In addition, progress in many developing countries is placing increasing numbers of people and property at risk as a consequence of unplanned occupancy and use of marginal and high risk land. In particular, overcultivation and deforestation aggravate the situation. As geological hazards are widespread and fairly common events they obviously give cause for concern. Many such geohazards impose notable constraints on development or exert a substantial influence on society. Hence, it is important that geological hazards are understood in order that their occurrence and behaviour can be predicted, and that measures can be taken to reduce their impact.

As individual types of geohazard tend to occur again and again, the frequency of a hazard event can be regarded as the number of events of a given magnitude that take place in a particular period of time. As such, a recurrence interval for such an event can be determined in terms of the average length of time between events of a certain size. Although the most catastrophic events are highly infrequent, their impacts usually have the greatest overall significance. Similarly, frequent events normally are too small to have significant aggregate effect.

A hazard involves a degree of risk, the elements at risk being life, property, possessions and the environment. Risk involves quantification of the probability that a hazard will be harmful and the tolerable degree of risk depends upon what is being risked – life being much more important than property. The risk to society can be regarded as the magnitude of a hazard multiplied by the probability of its occurrence. If there are no mitigation measures, no warning systems and no evacuation plans for an area that is subjected to a recurring hazard, then such an area has the highest vulnerability. Vulnerability, V, was defined by UNESCO as the degree of loss to a given element or set of elements at risk resulting from the occurrence of a natural phenomenon of a given magnitude. It is expressed on a scale from 0 (no damage) to 1 (total loss). The factors that usually influence loss are population distribution, social infrastructure, and socio-cultural and socio-political differentiation and diversity.

The effects of a disaster may be lessened by reduction of vulnerability. Short-term forecasts a few days ahead of the event may be possible and complement relief and rehabilitation planning. In addition, it is possible to reduce the risk of disaster by a combination of preventative and mitigative measures. To do this successfully, the patterns of behaviour of the geological phenomena posing the hazards need to be understood and the areas at risk identified. Then, the level of potential risk may be decreased and the consequence of disastrous events mitigated against by introducing regulatory measures or other inducements into the physical planning process. The impact of disasters may be reduced further by incorporating appropriate measures into building codes and other regulations so that structures will withstand or accommodate potentially devastating events.

UNESCO went on to define specific risk, elements at risk and total risk. Specific risk, R, refers to the expected degree of loss associated with a particular hazard. It may be expressed as the product of hazard, H, and vulnerability ($R = H \times V$). Elements at risk, E, refers to the population, property and economic activity at risk in a given area. Lastly, total risk, R_T, refers to the expected number of lives lost, of injuries, property damage and disruption to economic activity brought about by a given hazard. It can be regarded as the product of specific risk and the elements at risk ($R_T = R \times E$ or $R_T = E \times H \times V$).

The net impact of a hazard can be looked upon in terms of the benefits derived from inhabiting a zone in which a hazard occurs minus the costs involved in mitigation measures. Mitigation measures may be structural or technological on the one hand or regulatory on the other (Table 2.3).

The response to geohazards by risk evaluation leading to land-use planning and appropriate engineering design of structures is lacking in many parts of the world, yet most geohazards are amenable to avoidance or preventative measures. In addition, the causes of geological hazards are often reasonably well understood, and so most can be identified and even predicted with a greater or lesser degree of accuracy. Nonetheless, geological hazards can represent constraints that impede economic development, especially in developing countries. Consequently, measures should be included in planning processes in order to avoid severe problems attributable to hazards that lead to economic and social disruption.

Table 2.3 Non-structural and structural methods of disaster mitigation.

<u>**Non-structural methods**</u>

(a) Short-term:
    i. Emergency plans: Civil and military forces involvement.
    ii. Evacuation plans: Routes and reception centres for the general public, for vulnerable groups, e.g. the very young, elderly, sick or disabled
    iii. Prediction of impact: Monitoring equipment, forecasting methods and models
    iv. Warning processes: General and specialized warning

(b) Long-term
    i. Building codes and construction norms
    ii. Hazard microzonation: Selected risks; All risks
    iii. Land-use control: Regulations, prohibitions, moratoria, compulsory purchase
    iv. Probabilistic risk analysis
    v. Insurance
    vi. Taxation
    vii. Education and training

<u>**Structural methods**</u>

    i. Retrofitting of existing structures
    ii. Reinforcement of new structures
    iii. Safety features: Structural safeguards, fail safe design
    iv. Probabilistic prediction of impact strength

Unfortunately, in many developing countries the capacity may not exist to effect planning processes or to establish mitigation measures, even though in some instances economic losses due to natural disasters could be reduced with only small investment.

Geological hazards vary in their nature and can be complex. One type of hazard, for example an earthquake, can be responsible for the generation of others such as liquefaction of sandy soils, landslides or tsunamis. Certain hazards such as earthquakes and landslides are rapid-onset hazards and so give rise to sudden impacts. Others such as soil erosion and subsidence due to the abstraction of groundwater may take place gradually over an appreciable period of time. Furthermore, the effects of natural geohazards may be difficult to separate from those attributable to human influence. For example, although soil erosion is a natural process it frequently has been seriously aggravated by people. In fact, modification of nature by people often increases the frequency and severity of natural geohazards which then increase the threat to human occupancy. Yet other geohazards are primarily attributable to human activity such as pollution of groundwater, contamination of land and subsidence due to mining.

2.5 Risk assessment

Land-use planning for the prevention and mitigation of geological hazards should be based on criteria that establish the nature and degree of the risks present and their potential impact. Risk, as mentioned, should take account of the magnitude of the hazard and the probability of its occurrence. Risk arises out of uncertainty due to insufficient information being available about a hazard and to incomplete understanding of the mechanisms involved. The uncertainties prevent accurate predictions of hazard occurrence. Risk analysis involves identifying the degree of risk, then estimating and evaluating it. An objective assessment of risk is obtained from a statistical assessment of instrumentally obtained data and/or data gathered from past events. In the conventional statistical assessment, probability expresses the relative frequency of occurrence of an event based upon data collected from various sources. A number of computer programs are available to carry out such analyses.

The risk to society can be quantified in terms of the number of deaths attributable to a particular hazard in a given period of time and the resultant damage to property. The costs involved can then be compared with the costs of hazard mitigation. Unfortunately, not all risks and benefits are readily amenable to measurement and assessment in financial terms. Even so, risk assessment is an important objective in decision making by planners because it involves the vulnerability of people and property on the basis of probability of occurrence of an event. The aims of risk analysis are to improve the planning process, reduce vulnerability and mitigate damage. The acceptable level of risk is inversely proportional to the number of people exposed to a hazard. Acceptable risk, however, is complicated by cultural factors and political attitudes. It is the latter two factors that are likely to determine public policy in relation to hazard mitigation and regulatory control.

The occurrence of a given risk in a particular period of time can be expressed in terms of probability as follows:

$$R = \Sigma P(E)Ct \qquad (2.1)$$

where R is the level of risk per unit time; E is events expressed in terms of probability, P; C refers to the consequences of the events; and t is a unit of time. Risk analysis should include the magnitude of the hazard, as well as its probability. The magnitude of the hazard is related to the size of the population at risk. The assessment should mention all the assumptions and conditions on which it is based and, if these are not constant, the conclusions should vary according to the degree of uncertainty, a range of values representing the probability distribution. The confidence limits of the predictions should be provided in the assessment.

Although it can be assumed that there is a point where the cost involved in the reduction of risk equals the savings earned by the reduction of risk, society tends to set arbitrary tolerance levels based largely on the perception of risk and the priorities for their management. Some hazards are far more emotive than others, notably the disposal of radioactive waste, and therefore may attract more funds from politically sensitive authorities to deal with associated problems. In essence, something is safe if its risks are judged to be acceptable but there is little consensus as to what is acceptable. Tolerability of risk may be regarded as a willingness to live with a risk in order to obtain certain benefits in the confidence that the risks are properly controlled. To tolerate a risk does not mean that it can be regarded as negligible or that it can be ignored. A risk needs to be kept under review, and further reduced if and when necessary. The use of tolerability to guide public policy involves both technical and social considerations.

A number of problems are inherent in risk management. For instance, the degree of risk does not increase linearly with the length of exposure to a hazard. Moreover, with time the response to risk can change so that mitigation and risk reduction can change. Unfortunately, the dichotomy between actual and perceived risk does not help attempts to reduce risk and promotes a conflict of objectives. Nonetheless, risk management has to attempt to determine the level of risk that is acceptable. This has been referred to as risk balancing. Public and private resources then may be allocated to meet a level of safety that is acceptable to the public. Cost-benefit analyses can be made use of in order to develop a rational economic means for risk reduction expenditure.

Risk management requires a value system against which decisions are made, that is, the risk basis. In order to effect risk control certain actions need to be taken. Risk assessment and control is made in terms of either monetary value or in terms of potential loss of life. Risk management decisions that do not involve health risk can use monetary value for assessment of costs and benefits. If a health risk is used as a value system, then the average incremental mortality rate of a population affected by a hazard may be used as the value criterion. This mortality rate is the expected additional deaths per year divided by the total affected population. Other health risk measures can be used, such as average loss of life expectancy.

2.6   Hazard maps

Any spatial aspect of hazard can be mapped, providing there is sufficient information on its distribution. Hence, when hazard and risk assessments are made over a large area, the results can be expressed in the form of hazard and risk maps (Fig. 2.5). An ideal hazard map should provide information relating to the spatial and temporal probabilities of the hazard mapped. In addition to data gathered from surveys, hazard maps frequently are compiled from historical data related to past hazards that have occurred in the area concerned. The concept is based on the view that past hazards provide some guide to the nature of future hazards. Hazard zoning maps usually provide some indication of the degree of risk involved with a particular geological hazard. The hazard often is expressed in qualitative terms such as high, medium and low. These terms, however, must be adequately described so that their meaning is understood. The variation in intensity of a hazard from one location to another can be depicted by risk mapping. This attempts to quantify the hazard in terms of potential victims or damage. Risk mapping therefore attempts to estimate the location, probability and relative severity of future, probable hazardous events so that potential losses can be estimated, mitigated against or avoided. Specific risk zoning maps divide a region into zones indicating exposure to a specific hazard. For example, a method of mapping areas prone to geological hazards has been developed that uses map units based primarily on the nature of the potential hazards associated with them. The resultant maps, together with their explanation, can be combined with a land-use/geological hazard area matrix to provide some idea of the problems that may arise in the area represented by the individual map. For instance, the matrix can indicate the effects of any changes in slope or the mechanical properties of rocks or soils, and so attempt to evaluate the severity of hazard(s) for various land uses. The landslide hazard map of the Crested Butte-Gunniston area, Colorado, provides an illustration of this method (Fig. 2.2). This map attempts to show which factors within individual map units have the most significance as far as potential hazard is concerned. The accompanying matrix outlines the problems likely to be encountered as a result of human activity.

However, hazard maps, like other maps, do have disadvantages. For instance, they are highly generalized and represent a static view of reality. They need to be updated periodically as new data becomes available. For example, the hazard map of the Mount St Helens area had to be completely

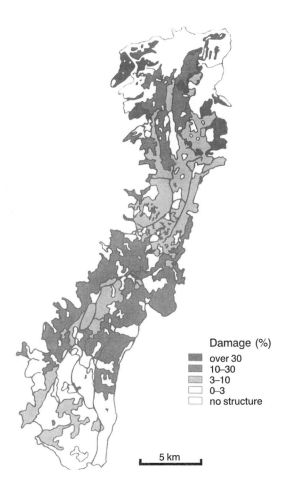

Damage (%)
◼ over 30
◼ 10–30
▨ 3–10
☐ 0–3
☐ no structure

5 km

Fig. 2.5 Distribution of structural damage resulting from a local earthquake in Quito, Equador (courtesy of Elsevier).

redrawn after the eruption of 1980. As catastrophic events occur infrequently, information initially may be completely lacking in an area where such an event subsequently occurs.

Microzonation has been used to depict the spatial variation of risk in relation to particular areas. Ideally, land use should adjust to the recommendations suggested by such microzoning so that the impacts of the hazard event(s) depicted have a reduced or minimum influence on people, buildings and infrastructure. Single-hazard/single-purpose, single-hazard/multiple-purpose and multiple-hazard/multiple-purpose maps have been distinguished. The first type of map obviously is appropriate where one type of hazard occurs and the effects are relatively straightforward. Single-hazard/multiple-purpose maps are prepared where the hazard is likely to affect more than one activity and shows the varying intensity of and the response to the hazard. The most useful map as far as planners are concerned is the multiple-hazard/multiple-purpose map (Fig. 3.9). Such a map can be used for risk analysis and assessment of the spatial variation in potential loss and damage. Both the hazard and the consequences should be quantified, and the relative significance of each hazard compared. The type and amount of data involved lends itself to processing by a geographical information system.

2.7 Land restoration and conservation

Obviously, societies make demands upon land that frequently mean that it becomes degraded. Derelict land can be regarded as land that has been damaged by industrial use or other means of exploitation to the extent that it has to undergo some form of remedial treatment before it can be of beneficial use. Such land usually has been abandoned in an unsightly condition and often is located in urban areas where land for development is scarce. Consequently, not only is derelict land a wasted resource but it also has a blighting effect on the surrounding area and can deter new development. Its restoration therefore is highly desirable, not only to improve the appearance of an area but also to enhance its economy by bringing derelict land back into worthwhile use. Accordingly, there is both economic and environmental advantage in the regeneration of derelict land. Land recycling in urban areas also can be advantageous since the infrastructure generally is still in place. Moreover, its regeneration should help prevent the exploitation of greenfield sites and encourage the use of brownfield sites. The term brownfield normally is applied in a broad sense to land that has been developed previously and as such includes vacant land, with or without buildings, together with derelict and contaminated land. The environmental geotechnics of a brownfield site influence its subsequent re-use and affect the costs of bringing the site into a state suitable for redevelopment.

The existence of abandoned derelict sites runs counter to the concept of sustainable development. Sustainable land use is essential if present day development needs are to be met without compromising the ability of future generations to meet their needs. A sustainable land use therefore requires derelict sites to be re-used, thereby recovering such sites as a land resource. It has been estimated that two-thirds of the world population will live in cities by 2015, which emphatically highlights the need for more efficient use of urban space, especially in those cities that are growing at a rapid rate.

Whatever the ultimate use to which the land is put after restoration, it is imperative that it should fit the needs of the surrounding area and be compatible with other forms of land use that occur in the neighbourhood. Accordingly, planning of the eventual land use must take account of overall plans for the area and must endeavour to include the integration of the restored area into the surrounding landscape. Restoring a site that represents an intrusion in the landscape to a condition that is well integrated into its surroundings upgrades the character of the environment far beyond the confines of the site.

Derelict land may present hazards, for example, disposal of industrial wastes may have contaminated land (in some cases so badly that earth has to be removed) or the ground may be severely disturbed by the presence of massive old foundations and subsurface structures such as tanks, pits and conduits for services. Contaminated land also may emit gases or may represent a fire hazard. Details relating to such hazards should be determined during the site investigation. Site hazards result in constraints on the freedom of action, necessitate following stringent safety requirements, may involve time-consuming and costly working procedures, and affect the type of development. Indeed, such constraints may mean that the development plan has to be changed so that the more sensitive land uses are located in areas of reduced hazard. Alternatively, where notable hazards exist, then a change to a less sensitive end use may be advisable.

Any project involving the restoration of derelict land requires a feasibility study. This needs to consider accessibility to the site; land use and market value; land ownership and legal issues; topography and geological conditions; site history and contamination potential; and the local environment and existing infrastructure. The results of the feasibility study allow an initial assessment to be made of possible ways to develop a site and the costs involved. Aerial photographs may prove of value, especially for large sites. A preliminary reconnaissance of a derelict site is desirable to determine the sequence of work for the following site investigation. The exact boundaries of the site and the various physical features it contains are recorded during the preliminary reconnaissance. The site investigation provides essential input for the design of remedial measures and indicates whether or not demolition debris can be used in the rehabilitation scheme.

Derelict sites may require varying amounts of filling, levelling and regrading. As far as fill is concerned, this should be obtained from on site if possible, otherwise from an area nearby. Once regrading has been completed the actual surface needs restoring. This is not so important if the area is to be built over (e.g. if it is to be used for an industrial estate), as it is if it is to be used for amenity or recreational purposes. In the case where buildings are to be erected, however, the ground must be adequately compacted so that the buildings are not subjected to adverse settlement. Settlement frequently presents a problem when derelict industrial land, which consists of a substantial thickness of rubble fill, is to be built on. If this is not contaminated, then dynamic compaction or vibro-compaction can be used to minimize the amount of settlement (see Chapter 11). Where the soils are soft, then they can be preloaded, with or without vertical drains, or stabilized by deep mixing whereby cement, lime or cementitious fly ash is mixed into the ground by auger rigs. On the other hand, where the land is to be used for amenity or recreational purposes, then soil fertility must be restored so that the land can be grassed and trees planted. This involves laying top-soil (where appropriate) or substitute materials, the application of fertilizers and seeding (frequently by hydraulic methods). Adequate subsoil drainage also needs to be installed.

Derelict land caused by some earlier industrial operation may contain structures or machinery of historical interest and worthy of preservation. Any assessment of the industrial archaeological value of a site must consider its scientific and engineering interest, its state of preservation and completeness, and its representativeness and rarity. The amenity, recreation and tourism aspects also need to be considered.

Conservation is concerned with safeguarding natural phenomena and the preservation and improvement of the quality of the environment. As such, it is closely associated with geology as geology and environment are intimately interrelated. However, the interests of conservation frequently are in conflict with social and economic pressures for development and so one of the roles of planning authorities is to balance the demands of urban, industrial, and infra-structural development with the need to preserve the countryside, and areas of scientific and cultural interest. Hence, geologists may find themselves in conflict with conservationists, especially in the case of the mineral extraction industry. Be that as it may, geological knowledge should be made use of in conservation programmes since these should seek to make the wisest use of natural resources.

Conservation is not simply preservation, it seeks to improve existing conditions rather than simply maintaining the status quo. Hence, conservation does involve the reconciliation of differing views so that the best compromise can be reached. Obviously, in this context land use is important. In some instances the same land can be used simultaneously to cater for several needs, whilst in others the uses to which it can be put are consecutive rather than concurrent and the final effect can be either to restore the land to its original use or to a new use that forms an acceptable part of the landscape. Thus, a sequence of events must be planned to ensure the greatest efficiency in the use of resources and to achieve the most acceptable final state. Hence, in such situations, the geologist must be involved with planning from the onset. If this is not done, then natural resources may be sterilized, as for example, when urban development takes place over valuable surface mineral deposits, which could have been extracted and the area then developed.

3

Natural Geohazards

Natural geohazards are geological processes that become a hazard when they have an adverse impact on the environment. Their hazardous nature depends upon their reaction with the environment, in particular when the geological process or processes cause destruction of property, or worse, loss of life. The damage to property brought about by a natural geohazard is dependent not just on the geohazard but also on the vulnerability of the property. Also important are the magnitude of the event, its duration, the area affected and the frequency with which the event occurs. Furthermore, one type of hazard may trigger another, as when landslides are caused by an earthquake event. The most spectacular natural geohazards are rapidly occurring events such as volcanic eruptions, earthquakes and landslides but there are other geohazards that are not spectacular and may occur over a longer time. Examples are provided by soil erosion and by the occurrence of subsidence depressions formed by dewatering. Similarly, ground movements attributable to expansive/shrinkable clays could not be described as spectacular but may cause more damage in terms of cost to a nation than more dramatic events.

3.1   Volcanic activity

Volcanic eruptions and other manifestations of volcanic activity are variable in type, magnitude and duration. As pointed out in Chapter 1, volcanic zones are associated with the boundaries of the crustal plates, the type of plate boundary offering some indication of the type of volcano that is likely to develop. In other words, andesitic magmas are associated with destructive plate margins whereas basaltic magmas are associated with constructive plate margins. Andesitic magmas are often responsible for violent eruptions while basaltic magmas are associated with less explosive activity.

3.1.1   Volcanic activity and hazards

When volcanic activity occurs near populated areas, it produces a variety of hazards for the people living in the vicinity. It is impossible to restrict people from all hazardous areas around volcanoes, especially those that are active only intermittently. Therefore, it is important to recognize the various types of hazard that may occur in order to prevent or mitigate against disasters.

Assessment of volcanic hazard is a complex function of the probability of eruptions of various intensities at a given volcano and of the location of the site in question with respect to the volcano. In addition, assessment of risk due to volcanic activity has to take account of the number of lives at stake, the capital value of the property and the productive capacity of the area concerned. Although evacuation from danger areas is possible if enough time is available, the vulnerability of property frequently is close to 100% for most violent volcanic eruptions. Such eruptions are rare events so there are insufficient observational data for effective analysis and the associated hazard is difficult to estimate. For the foreseeable future humans are unlikely to be able influence the degree of hazard linked to volcanoes so reduction of risk can only be achieved by reducing the exposure of life and property. This can be assessed by balancing the loss of income resulting from non-exploitation of a particular area against the risk of loss in the event of an eruption.

The active lifespan of most volcanoes is probably between one and two million years. Since the activity frequently follows a broadly cyclical pattern some benefit in terms of hazard assessment may accrue from the determination of the recurrence interval of particular types of eruption, the distribution of the resulting deposits, the magnitude of events and the recognition of any short-term patterns of activity. Four categories of hazard have been distinguished in Italy:

1. Very high frequency events with mean recurrence intervals (MRI) of less than two years. The area affected by such events is usually less than 1 km^2.

2. High frequency events with MRI values of 2 to 200 years. In this category damage may extend up to 10 km^2.

3. Low frequency events with MRI values of 200 to 2000 years. Areal damage may cover 1000 km^2.

4. Very low frequency events are associated with the most destructive eruptions and have MRI values in excess of 2000 years. The area affected may be greater than 10 000 km^2.

The return periods of particular types of activity of individual volcanoes or centres of volcanic activity can be obtained by thorough stratigraphic study and dating of the associated deposits so as to reconstruct past events. Used in conjunction with any available historical records, these data form the bases for assessing the degree of risk involved.

3.1.2 Types of volcanic hazard

Several categories of volcanic hazards exist, namely, premonitory earthquakes, pyroclast falls, lateral blasts, pyroclast flows and surges, lava flows, structural collapse, and associated hazards. Each type represents a specific phase of activity during a major eruptive cycle of a polygenetic volcano and may occur singly or in combination with other types. Damage resulting from volcano-seismic activity is rare. However, intensities on the Mercalli scale (see Section 3.2) varying from 6 to 9 have been recorded over limited areas.

Pyroclastic fall deposits may consist of bombs, scoria, lapilli, pumice, dense lithic material, crystals or any combination of these. In violent eruptions intense falls of ash interrupt human activities and cause serious damage. The size of the area affected by pyroclastic falls depends on the amount of material ejected and the height to which it is thrown, as well as the wind speed and direction. They can affect areas up to several tens of kilometres from a volcano within a few hours from the commencement of an eruption Ash-fall during and after an eruption may contain a small amount of free silica, which if inhaled over long periods could cause silicosis of the lungs. The principal hazards to property resulting from pyroclastic falls are burial, impact damage, and fire if the material has a high temperature. The latter hazard is most dangerous within a few kilometres of the vent. The weight of the material collected on roofs may cause them to collapse.

Laterally directed blasts are among the most destructive of volcanic phenomena. They travel at high speeds, for example, the lateral blast associated with the Mount St Helens eruption initially travelled at 600 km h^{-1}, slowing to 100 km h^{-1} some 25 km from the volcano. They can occur with little or no warning in a period of a few minutes and can affect hundreds of square kilometres. The material carried by lateral blasts can vary from cold to temperatures high enough to scorch vegetation and start fires. Such blasts kill virtually all life by impact, abrasion, burial and heat. High concentration of ash contaminates the air. Laterally directed blasts are the result of sudden decompression of magmatic gases or explosion of a high pressure hydrothermal system. Release of gases may be caused by volatile pressures exceeding the weight of the overlying rocks, so giving rise to explosion.

Pyroclastic flows are hot dry masses of clastic volcanic material that move over the ground surface. Most pyroclastic flows consist of a dense basal flow, the pyroclastic flow proper, one or more pyroclastic surges and clouds of ash. Two major types of pyroclastic flows may be recognized. Pumiceous pyroclastic flows are concentrated mixtures of hot to incandescent pumice, mainly of ash and lapilli size. Ash flow tuffs and ignimbrites are associated with these flows. Individual flows vary in length from less than 1 km up to 200 km, covering areas ranging up to 20 000 km^2 with volumes from less than 0.001 to over 1000 km^3. Pyroclastic flows formed primarily of scoriaceous or lithic volcanic debris are known as hot avalanches, glowing avalanches, nuées ardentes or block and ash flows (Fig. 3.1). They generally affect a narrow

Fig. 3.1   Eruption of Mt St Helen's volcano in May 1980, Washington State, USA. The associated ash cloud travelled some 400 km downwind in the first six hours after eruption.

sector of a volcano, perhaps only a single valley. Maximum temperatures of pyroclastic flow material soon after it has been deposited may range from 350° to 550°C. Hence, they are hot enough to kill anything in their path. Because of their high mobility (up to 160 km h^{-1} on the steeper slopes of volcanoes) they constitute a great potential danger to many populated areas. Other hazards associated with pyroclastic flows, apart from incineration, include burial, impact damage, and asphyxiation.

A pyroclastic surge is a turbulent, low density cloud of gases and rock debris, which hugs the ground over which it moves. Hot pyroclastic surges can originate by explosive disruption of volcanic domes caused by rapidly escaping gases under high pressure or by collapse of the flank of a dome. They also can be caused by lateral explosive blast. Hot pyroclastic surges can occur together with pyroclastic flows. Generally, surges are confined to a narrow valley of a volcano but they may reach speeds up to 300 km h^{-1} so that escape from them is impossible. They give rise to similar hazards to pyroclastic flows. Cold pyroclastic surges are produced by phreatic and phreatomagmatic explosions.

The distance that a lahar (i.e. a mudflow associated with a volcanic event) may travel depends on its volume and water content, and the gradient of the slope of the volcano down which it moves (Fig. 3.2). Some may travel for more than 100 km. The speed of a lahar also is influenced by the water-sediment ratio and its volume, as well as the gradient and shape of the channel it moves along, with speeds of up to 165 km h^{-1} being recorded at times. However, the average speed over several tens of kilometres is generally less than 25 km h^{-1}. Because of their high bulk density lahars can destroy structures in their path and block highways. They can reduce the channel capacity of a river and so cause flooding, as well as adding to the sediment load of a river.

Because the rate of flow of lava usually is sufficiently slow and along courses that are predetermined by topography, they rarely pose a serious threat to life. The arrival of a lava flow along its likely course can be predicted if the rate of lava emission and movement can be determined. Damage to property, however, may be complete, destruction occurring by burning, crushing or burial of structures in the lava path. Lava flood eruptions are the most serious and may cover large areas with immense volumes of lava. The length of a lava flow is dependent upon the rate of eruption, the viscosity of the lava and the topography of the area involved. Given the rate of eruption it may be possible to estimate the length of flow. The possibility of a given location being inundated with lava at some future time can be estimated

Fig. 3.2 Lahar on the side of Lassen Peak, California, formed after the eruption of 19 May, 1915.

from information relating to the periodicity of eruptions in time, the distribution of rift zones on the flanks of a volcano, the topographic constraints on the directions of flow of lavas, and the rate of covering by lava. However, each new eruption of lava alters the topography of the slopes of a volcano to a certain extent and therefore flow paths may change. What is more, prolonged eruptions of lava may eventually surmount obstacles that lie in their path that act as temporary dams. This then may mean that the lava invades areas that formerly were considered safe.

The formation of calderas and landslip scars due to the structural collapse of large volcanoes are rare events. They frequently are caused by magma reservoirs being evacuated during violent Plinian eruptions. Because calderas develop near the summits of volcanoes and subside progressively as evacuation of their magma chambers takes place, caldera collapses do not offer such a threat to life and property as does sector collapse. Sector collapse involves subsidence of a large area of a volcano. It takes place over a comparatively short period of time and may involve volumes up to tens of cubic kilometres. Collapses that give rise to landslides may be triggered by volcanic explosions or associated earthquakes.

Hazards associated with volcanic activity include destructive floods caused by sudden melting of the snow and ice that cap high volcanoes, or by heavy downfalls of rain (vast quantities of steam may be given off during an eruption), or the rapid collapse of a crater lake. Far more dangerous are the tsunamis generated by violent explosive eruptions and sector collapse. Tsunamis may decimate coastal areas. Dense poisonous gases normally offer a greater threat to livestock than to humans. Nonetheless, the large amounts of carbon dioxide released at Lake Manoun in 1984 and neighbouring Lake Nyos in 1986, in Cameroon, led to the deaths of 37 and 1887 people respectively. In addition, gases can be injurious to the person mainly because of the effects of acid compounds on the eyes, skin and respiratory system. They can kill crops. Acid rain can form as a result of rain mixing with aerosols and gases adhering to tephra. Such rain can cause severe damage to natural vegetation and crops, and skin irritation to people. Air-blasts, shock waves and counter-blasts are relatively minor hazards, although they can break windows several tens of kilometres away from major eruptions.

3.1.3 Volcanic activity and hazard zoning

Hazard zoning involves mapping deposits that have formed during particular phases of volcanic activity and extrapolating to identify areas that would be likely to suffer a similar fate at some future time. The

63

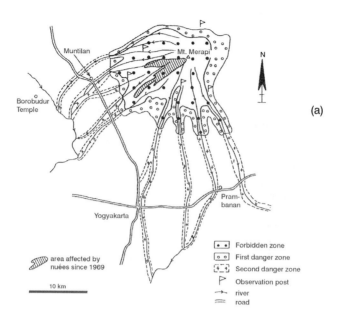

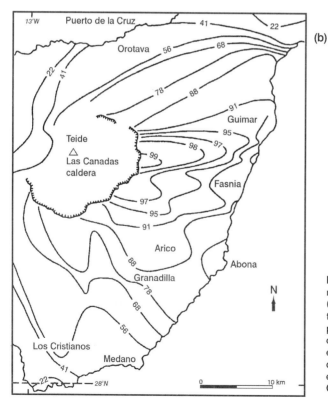

Fig. 3.3 (a) A volcanic hazard map of Merapi volcano, Java, (b) South eastern Tenerife, showing the extent and percentage probability of burial by more than one metre of airfall ash if a large eruption occurs, based on the deposits from 27 previous such eruptions. (Courtesy of the Geological Society of London.)

zone limits on such maps normally assume that future volcanic activity will be similar to that recorded in the past. Unfortunately, this is not always the case. Volcanic hazard maps have been produced in Indonesia that define three zones, namely, the forbidden zone, the first danger zone and the second danger zone. The forbidden zone is meant to be abandoned permanently since it is affected by nuées ardentes (Fig. 3.3a). The first danger zone is not affected by nuées ardentes but may be affected by bombs. Lastly, the second danger zone is that likely to be affected by lahars. This zone is subdivided into an abandoned zone from which there is no escape from lahars and an alert zone where people are warned and from where evacuation may be necessary.

Volcanic risk maps indicate the specified maximum extents of particular hazards such as lava and pyroclastic flow paths, expected ash-fall depths and the areal extent of lithic missile fall-out. They are needed by local and national governments so that appropriate land uses, building codes and civil defence responses can be incorporated into planning procedures. Events with MRI of less than 5000 years should be taken into account in the production of maps of volcanic hazard zoning and data on any events that have taken place in the last 50 000 years are probably significant. Two types of map are of use for economic and social planning. One type indicates areas liable to suffer total destruction by lava flows, nuées ardentes and/or lahars. The other shows areas likely to be affected temporarily by damaging but not destructive phenomena, such as heavy falls of ash, toxic emissions, or pollution of surface or underground waters (Fig. 3.3b).

3.1.4 Prediction of volcanic activity

Losses caused by volcanic eruptions can be reduced by a combination of prediction, preparedness and land-use control. Emergency measures include alerting the public to the hazard, followed by evacuation. Appropriate structural measures for hazard reduction are not numerous, but include building steeply-pitched reinforced roofs that are unlikely to be damaged by ash-fall, and constructing walls and channels to deflect lava flows. Hazard zoning, insurance, local taxation and evacuation plans are appropriate non-structural measures. Risk management depends on identifying hazard zones and forecasting eruptions. The levels of risk must be defined and linked to appropriate social responses. In the long term, appropriate controls should be placed on land use and the location of settlements.

Prediction studies are based on the detection of reliable premonitory symptoms of eruptions. Such forerunners may be geophysical or geochemical in nature. Geophysical observations, principally tiltmetry, seismography and thermometry, provide the basis for forecasting volcanic eruptions. Data from satellites can be used to monitor volcanic activity. Three aspects of volcanism, namely, the movement of eruption plumes, changes in thermal radiation of the ground, and output of gases can be measured using remote sensing.

A volcanic eruption involves the transfer towards the surface of millions of tonnes of magma. This leads to the volcano concerned undergoing changes in elevation, notably uplift. The uplift usually is measured by a network of geodimeters and tiltmeters set up around the volcano. Gravity meters can be used to detect any vertical swelling. Such uplift also means that rocks are fractured so that volcanic eruptions may be preceded by seismic activity. Where earthquake swarms do occur, the number of tremors increases as the time of eruption approaches. However, tremors may continue for a few days on the one hand to a year or more on the other. A network of seismic stations can be set up to monitor the tremors and, from the data obtained, the position and depth of origin of the tremors ascertained. More recently, long-period events have been analysed. These shock waves are believed to be generated by resonating magma that is coming under pressures inside the volcano. Long-period events increase in number as the volcano nears its climax. In 2001 long-period events were used to predict an eruption of Popocatepetl in Mexico.

Infrared techniques have been used in the prediction of volcanic eruptions since the volcano area usually becomes hotter than its surroundings due to the rising magma. Thermal maps of volcanoes can be produced quickly by ground-based surveys using infrared telescopes. However, consistent monitoring is necessary in order to distinguish between real and apparent thermal anomalies. Aerial surveys provide better data but are too expensive to be used for routine monitoring. However, basaltic magmas of low viscosity may prove an exception if their ascent is more rapid than the rate at which heat is conducted from them. On the other hand, thermal techniques may not prove satisfactory when monitoring explosive andesitic and dacitic volcanoes, since the ascent of highly viscous acid magmas may be too slow to give rise to easily detectable temperature changes. Detectable anomalies are more likely to develop when heat is transferred by circulating groundwater rather than by conduction from a magma. The temperatures of hot springs may rise, as may the water in crater lakes, with the approach of a volcanic eruption. Hot groundwater also gives rise to the appearance of new fumaroles or to an increase in the temperatures at existing fumaroles. Gauges are used to monitor changes in gas temperatures, pH values and amounts of suspended mineral matter. An increase in temperature also leads to demagnetization of rock, as the magnetic minerals are heated above their Curie points. This can be monitored by magnetic surveying. Changes also occur in the gravitational and electrical properties of rocks.

The evolution of an eruption involves changes in matter and energy. The most significant variations take place in the gas phase. Gas sensors can be stationed on the ground or carried by aircraft to record the changes in the composition of gases. Gas emitted from fumaroles on the flanks of a volcano may contain increases in HCl, HF and SO_2, or the ratios of, for example, S to Cl may alter. These changes may be related to an impending eruption since the proximity of magma tends to emit more sulphur, chlorine and fluorine. Moreover, increase of the dissolution of acidic volcanic gases in the hydrothermal system often results in small decreases in the pH value. In addition, lake water chemistry may alter.

Monitoring rainfall and storm build-up is necessary for lahar prediction. Telemetered rain gauges may be used that are activated when a certain amount of rain falls in a given time.

3.2 Earthquakes

Although some seismicity is caused by volcanic activity, most is due to movements along faults within the Earth's crust. In fact, earthquakes have been reported from all parts of the world but they are primarily associated with the margins of the crustal plates that move with respect to each other. Earthquakes are a manifestation of this movement – differential displacements giving rise to elastic strains, which eventually exceed the strength of the rocks involved. The strained rocks rebound along the fault, elastic strain energy being released in the form of seismic waves.

Initially, movement may occur over a small area of a fault plane, to be followed by slippage over a much larger surface. The initial movements give rise to the foreshocks that precede an earthquake. These are followed by the principal movements, but complete stability is not restored immediately. The shift of the rock masses involved in faulting relieves the main stress but new stresses develop in adjacent areas. Because stress is not relieved evenly everywhere, minor adjustments arise along a fault plane generating aftershocks. The decrease in strength of the aftershocks is irregular and occasionally they may continue for a year or more.

3.2.1 Earthquake magnitude, frequency and duration

The energy generated by an earthquake may be expressed in terms of intensity or magnitude. Earthquake intensity scales are qualitative expressions of the damage caused by an event. The most widely accepted intensity grading is the Mercalli scale (Table 3.1). The magnitude of an earthquake is an instrumentally measured expression of the energy liberated during the event, the maximum amplitude of the resulting seismic waves being expressed on a logarithmic scale. An earthquake of magnitude 2 is the smallest likely to be felt by humans while earthquakes of magnitude 5 or less are unlikely to cause damage to well constructed buildings. The maximum magnitude of earthquakes is limited by the amount of strain energy that the rock masses involved can sustain before failure occurs, hence the largest tremors have had a magnitude of around 8.9. Such an event causes severe damage over a wide area.

The magnitude of an earthquake depends on the length of the fault break and the amount of displacement that occurs. Generally, movement only occurs over a limited length of a fault during one event. A magnitude 7 earthquake would be produced if a 150 km length of a fault underwent a

Table 3.1 Modified Mercalli Scale, 1956 version, with Cancani's equivalent acceleration. (These are not peak accelerations as instrumentally recorded).

Degrees	Description	Acceleration mm s^{-2}
I	Not felt. Only detected by seismographs.	Less than 2.5
II	Feeble. Felt by persons at rest, on upper floors, or favourably placed.	2.5 to 5.0
III	Slightly felt indoors. Hanging objects swing. Vibration like passing of light trucks. Duration estimated. May not be recognized as earthquake.	5.0 to 10
IV	Moderate. Hanging objects swing. Vibration like passing of heavy trucks, or sensation of a jolt like a heavy ball striking the walls. Standing motor cars rock. Windows, dishes, doors rattle. Glasses clink. Crockery clashes. In the upper range of IV wooden walls and frame creak.	10 to 25
V	Rather strong. Felt outdoors, direction estimated. Sleepers wakened. Liquids disturbed, some spilled. Small unstable objects displaced or upset. Doors swing, close, open. Shutters and pictures move. Pendulum clocks stop, start, change rate.	25 to 50
VI	Strong. Felt by all. Many frightened and run outdoors. Persons walk unsteadily. Windows, dishes, glassware broken. Ornaments books, etc, fall off shelves. Pictures fall off walls. Furniture moved or overturned. Weak plaster and masonry cracked. Small bells ring (church, school). Trees, bushes shaken visibly or heard to rustle.	50 to 100
VII	Very strong. Difficult to stand. Noticed by drivers of motor cars. Hanging objects quiver. Furniture broken. Damage to masonry D, including cracks. Weak chimneys broken at roof line. Fall of plaster, loose bricks, stones, tiles, cornices, also unbraced parapets and architectural ornaments. Some cracks in masonry C. Waves on ponds, water turbid with mud. Small slides and caving in along sand or gravel banks. Large bells ring. Concrete irrigation ditches damaged.	100 to 250
VIII	Destructive. Steering of motor cars affected. Damage to masonry C, partial collapse. Some damage to masonry B, none to masonry A. Fall of stucco and some masonry walls. Twisting, fall of chimneys, factory stacks, monuments, towers, elevated tanks. Frame houses moved on foundations if not bolted down, loose panel walls thrown out. Decayed piling broken off. Branches broken from trees. Changes in flow or temperature of springs and wells. Cracks in wet ground and on steep slopes.	250 to 500
IX	Ruinous. General panic. Masonry D destroyed, masonry C heavily damaged, sometimes with complete collapse, masonry B seriously damaged. General damage to foundations. Frame structures, if not bolted, shifted off foundations. Frames cracked, serious damage to reservoirs. Underground pipes broken. Conspicuous cracks in ground. In alluviated areas sand and mud ejected, earthquake fountains, sand craters.	500 to 1000
X	Disastrous. Most masonry and frame structures destroyed with their foundations. Some well built wooden structures and bridges destroyed. Serious damage to dams, dykes, embankments. Large landslides. Water thrown on banks of canals, rivers, lakes, etc. Sand and mud shifted horizontally on beaches and flat land. Railtracks bent slightly.	1000 to 2500
XI	Very disastrous. Railtracks bent greatly. Underground pipelines completely out of service.	2500 to 5000
XII	Catastrophic. Damage nearly total. Large rock masses displaced. Lines of sight and level distorted. Objects thrown into the air.	Over 5000

Masonry types. A: Good workmanship, mortar, and designed to resist lateral forces; reinforced and bound together with steel, concrete, etc. **B:** Good workmanship and mortar; reinforced, but not designed in detail to resist lateral forces. **C:** Ordinary workmanship and mortar; no extreme weaknesses but not reinforced or designed against horizontal forces. **D:** Weak materials, e.g. adobe, poor mortar, poor workmanship, weak horizontally.

displacement of about 1.0 m. The length of the fault break during a particular earthquake is usually only a fraction of the true length of the fault. Individual fault breaks during simple earthquakes have ranged in length from less than a kilometre to several hundred kilometres. What is more, fault breaks do not only occur in association with large and infrequent earthquakes but also occur in association with small shocks and continuous slow slippage known as fault creep. Fault creep may amount to several millimetres per year and progressively deforms buildings located across such faults.

There is little information available on the frequency of breakage along active faults. Some master faults have suffered repeated movements, in some cases recurring in less than 100 years. By contrast, much longer intervals, totalling many thousands of years, have occurred between successive breaks. Therefore, because movement has not been recorded in association with a particular fault in an active area, it cannot be concluded that the fault is inactive.

The duration of an earthquake is one of the most important factors as far as damage or failure of structures, soils and slopes are concerned. What is important in hazard assessment is the prediction of the duration of seismic shaking above a critical ground acceleration threshold. The magnitude of an earthquake affects the duration much more than it affects the maximum acceleration, since the larger the magnitude the greater the length of ruptured fault. Hence, the more extended the area from which the seismic waves are emitted. With increasing distance from a fault, the duration of shaking is longer but the intensity of shaking is less, the higher frequency waves being attenuated more than those with lower frequencies.

3.2.2 Ground behaviour and earthquakes

Surface ground motion is influenced by the physical properties of the soil and rock masses through which seismic waves travel, and by the geological structure. Maximum acceleration within an earthquake source area may exceed $2\,g$ for competent bedrock. On the other hand, normally consolidated clay soils of low plasticity are incapable of transmitting accelerations greater than 0.1 to $0.15\,g$ to the surface. Clay soils with high plasticity allow accelerations of 0.25 to $0.35\,g$ to pass through. Saturated sandy clay soils and medium dense sands may transmit 0.5 to $0.6\,g$, and in clean gravel and dry dense sand accelerations may reach much higher values.

In general, structures not specifically designed for earthquake loadings have fared far worse on soft saturated alluvium than on hard rock. This is because motions and accelerations are much greater in deep alluvium than rock. By contrast, a rigid building may suffer less on alluvium than on rock. The explanation is attributable to the alluvium having a cushioning effect and the motion may be changed to a gentle rocking. This is easier on such a building than the direct effect of earthquake motions experienced on harder ground. Intensity attenuation on rock is very rapid whereas it is extremely slow on soft formations increasing only in the fringe area of the shock. Hence, the character of intensity attenuation in any earthquake depends largely on the surface geology of the area shaken. It therefore is important to relate the dynamic characteristics of a building to those of the subsoil in which it is founded. The vulnerability of a structure to damage is enhanced considerably if the natural frequency of vibration of the structure and the subsoil are the same.

Ground vibrations caused by earthquakes often lead to compaction of sandy soil and associated settlement of the ground surface. Loosely packed saturated sands and silts tend to lose all strength and behave like fluids during strong earthquakes. When such materials are subjected to shock, densification occurs. During the relatively short time of an earthquake, drainage cannot be achieved and this densification therefore leads to the development of excessive pore water pressures that cause the soil mass to act as a heavy fluid with practically no shear strength, that is, a quick condition develops (Fig. 3.4). Water moves upward from the voids in the soil to the ground surface where it emerges to form sand boils. If liquefaction occurs in a sloping soil mass, the entire mass begins to move as a flow slide. Such slides develop in loose saturated sandy materials during earthquakes. Loose saturated silts and sands often occur as thin layers underlying firmer materials. In such instances, liquefaction of the silt or sand during an earthquake may cause the overlying material to slide over the liquefied layer. Structures on

Fig. 3.4 Overturning and tilting of apartment blocks in Niigata due to liquefaction of ground following the Niiagata earthquake, Japan, 1964. (Courtesy of the late Professor H. Bolton Seed.)

the main slide frequently are moved without suffering damage. However, a graben-like feature often forms at the head of the slide and buildings located in this area are subjected to large differential settlements and often are destroyed (Fig. 3.5). Buildings near the toe of the slide are commonly heaved upwards or are even pushed over by the lateral thrust.

Clay soils do not undergo liquefaction when subjected to earthquake activity. However, under repeated cycles of loading large deformations can develop, although the peak strength remains about the same. Nonetheless, these deformations can reach the point where, for all practical purposes, the soil has failed. Major slides can result from failure in deposits of clay.

Fig. 3.5 Graben collapse in Fairbanks after the 1964 Alaskan earthquake. (Courtesy of the late Professor H. Bolton Seed.)

3.2.3   Earthquake forecasting

The purpose of earthquake forecasting is reduction of loss of life and damage to property. Earthquake prediction involves using studies of historical seismicity and the results of intensive monitoring of seismic and geological phenomena to establish the probability that a given magnitude of earthquake or intensity of damage will recur during a given period of time. Such predictions help the development of safety measures to be taken in relation to the degree of risk.

Earthquake precursors are phenomena that precede major ground tremors, more specifically they are anomalous values of such phenomena that may indicate the onset of an earthquake. Most important is the variation in seismic wave velocity. The P-waves slow down (e.g. by up to 10% in the focal region) due to the minute fracturing in rocks during dilation but then increase as water occupies microfractures prior to an earthquake. The period of time covered by these changes indicates the size of earthquake that can be expected. For example, an earthquake of magnitude 5.4 may be preceded by a period of four months of lowered velocities, whereas if the period of changes lasts for 14 years it suggests a potentially violent earthquake of magnitude 7 or over. Furthermore, the ratio of the P- and S-wave velocities decreases by up to 20% a few hours or days before a major earthquake. This is primarily due to the reduction in the compressional wave velocity. The duration, but not the amount of the decrease in velocity ratio appears to be proportional to the earthquake magnitude. The velocity ratio returns to normal at the beginning of the 'critical' period immediately before an earthquake.

Dilation of rocks due to crustal deformation prior to an earthquake leads to an increase in their volume and results in ground shortening or lengthening, minor tilting, or uplift or subsidence of the ground surface near an active fault. This ground movement can be measured by very accurate surveying or by high-precision tiltmeters and extensometers. Unfortunately, however, the point at which movements become critical is not always easy to detect. Other features brought about by rock dilation include a decrease in electrical resistivity and a change in magnetic susceptibility of the rocks concerned, that is, the local magnetic field becomes stronger shortly before an earthquake. Changes in density are recorded by sensitive gravity meters.

The release of small quantities of the inert gases, notably radon, but also argon, helium, neon and xenon takes place prior to an earthquake. The increase in radon has been noted in well waters and has been attributed to increasing water flow that carries radon with it into wells after dilation and cracking in neighbouring rock masses. Stresses in saturated strata may cause spring discharge to increase, or the water in wells to alter level or become turbid. Stress in the rocks may speed the movement of water in the phreatic zone and hence the migration of radon and halogens to the surface. Moreover, many active faults are zones of geothermal energy release, where it may be possible to relate the temperature of hot springs to incipient seismic activity.

As strain increases, the more likely it is that sudden slip will take place along a fault and that an earthquake will be generated. Recording geological and geodetic measurements should help to determine which segments of a fault are likely to slip in future. Hence, the importance of mapping fault segments accurately. It can be assumed that the earthquake that generates the largest magnitude is that where movement occurs over the complete segment. Accordingly, it is necessary to determine which fault segments along an active fault have slipped in the past and to calculate the rate at which strain is accumulating in the area. Each time slip has taken place can be related to the magnitude of the resultant earthquake so that the recurrence intervals between earthquakes of a given magnitude can be determined. From this, the range of probable occurrence of an earthquake of a particular size in a specific time period can be made.

Most major fault zones do not consist of a single break but are made up of a great number of parallel and interfingering breaks that may range over a kilometre apart. This indicates that displacements associated with intermittent great earthquakes have tended to migrate back and forth throughout the fault zone concerned. In many areas the more abundant recurrent displacements have taken place along the principal fault breaks, which suggests that these breaks extend to master faults located at depth and so would have a higher probability of future movements.

Rock bending at faults means that strain is accumulating in rocks, possibly foreshadowing an earthquake and sudden fault displacement. Such tectonic movement can be measured with electronic distance measurement (EDM) equipment, by triangulation, by fault movement quadrilaterals and by tiltmeters. It seems that sudden strain release is more likely along a segment of a fault where the rate of movement differs from that of adjacent areas or from its past rate, so that this may aid earthquake prediction. Triangulation in critical fault zone areas provides precise survey control and a basis for determination of ground movements in the future. Geodimeter measurement does not distinguish between movement due to fault slippage and that due to strain, primarily because the lines measured are long, from 13 to 32 km. Measurements over much shorter distances, 250 to 500 m, provide indication of slippage. These lines are arranged in quadrilaterals across a fault. Gradual vertical movements result in some tilting at the surface. Tiltmeters can detect tilting of 0.0001 mm along a 30.5 m horizontal line. Arrays of seismographs, spread over large areas, are buried at various depths and arranged in two lines at right angles to each other in order to detect shock waves. The arrays also can include electrical resistivity meters, magnetometers, gravity meters and strain gauges.

It may be possible to assess the relative activity of a fault using geological, seismological and historical data, and so classify it as active, potentially active, of uncertain activity, or inactive. Seismological data used to recognize active faults include historical or recent surface faulting associated with strong earthquakes, and tectonic fault creep or geodetic indications of fault movement. A lack of known earthquakes, however, does not necessarily mean that a fault is not active. An active fault is indicated by geological features such as young deposits displaced or cut by faulting, fault scarps, fault rifts, pressure ridges, offset streams, enclosed depressions, fault valleys, and ground features such as open fissures, rejuvenated streams, folding or warping of young deposits, groundwater barriers in recent alluvium, *en echelon* faults and fault paths on recent surfaces (Fig. 3.6). Historical sources of displacement of man-made structures such as roads, railways, etc. may provide further evidence of active faults. Faults can be regarded as potentially active faults if there is no reliable report of historic surface faulting; faults are not known to cut or displace the most recent alluvial deposits, but may be found in older alluvial deposits; geological features characteristic of active fault zones are subdued, eroded and discontinuous; water

Fig. 3.6 Fault scarps formed by earthquake in gravels of Recent age, near Springfield, South Island, New Zealand.

barriers may be present in older materials; and the geological setting in which the geometrical relationship to active or potentially active faults suggests similar levels of activity. Faults are of uncertain activity if data is insufficient to provide criteria definitive enough to establish fault activity. Additional studies are necessary if the fault is considered critical to a project. In the case of inactive faults, a thorough study of local sources will show no historical activity. Geologically, features characteristic of active fault zones are not present in areas where faults are inactive, and geological evidence is available to indicate that faults have not moved in the recent past. Although not a sufficient condition for inactivity, seismologically the fault concerned should not have been recognized as a source of earthquakes.

Data relating to ground motion is essential for an understanding of the behaviour of structures during earthquakes. From the engineering point of view, strong motion earthquakes are the most important, since they damage or even destroy man-made structures. Records of shocks produced by such earthquakes are obtained by using ruggedly constructed seismometers called accelerometers. Strong motion records provide information concerning the acceleration, displacement periods and duration of earthquakes. Accelerometers are placed on the surface and below ground level. They also have been installed in dams and at different levels in multi-storey buildings to compare structural motion at different heights with ground motion.

3.2.4 Seismic hazards and zoning

Earthquake hazard relates to the probability of ground motion occurring in a certain area, during a given period of time, that is capable of causing significant loss of property or life. It may be expressed by indicating the probability of occurrence of certain ground accelerations, velocities and displacements; of ground movements of various duration; or any other physical parameter that adversely affects a structure. Earthquake risk may be defined in terms of the probability of the loss of life or property, or loss of function of structures or utilities. The assessment of that risk must take into consideration the probability of occurrence of ground motions due to an earthquake, the value of the property and lives exposed to the hazard, and their vulnerability to damage or destruction, injury or death by ground motions associated with the hazard.

Although the hazards due to seismicity include the possibility of a structure being severed by fault displacement, it is much more likely that damage will result from shaking (Fig. 3.7). The destruction caused by an earthquake depends on many factors. Of prime importance are the magnitude of the event

Fig. 3.7 Loma Prieta earthquake of October, 1989. The view along Jefferson Street in the Marina district, San Francisco.

and its duration on the one hand, and the response of buildings and other elements of the infrastructure on the other. In addition, other hazards such as landslides, floods, subsidence and secondary earthquakes may be triggered by a seismic event.

Seismic zoning and micro-zoning provide a means whereby regional and local planning departments can attempt to reduce the impact of seismic hazard and are also of use in relation to earthquake-resistant design. While seismic zoning takes into account the distribution of earthquake hazard within a region, seismic micro-zoning defines the distribution of earthquake risk in a seismic zone. A seismic zoning map shows the zones of different seismic hazard in a particular area. Seismic evidence obtained instrumentally and from the historical record can be used to produce such maps. Maximum hazard levels can be based on the assumption that future earthquakes will occur with the same maximum magnitudes and intensities recorded at a given location as in the past. Hence, seismic zoning provides a broad picture of the earthquake hazard that can be present in seismic regions and so has led to a reduction of earthquake risk. Detailed seismic zoning maps should take account of local engineering and geological characteristics, as well as the differences in the spectrum of seismic vibrations and, most important of all, the probability of the occurrence of earthquakes of various intensities. Expected intensity maps define the source areas of earthquakes and either the maximum intensity that can be expected at each location or the maximum intensity to be expected in a given time interval. Quantities related to earthquake-resistant design also should be taken into account, such as maximum acceleration or peak particle velocity, predominant period of shaking or probability of occurrence (Fig. 3.8).

Most studies of the distribution of damage attributable to earthquakes indicate that areas of severe damage are highly localized and that the degree of damage can change abruptly over short distances. These differences frequently are due to changes in soil conditions or local geology. Such behaviour has an important bearing on seismic micro-zoning so that such maps may take account of active faults likely to give rise to ground rupture, and areas of potential landslide and liquefaction risk (Fig. 3.9). Hence, seismic micro-zoning maps can be used for detailed land-use planning and for insurance risk evaluation. These maps are most detailed and more accurate where earthquakes have occurred quite frequently, and the local variations in intensity have been recorded.

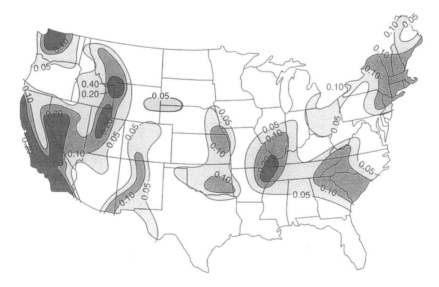

Fig. 3.8 Seismic risk map of the United States. The contours indicate the effective peak, or maximum, acceleration levels (values are in decimal fractions of gravity) that might be expected (with a probability of 1 in 10) to be exceeded in a 50 year period.

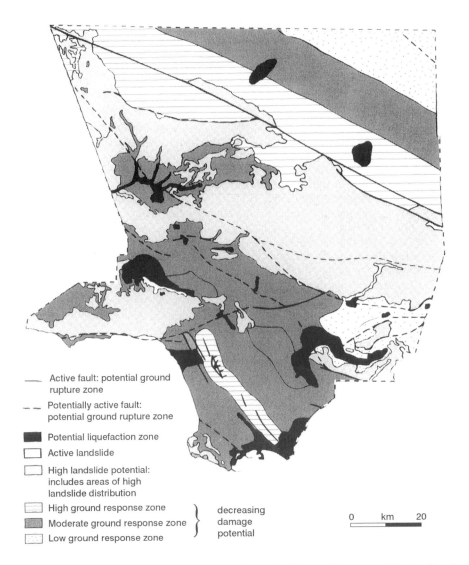

Fig. 3.9 Micro-seismic zoning map of Los Angeles County, California. (Courtesy of the Los Angeles County Department of Regional Planning.)

3.3 Slope movements

3.3.1 Soil creep

Movements on slopes can range in magnitude from soil creep on the one hand to instantaneous and colossal landslides on the other. Soil creep is the slow downslope movement of superficial rock or soil debris, which usually is imperceptible except by observations of long duration. It is a more or less continuous process, which is a distinctly surface phenomenon and occurs on slopes with gradients somewhat in excess of the angle of repose of the material involved. Like landslip, its principal cause is gravity. Evidence of soil creep may be found on almost every soil-covered slope. For example, it occurs in the form of small terracettes, downslope tilting of poles, the curving downslope of trees, and soil accumulation on the uphill sides of walls. Indeed, walls may be displaced or broken and sometimes roads may be moved out of alignment by soil creep.

Solifluction is a form of creep that occurs in cold climates or high altitudes where masses of saturated rock waste move downslope. Generally, the bulk of the moving mass consists of fine debris but blocks of appreciable size also may be moved. Saturation may be due to either water from rain or melting snow. Moreover, in many periglacial regions water cannot drain into the ground since it is frozen permanently. Solifluction differs from mudflow in that it moves much more slowly, the movement is continuous and it occurs over the whole slope. Solifluction processes vary at different altitudes. At higher elevations surfaces are irregular and terraces are common, but at lower elevations the most common solifluction phenomena are continuous aprons of detritus that skirt the bases of all the more prominent relief features.

3.3.2   Causes of landslides

Landslides are rapid downward and outward movement of slope-forming materials, the displaced material having well defined boundaries. Movement may take place by falling, sliding or flowing, or some combination of these factors. This movement generally involves the development of a slip surface between the separating and remaining masses. However, rock falls, topples and debris flows involve little or no true sliding on a slide surface. In most landslides a number of causes contribute towards movement. Often the final factor is nothing more than a trigger mechanism that sets in motion a mass that was already on the verge of failure. Although many factors can influence slope stability, basically landslides occur because the forces creating movement, the disturbing forces (M_D), exceed those resisting it, the resisting forces (M_R), that is, the shear strength of the material concerned. In general terms, therefore, the stability of a slope may be defined by a factor of safety (F) where:

$$F = M_R/M_D \qquad (3.1)$$

If the factor of safety exceeds one, then the slope is stable, whereas if it is less than one, the slope is unstable.

The common force tending to generate movements on slopes is gravity. Generally, the steeper the slope, the greater is the likelihood that landslides will occur. However, stability must be related to the ground conditions. Indeed, many steep slopes on competent rock are more stable than comparatively gentle slopes on weak material. Other causes of landslides can be grouped into two categories, namely, external causes and internal causes. External mechanisms are those outside the mass involved, which are responsible for overcoming its internal shear strength, thereby causing it to fail. By constrast, internal causes include those mechanisms within the mass that help to reduce its shear strength to a point below the external forces imposed on the mass by its environment, thus inducing failure.

An increase in the weight of slope material means that shearing stresses are increased, leading to a decrease in the stability of a slope, which may ultimately give rise to a slide. This can be brought about by natural or artificial (man-made) activity. For instance, removal of support from the toe of a slope, either by erosion or excavation, is a frequent cause of slides, as is overloading the top of a slope. Such slides are external slides in that an external factor causes failure. Other external mechanisms include earthquakes or other shocks and vibrations.

Internal slides generally are caused by an increase of pore water pressures within the slope material, which causes a reduction in effective shear strength. Indeed, groundwater constitutes the most important single contributory cause in most landslides. An increase in water content also means an increase in the weight of the slope material or its bulk density, which can induce slope failure. Significant volume changes may occur in some materials, notably expansive clay soils, on wetting and drying. Not only does this weaken the clay soil by developing desiccation cracks within it, but any enclosing strata also may be affected adversely. Weathering also can effect a reduction in strength of slope material, leading to sliding.

A slope in dry coarse-grained soil should be stable provided its inclination is less than the angle of repose. Slope failure tends to be caused by the influence of water. For instance, seepage of groundwater through a deposit of sand in which slopes exist can cause them to fail. Failure on a slope composed of granular soil involves the translational movement of a shallow surface layer. The slip often is appreciably longer than it is in depth, which is due to the strength of granular soils increasing rapidly with depth. Although shallow slips are common, deep-seated shear slides can occur in coarse-grained soils.

Slope and height are interdependent in fine-grained soils, notably clay soils, and can be determined when the shear strength characteristics of the material are known. Because of their water-retaining capacity, due to their low permeability, pore water pressures (u) are developed in such soils. These pore water pressures reduce the strength of the soil. Thus, in order to derive the strength of an element of the failure surface within a slope in fine-grained soil, the pore water pressure at that point needs to be determined to obtain the total (σ) and effective stresses (σ'), ($\sigma - u = \sigma'$). This effective stress then is used as the normal stress in a shear box or triaxial test to assess the shear strength of the soil concerned. The resistance offered along a slip surface in a slope in fine-grained soil, that is, its shear strength (s) is given by

$$s = c' + (\sigma - u)\tan\phi' \tag{3.2}$$

where c' is the cohesion intercept, ϕ' is the angle of shearing resistance (these are average values around the slip surface and are expressed in terms of effective stress), σ is the total overburden pressure, and u is the pore water pressure. In a stable slope only part of the total available shear resistance along a potential slip surface is mobilized to balance the total shear force (τ), hence

$$\Sigma\tau = \Sigma c'/F + \Sigma(\sigma - u)\tan\phi'/F \tag{3.3}$$

If the total shear force equals the total shear strength a slip may occur at some point in the future (that is, $F = 1.0$). Fine-grained soils, especially in the short-term, may exhibit relatively uniform strength with increasing depth. As a result slope failures, particularly short-term failures, may be comparatively deep-seated, with roughly circular slip surfaces. This type of failure is typical of relatively small slopes. Landslides on larger slopes are often non-circular failure surfaces following bedding planes or other weak horizons.

Landsliding in hard unweathered rock is largely dependent on the incidence, orientation and nature of the discontinuities present. The value of the angle of shearing resistance (ϕ) depends on the type and degree of interlock between the blocks on either side of the surface of sliding, but in such rock masses interlocking is independent of the orientation of that surface.

In a bedded and jointed rock mass, if the bedding planes are inclined, the critical slope angle depends upon their orientation in relation to the slope and the orientation of the joints. The relation between the angle of shearing resistance (ϕ) along a discontinuity, at which sliding will occur under gravity, and the inclination of the discontinuity (α) is important. If $\alpha < \phi$, then the slope is stable at any angle, whilst if $\phi < \alpha$, then gravity will induce movement along the discontinuity surface and the slope cannot exceed a critical angle, which will have a maximum value equal to the inclination of the discontinuities. It must be borne in mind, however, that rock masses generally are interrupted by more than one set of discontinuities.

3.3.3 Classification of landslides
Landslides can be classified according to the type of movement undergone on the one hand and the type of materials involved on the other. Three types of movement are recognized, namely, falls, slides and flows, but one can grade into another. The materials concerned are simply rocks and soils.

In rock falls, the moving mass travels mostly through the air by free-fall, saltation or rolling, with little or no interaction between the moving fragments. Movements are very rapid and may not be preceded by minor movements. The fragments are of various sizes in rock falls, generally are broken in the fall and mainly accumulate at the bottom of a slope. Toppling failure is a special type of rock fall that involves overturning and failure of steeply dipping strata (Fig. 3.10).

In true slides the movement results from shear failure along one or several surfaces, such surfaces offering the least resistance to movement. One of the most common types of slide occurs in clay soils where the slip surface is approximately spoon-shaped. Such slides are referred to as rotational slides

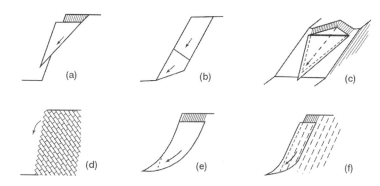

Fig. 3.10 Idealized failure mechanisms in rock slopes (a) plane (b) active and passive blocks (c) wedge (d) toppling (e) circular (f) non-circular.

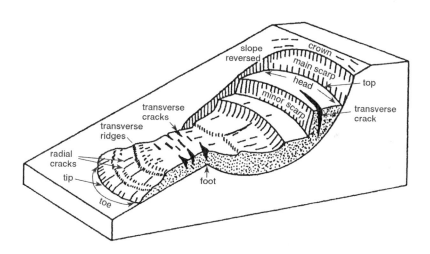

Fig. 3.11 Main features of rotational sliding, showing a series of retrogressive slides.

(Fig. 3.11). Rotational slides usually develop from tension scars in the upper part of a slope, the movement being more or less rotational about an axis located above the slope. When the scar at the head of a rotational slide is almost vertical and unsupported, then further failure usually is just a matter of time. Consequently, successive rotational slides occur until the slope is stabilized. These are retrogressive slides and they develop in a headward direction.

Translational slides occur in inclined stratified deposits, the movement occurring along a planar surface (Fig. 3.10), frequently a bedding plane. The mass involved in the movement becomes dislodged because the force of gravity overcomes the frictional resistance along the potential slip surface, the mass having been detached from the parent rock by a prominent discontinuity such as a major joint. A wedge

Fig. 3.12   A mudflow invading a house in Westville, South Africa.

is a type of translational slide in which two discontinuities intersect (Fig. 3.10). Slab slides, in which the slip surface is roughly parallel to the ground surface, are a common type of translational slide.

Rock slides and debris slides are usually the result of a gradual weakening of the bonds within a rock mass and generally are translational in character. Most rock slides are controlled by the discontinuity patterns within the parent rock. Rock slides commonly occur on steep slopes and most of them are of single rather than multiple occurrence. They are composed of rock boulders. Debris slides usually are restricted to the weathered zone or to surficial talus. With increasing water content debris slides grade into mudflows.

Some water content is necessary for most types of flow movement but dry flows can occur. Dry flows are very rapid and short-lived, and frequently are composed mainly of silt or sand. Debris flows are distinguished from mudflows on the basis of particle size, the former containing a high percentage of coarse fragments, whilst the latter consist of at least 50% sand-size particles or less. These flows follow unusually heavy rainfall or sudden thaw of frozen ground. They have a high density, perhaps 60 to 70% solids by weight, and are capable of carrying large boulders. Mudflows may develop when a rapidly moving stream of storm water mixes with a sufficient quantity of debris to form a pasty mass (Fig. 3.12). They frequently move at rates ranging between 10 and 100 m min^{-1} and can travel down shallow slopes. Both debris flows and mudflows may move over many kilometres. An earthflow involves mostly fine-grained material that may move slowly or rapidly, depending on its water content in that the higher the content, the faster the movement. If the material is saturated, a bulging frontal lobe is formed and this may split into a number of tongues that advance with a steady rolling motion. Earthflows frequently form the spreading toes of rotational slides due to the material being softened by the ingress of water.

3.3.4   Monitoring movement

Small movements usually precede slope failure, particularly a catastrophic failure, and accelerating displacement frequently precedes collapse. If these initial movements are detected in sufficient time, remedial action can be taken to prevent or control further movement. A slope-monitoring system provides a means of early warning. One of the first steps in the planning and design of a monitoring system is to assess the extent and depth of potentially unstable material, and to determine the factors of safety against sliding for various modes of failure. This indicates whether there is a problem of slope stability, and aids the choice of instrumentation and its location within the slope to be made.

Monitoring of movement provides a direct check on the stability of a slope. Instruments indicate the location, direction and maximum depth of movements, and the results help determine the extent and depth of treatment that is necessary. What is more, the same instruments can be used to determine the effect of this treatment. Monitoring of surface movement can be done by conventional surveying techniques, the use of electronic distance measurement or laser equipment providing accurate results. Automated slope-monitoring instruments can be programmed to take various measurements across a slope. Movements also may be revealed by an examination of a sequence of photographs taken at suitable time intervals. The appearance of tension cracks at the crest of a slope may provide the first indication of instability. Crack measurements, that is, their width and vertical offsets, should be taken since they may provide an indication of slope behaviour.

Single point tiltmeters also can be used to monitor surface movements, when the surface of the landslide has a rotational component. Portable tiltmeters can be used with a series of reference places or tiltmeters can be left in place and connected to a datalogger.

Measurements of subsurface horizontal movements are more important than measurements of subsurface vertical movements. Subsurface movements can be recorded by using extensometers, inclinometers and deflectometers. Borehole extensometers are used to measure vertical displacement of the ground at different depths. An inclinometer is used to measure horizontal movements below ground. High accuracy inclinometer measurements frequently represent the initial data relating to subsurface movement. Multiple deflectometers operate on a similar principle to in-place inclinometers but rotation is measured by angle transducers instead of tilt transducers. These deflectometers can detect shear deformation across a borehole at any inclination.

Groundwater is one of the most influential factors governing the stability of slopes. Instability problems may be associated with either excessive discharge or excessive pore water pressure. Pore water pressures are recorded by a piezometer installed in a borehole (see Chapter 8). There are important differences between the monitoring of water pressures in rock and in soil. Usually, in rock the majority of flow takes place via discontinuities rather than through intergranular pore space. The predominance of fissure flow means that piezometer heads in rock slopes often vary considerably from point to point and therefore a sufficient number of piezometers must be installed to define the overall conditions. They should be located with reference to the geological conditions, especially with regard to the intersection of major discontinuities in rock masses.

Movements in rock or soil masses are accompanied by generation of acoustic emissions or noise. The detection of acoustic emissions is most effective when the amplitude of the signals is high. Hence, detection is more likely in rock masses or coarse-grained soils than fine-grained soils. Obviously, when a slope collapses noise is audible, but sub-audible noises are produced at earlier stages in the development of instability. Normally, the rate of these microseismic occurrences increases rapidly with the development of instability. Such noises can be picked up by an array of geophones located in the vicinity of a slope or in shallow boreholes. Most movements generating noise originate near or along the plane of failure, so that seismic detection helps locate the depth and extent of the surface of sliding.

3.3.5 Landslide hazard maps

A checklist is a useful starting point in landslide investigations. Each separate slope unit can be classified according to a stability rating and, furthermore, a checklist provides for systematic examination of the main factors influencing mass movement. Aerial photographs, precise surveying and geophysical methods have proved of value in landslide investigation. Computers can be used to produce maps of slope stability from the data obtained from aerial photographs and field investigations. Landslide hazard maps delineate areas that probably will be affected by slope instability within a given period of time. Landslide risk maps attempt to quantify the vulnerability of an area either in relation to the probability of occurrence of a landslide or to the likelihood of damage to property or injury or death to the person.

Generally, the purpose of mapping landslide hazards is to locate problem areas and to help understand why, when and where landslides are likely to occur. Three basic principles have guided the production of landslide hazard zonation maps. The first involves the concept that slope failures in the future will take place for reasons similar to those that gave rise to failures in the past. Secondly, the basic causes of slope instability are fairly well known, thereby aiding recognition and mapping of slides. Also, it often is possible to estimate the relative contribution to slope instability of the conditions and processes that are responsible for slope instability once they have been identified. Thirdly, in this way the degree of potential hazard can be assessed, depending on the correct recognition of the number of failure-inducing factors present, their severity and interaction.

Many landslide maps only show the hazards known at a particular time whereas others provide an indication of the possibility of landslide occurrence (Fig. 3.13). The latter are landslide susceptibility maps and involve some estimation of relative risk. An assessment of the level of risk attributable to landslide occurrence for a particular area involves classification of the data obtained into risk groups. Most risk classifications recognize low, medium and high levels of risk but the categories usually are poorly defined. Multivariate landslide susceptibility maps produced by the statistical evaluation of the physical factors that influence slope instability, generally are based on grid cells and produced by computer. However, grid cells of fixed size have the disadvantage of often relating poorly to geomorphological slope units. Consequently, later work, facilitated by the use of geographical information systems (GIS), has associated land characteristics with geomorphological units.

Effective landslide hazard management has done much to reduce economic and social losses due to slope failure by avoiding the hazards or by reducing the damage potential. This has been accomplished by restrictions placed on development in landslide-prone areas; application of excavation grading, landscaping and construction codes; use of remedial measures to prevent or control slope failures; and landslide warning systems. Future policy should develop and promote techniques, and initiate hazard recognition and reduction schemes as preventative measures. Risk reduction can be achieved with reference to either the process system or the degree of vulnerability of the land use. The final choice of risk reduction measures depends on the type of development, either in existence or proposed, and the type, magnitude and time scale of the hazardous process.

3.3.6 Landslide prevention and remediation (see also Section 11.1.3)

If landslides are to be prevented, then areas of potential landsliding first must be identified, as must their type and possible amount of movement. Then, if the hazard is sufficiently real the engineer can devise a method of preventative treatment. Economic considerations, however, cannot be disregarded. In this respect it is seldom economical to design cut slopes sufficiently flat to preclude the possibility of landslides. Indeed, many roads in upland terrain could not be constructed with the finance available without accepting some risk of landslides.

Rock slopes normally cannot be designed economically so that no rock falls occur, hence unless absolute security is necessary, small falls of rock under controlled conditions are acceptable. Subsequent slope treatment may take the form of a reduction in the overall slope angle so as to increase the factor of safety. Care must be taken to avoid damaging a slope when it is being trimmed by blasting and to maintain a constant slope line. One of the prerequisites for safe rock slopes involves sealing of loose blocks. Single-mesh fencing supported by rigid posts will contain small rock falls. Larger, heavy-duty catch fences or nets are required for larger rock falls. Yet other types of fences engage energy absorbing friction brakes that extend the time of collision and in this way increase the capacity of the nets to restrain the falling rocks. Wire meshing bolted into rock slopes is one of the most effective methods of preventing rock falls from steep slopes. Wire mesh also can be suspended from the top of a face in order to control rock fall. Cable lashing and cable nets can be used to restrain loose rock blocks. Rock traps in the form of a ditch and/or barrier can be installed at the foot of a slope. Benches also may act as traps to retain rock fall, especially if a barrier is placed at their edge. Where a road or railway passes along the foot of a steep slope, then protection from rock fall can be afforded by the construction of a rigid canopy from the rock face.

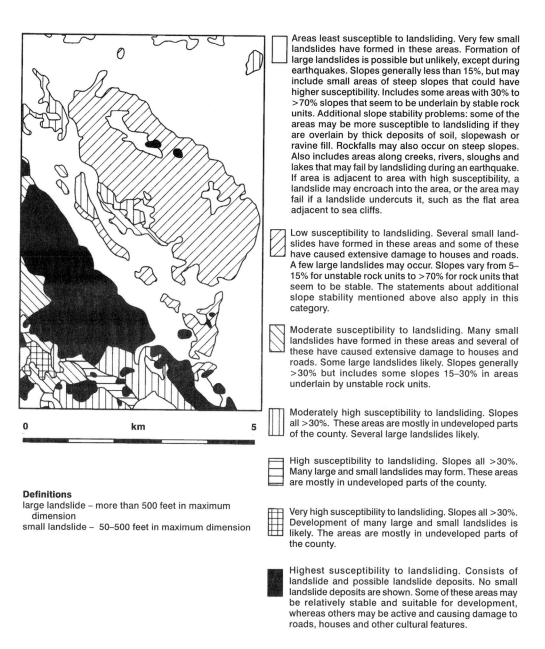

Areas least susceptible to landsliding. Very few small landslides have formed in these areas. Formation of large landslides is possible but unlikely, except during earthquakes. Slopes generally less than 15%, but may include small areas of steep slopes that could have higher susceptibility. Includes some areas with 30% to >70% slopes that seem to be underlain by stable rock units. Additional slope stability problems: some of the areas may be more susceptible to landsliding if they are overlain by thick deposits of soil, slopewash or ravine fill. Rockfalls may also occur on steep slopes. Also includes areas along creeks, rivers, sloughs and lakes that may fail by landsliding during an earthquake. If area is adjacent to area with high susceptibility, a landslide may encroach into the area, or the area may fail if a landslide undercuts it, such as the flat area adjacent to sea cliffs.

Low susceptibility to landsliding. Several small landslides have formed in these areas and some of these have caused extensive damage to houses and roads. A few large landslides may occur. Slopes vary from 5–15% for unstable rock units to >70% for rock units that seem to be stable. The statements about additional slope stability mentioned above also apply in this category.

Moderate susceptibility to landsliding. Many small landslides have formed in these areas and several of these have caused extensive damage to houses and roads. Some large landslides likely. Slopes generally >30% but includes some slopes 15–30% in areas underlain by unstable rock units.

Moderately high susceptibility to landsliding. Slopes all >30%. These areas are mostly in undeveloped parts of the county. Several large landslides likely.

High susceptibility to landsliding. Slopes all >30%. Many large and small landslides may form. These areas are mostly in undeveloped parts of the county.

Very high susceptibility to landsliding. Slopes all >30%. Development of many large and small landslides is likely. The areas are mostly in undeveloped parts of the county.

Highest susceptibility to landsliding. Consists of landslide and possible landslide deposits. No small landslide deposits are shown. Some of these areas may be relatively stable and suitable for development, whereas others may be active and causing damage to roads, houses and other cultural features.

0   km   5

Definitions
large landslide – more than 500 feet in maximum dimension
small landslide – 50–500 feet in maximum dimension

Fig 3.13 Landslide susceptibility classification used in a study of San Mateo County, California. (Courtesy of US Geological Survey.)

Landslide prevention may be brought about by reducing the activating forces, by increasing the forces resisting movement or by avoiding or eliminating the slide. Reduction of the activating forces can be accomplished by removing material from that part of the slide that provides the force that gives rise to movement. Complete excavation of potentially unstable material from a slope may be feasible, however, there is an upper limit to the amount of material that can be removed economically. Although partial removal is suitable for dealing with most types of mass movement, for some types it is inappropriate. For example, removal of the head has little influence on flows or slab slides. On the other hand, this treatment is eminently suitable for rotational slips. Slope flattening, however, rarely is applicable to rotational or slab slides. Slope reduction may be necessary in order to stabilize the toe of a slope and so prevent successive undermining with consequent spread of failure upslope. Benching can be used on steeper slopes. It brings about stability by dividing a slope into segments.

If some form of reinforcement is required to provide support for a slope, then it is advisable to install it as quickly as possible. Dentition refers to masonry or concrete infill placed in fissures or cavities in a rock slope. Steel dowels, grouted into predrilled holes, and rock bolts may be used as reinforcement to enhance the stability of slopes in jointed rock masses. Rock anchors are used for major stabilization works. Gunite or shotcrete frequently is used to seal rock surfaces, and can be reinforced by wire mesh and/or rock bolts. In soil nailing the nails consist of steel bars, metal rods or metal tubes driven into the slope and covered with a layer of shotcrete reinforced with wire mesh.

Restraining structures, such as retaining walls, control sliding by increasing the resistance to movement. Similarly, reinforced earth can be used for retaining earth slopes. Cribs and gabions also are used to restrain slopes, the former are constructed of precast reinforced concrete or steel units and the latter consist of stones placed in wire mesh containers (Fig. 3.14).

Drainage is the most generally applicable method of improving slope stability or for the corrective treatment of slides, regardless of type, since it reduces the effectiveness of excess pore water pressure. In rock masses groundwater also tends to reduce the shear strength along discontinuities. Moreover, drainage is the only economic way of dealing with slides involving the movement of several million cubic metres. The surface of a landslide generally is uneven, hummocky and traversed by deep fissures. This is particularly the case when the slipped area consists of a number of slices. Water collects in depressions and fissures, and pools and boggy areas are formed. In such cases the first remedial measure to be carried out is surface drainage. Surface run-off should be prevented from flowing unrestrained over a slope by a drainage ditch at the top of a slope and herringbone ditch drainage, leading into an interceptor drain at the foot of the slope, which conveys water from its surface. Subsurface drainage galleries are designed to drain by gravity. Galleries may be essential in large slipped masses where drainage has to be carried out over lengths of 200 m or more. Drain holes may be drilled around the perimeter of a gallery to enhance drainage. Subhorizontal drainage holes are much cheaper than galleries and are satisfactory over short lengths but it is more difficult to intercept water-bearing layers with them. They can be inserted from the ground surface or by drilling from drainage galleries, large diameter wells or caissons.

3.4 River action and flooding

3.4.1 Basic stream hydraulics

All rivers form part of a drainage system, the form of which is influenced by rock type and structure, the nature of the vegetation cover and the climate. Rivers form part of the hydrological cycle in that they carry precipitation run-off. Although, due to heavy rainfall or in areas with few channels, the run-off may occur as a sheet, generally speaking it quickly becomes concentrated into channels.

The velocity of a stream depends upon channel gradient, volume and configuration. One of the most commonly used equations applicable to open channel hydraulics is the Chezy formula:

$$v = C\sqrt{(RS)} \qquad (3.4)$$

where v is the mean velocity, C is a coefficient that varies with the characteristics of the channel, R is the hydraulic radius and S is the slope. Numerous attempts have been made to find a generally acceptable

Fig. 3.14 Typical use of gabions. (a) Gabion wall, and drainage with stilling blocks,used to retain landslipped material, Hong Kong. (b) Woven mesh gabion wall at Langside College, Glasgow, Scotland. Stone fill was reused and reclaimed sandstone from demolition works elsewhere on the site (Courtesy, Maccaferri Ltd, Oxford, UK). (c) Woven mesh gabions (maximum height 7m) on the M3 motorway Winchester, Hants., England (Courtesy, Maccaferri Ltd, Oxford, UK).

expression for *C*. The Manning formula, based on field and experimental determinations of the value of the resistance coefficient (*n*) is widely used:

$$v = \frac{1.49}{n} R^{2/3} S^{1/2} \tag{3.5}$$

The experimental values of *n* vary from approximately 0.01 for smooth metal surfaces to 0.06 for irregular channels containing large stones.

The efficiency of a river channel is defined as the ratio between its cross-sectional area and the length of its wetted perimeter. This ratio is termed the hydraulic radius, and the higher its value, the more efficient is the river. The most efficient forms of channel are those that have approximately circular or rectangular sections with widths approaching twice their depths whereas the most inefficient channel forms are very broad and shallow with wide wetted perimeters.

The quantity of flow can be estimated from measurements of cross-sectional area and current speed of a river. Generally, channels become wider relative to their depth and adjusted to larger flows with increasing distance downstream. Bankfull discharges also increase downstream in proportion to the square of the width of the channel or of the length of individual meanders, and in proportion to the 0.75 power of the total drainage area focused at the point in question. Statistical methods are used to predict river flow and assume that recurrence intervals of extreme events bear a consistent relationship to their magnitudes. A recurrence interval, generally of 50 or 100 years, is chosen in accordance with given hydrological requirements. The concept of the unit hydrograph postulates that the most important

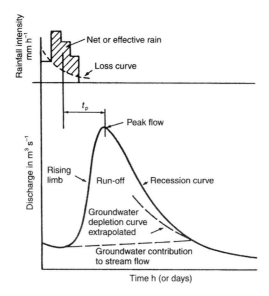

Fig. 3.15 Component parts of a hydrograph. There is an initial period of interception and infiltration when rainfall commences before any measurable run-off reaches the stream channels. During the period of rainfall, these losses continue in a reduced form, so the rainfall graph must be adjusted to show effective rain. When the initial losses are met, surface run-off begins and continues to a peak value, which occurs at time t_p, measured from the centre of gravity of the effective rain on the graph. Thereafter, surface run-off declines along the recession limb until it disappears. Baseflow represents the groundwater contribution along the banks of the river.

hydrological characteristics of any basin can be seen from the direct run-off hydrograph resulting from 25 mm of rainfall evenly distributed over 24 h. This is produced by drawing a graph of the total stream flow at a chosen point as it changes with time after such a storm, from which the normal base-flow caused by groundwater is subtracted (Fig. 3.15).

There is a highly significant relationship between mean annual discharge per unit area and drainage density. Drainage density is defined as the stream length per unit of area and varies inversely as the length of overland flow, so providing some indication of the efficiency of a drainage basin. It reflects the concentration or stream frequency of an area. Peak discharge and the lag time of discharge (the time that elapses between maximum precipitation and maximum run-off) also are influenced by drainage density, as well as by the shape and slope of the drainage basin. A relationship exists between drainage density and base-flow or groundwater discharge. This is related to the permeability of the rocks present in a drainage basin in that the greater the quantity of water that moves on the surface of the drainage system, the higher the drainage density, which in turn means that the base-flow is lower. River flow generally is most variable and flood discharges at a maximum per unit area in small basins. This is because storms tend to be local in occurrence.

3.4.2 Work of rivers

The work undertaken by a river is threefold, namely, it erodes soils and rocks, and transports the resultant products, which subsequently are deposited. In the early stages of river development erosion tends to be greatest in the lower part of a drainage basin. As the basin develops, however, the zone of maximum erosion moves upstream and in the later stages it is concentrated along the river basin divides. The

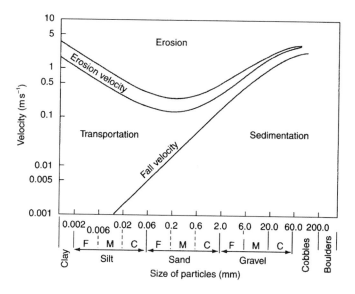

Fig. 3.16 Curves for erosion, transportation and deposition of uniform sediment. F = fine, M = medium, C = coarse-grained. Note that fine sand is the sediment eroded most easily.

amount of erosion accomplished by a river in a given time depends upon its volume and velocity of flow; the character and size of its load; the soil/rock type and geological structure over which it flows; the infiltration capacity of the area it drains; the vegetation that, in turn, affects soil stability; and the permeability of the soil. The volume and velocity of a river influences the quantity of energy it possesses. When flooding occurs the volume of a river is increased greatly, thereby leading to an increase in its velocity, capacity to erode and competence to transport material. The velocity required to initiate particle movement, that is, erosional velocity, is appreciably higher than that required to maintain movement. Figure 3.16 indicates the range of current velocity for erosion, transportation and deposition of well sorted sediments.

The competence of a river to transport its load is demonstrated by the largest boulder it is capable of moving and varies according to velocity and volume. In fact, the competence of a river varies as the sixth power of its velocity and is at a maximum during flood. The capacity of a river refers to the total amount of sediment that it carries and varies according to the size of the particles that form the load, and the velocity of the river. When the load consists of fine particles, the capacity is greater than when it is comprised of coarse material. Usually, the capacity of a river varies as the third power of its velocity. Both the competence and capacity of a river are influenced by changes in the weather, and the lithology and structure of the ground over which it flows, as well as by vegetative cover and land use. Because the discharge of a river varies, some sediments are not transported continuously, for instance, boulders may be moved only a few metres during a single flood. Sediments that are deposited over a flood-plain may be regarded as being stored there temporarily.

Deposition occurs where turbulence is at a minimum or where the region of turbulence is near the surface of a river. For example, lateral accretion occurs, with deposition of a point bar, on the inside of a meander bend. An individual point bar grows as a meander migrates downstream and new ones are formed as a river changes its course during flood. Consequently, old meander scars are common features of flood plains. The combination of point bar and filled slough or oxbow lake gives rise to ridge and

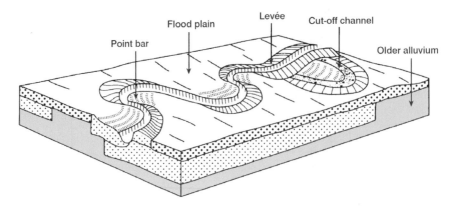

Fig. 3.17   The main depositional features of a meandering channel.

swale topography. The ridges consist of sandbars and the swales are sloughs filled with silt and clay. An alluvial flood-plain is the most common depositional feature of a river. The alluvium is made up of many kinds of deposits, laid down both in the channel and outside it. Vertical accretion of a flood-plain is accomplished by in-channel filling and the growth of overbank deposits during and immediately after floods. Gravel and coarse sand are moved chiefly at flood stages and deposited in the deeper parts of a river. As the river overtops its banks, its ability to transport material is lessened so that coarser particles are deposited near the banks to form levées (Fig. 3.17).

3.4.3   Floods

Floods represent one of the commonest types of geohazard (Fig. 3.18). Rivers may be considered in flood when their level has risen to an extent that damage occurs. Most disastrous floods are the result of excessive precipitation or snowmelt. However, the likelihood of flooding is more predictable than some other types of geohazard. Usually, it takes some time to accumulate enough run-off to cause a major disaster. This lag time is an important parameter in flood forecasting. For example, long-rain floods are associated with several days or even weeks of rainfall, which may be of low intensity. They are the most

Fig. 3.18   Flooding in Ladysmith, South Africa, February, 1994.

common cause of major flooding. As the physical characteristics of a river basin, together with those of the stream channel, affect the rate at which discharge downstream occurs, this means that calculation of the lag time is a complicated matter. Flash floods prove the exception. Flash floods are short-lived extreme events. They frequently occur as a result of slow moving or stationary thunderstorms. The resulting rainfall intensity exceeds infiltration capacity so that run-off takes place very rapidly. Flash floods generally are very destructive since the high energy flow can transport much sedimentary material. In most regions floods occur more frequently in certain seasons than others. Meltwater floods, for instance, are seasonal and are characterized by a substantial increase in river discharge so that there is a single flood wave. The latter may have several peaks. The two most important factors that govern the severity of floods due to snowmelt are the depth of snow and the rapidity with which it melts.

The influence of human activity can bring about changes in drainage basin characteristics, for example, removal of forest from parts of a river basin can lead to higher peak discharges that generate increased flood hazard. A most notable increase in flood hazard can arise as a result of urbanization, the impervious surfaces created mean that infiltration is reduced, and together with stormwater drains, they give rise to increased run-off. Not only does this produce higher discharges but lag-times also are reduced. The problem of flooding is particularly acute where rapid expansion has led to the development of urban sprawl without proper planning, or worse where informal settlements have sprung up.

An estimate of future flood conditions is required for either forecasting or design purposes. In terms of flood forecasting more immediate information is needed regarding the magnitude and timing of a flood so that appropriate evasive action can be taken. In terms of design, planners and engineers require data on magnitude and frequency of floods. Hence, there is a difference between flood forecasting for warning purposes and flood prediction for design purposes. A detailed understanding of the run-off processes involved in a catchment and stream channels is required for the development of flood forecasting but is less necessary for long-term prediction. The ability to provide enough advance warning of a flood means that it may be possible to evacuate people from an area likely to be affected, along with some of their possessions. The most reliable forecasts are based on data from rainfall or melt events that have just taken place or are still occurring. Hence, advance warning generally is measured only in hours or sometimes in a few days. Longer-term forecasts commonly are associated with snowmelt.

Basically, flood forecasting involves the determination of the amount of precipitation and resultant run-off within a given catchment area. Once enough data on rainfall and run-off versus time have been obtained and analysed, an estimate of where and when flooding will occur along a river system can be made. Analyses of discharge are related to the recurrence interval to produce a flood frequency curve. The recurrence interval is the period of years within which a flood of a given magnitude or greater occurs and is determined from field measurements at a gauging centre. It applies to that station only. The flood frequency curve can be used to determine the probability of the size of discharge that could be expected during any given time interval. As far as prediction for design purposes is concerned, the design flood, namely, the maximum flood against which protection is being designed, is the most important factor to determine.

3.4.4   River control and flood warning systems

The lower stage of a river, in particular, can be divided into a series of hazard zones based on flood stages and risk. A number of factors have to be taken into account when evaluating flood hazard such as the loss of life and property; erosion and structural damage; disruption of socio-economic activity including transport and communications; contamination of water, food and other materials; and damage to agricultural land and loss of livestock. Obviously, a floodplain management plan involves the determination of such zones. These are based on the historical evidence related to flooding, which includes the magnitude of each flood and the elevation it reached, as well as the recurrence intervals; the amount of damage involved; the effects of urbanization and any further development; and an engineering assessment of flood potential. Maps then are produced from such investigations that, for example, show the zones of most frequent flooding and the elevation of the flood waters. Flood hazard

Table 3.2 Generalized flood hazard zones and management.*

Flood hazard zone I (Active floodplain area)

Prohibit development (business and residential) within floodplain
Mountain area in a natural state as an open area or for recreational uses only.

Flood hazard zone II (Alluvial fans and plains with channels less than a metre deep, bifurcating, and intricately interconnected systems subject to inundation from overbank flooding)

Flood proofing to reduce or prevent loss to structures is highly recommended.
Residential development densities should be relatively low; development in obvious drainage channels should be prohibited.
Dry stream channels should be maintained in a natural state and/or the density of native vegetation should be increased to facilitate superior water drainage retention and infiltration capabilities.
Installation of upstream stormwater retention basins to reduce peak water discharges.Construction should be at the highest local elevation site where possible.

Flood hazard zone III (Dissected upland and lowland slopes; drainage channels where both erosional and depositional processes are operative along gradients generally less than 5%)

Similar to flood hazard zone II.
Roadways which traverse channels should be reinforced to withstand the erosive power of a channelled stream flow.

Flood hazard zone IV (Steep gradient drainages consisting of incised channels adjacent to outcrops and mountain fronts characterized by relatively coarse bedload material)

Bridges, roads, and culverts should be designed to allow unrestricted flow of boulders and debris up to a metre or more in diameter.
Abandon roadways which presently occupy the wash flood plains.
Restrict residential dwelling to relatively level building sites.
Provisions for subsurface and surface drainage on residential sites should be required.
Stormwater retention basins in relatively confined upstream channels to mitigate against high peak discharges.

---

*Kenny, R. 1990. Hydrogeomorphic flood hazard evaluation for semi-arid environments. *Quarterly Journal of Engineering Geology*, **23**, 333–336 (Courtesy of the Geological Society).

maps provide a basis for flood management schemes, for instance, four geomorphological flood hazard map units in central Arizona have formed the basis of a flood management plan (Table 3.2).

Land-use regulation seeks to obtain the beneficial use of floodplains with minimum flood damage and minimum expenditure on flood protection. In other words, the purpose of land-use regulation is to maintain an adequate floodway (i.e. the channel and those adjacent areas of the floodplain that are necessary for the passage of given flood discharge) and to regulate land-use development alongside it (i.e. in the floodway fringe that, for example, in the United States is the land between the floodway and the maximum elevation subject to flooding by the 100 year flood). Land use in the floodway in the United States now is severely restricted to agriculture and recreation. The floodway fringe also is a restricted zone in which land use is limited by zoning criteria based upon the degree of flood risk. However, land-use control involves the cooperation of the local population and authorities, and often central government. Any relocation of settled areas that are of high risk involves costly subsidization. Such cost must be balanced against the cost of alternative measures and the reluctance of some to move. In fact, change of land use in intensely developed areas usually is so difficult and so costly that the inconvenience of occasional flooding is preferred.

In some situations properties of high value that are likely to be threatened by floods can be flood-proofed. For instance, an industrial plant can be protected by a flood wall or buildings may have no windows below high water level and possess watertight doors, with valves and cut-offs on drains and sewers to prevent backing-up. Other structures may be raised above the most frequent flood levels. Most flood-proofing measures can be more readily and economically incorporated into buildings at the time of their construction rather than subsequently. Hence, the adoption and implementation of suitable design standards should be incorporated into building codes and planning regulations to ensure the integrity of buildings during flood events.

Reafforestation of slopes denuded of woodland tends to reduce run-off and thereby lowers the intensity of flooding. As a consequence, forests are commonly used as a watershed management technique. They are most effective in relation to small floods where the possibility exists of reducing flood volumes and delaying flood response. Nonetheless, if the soil is saturated differences in interception and soil moisture storage capacity due to forest cover will be ineffective in terms of flood response. Agricultural practices such as contour ploughing and strip cropping are designed to reduce soil erosion by reducing the rate of run-off. Accordingly, they can influence flood response.

Emergency action involves erection of temporary flood defences and possible evacuation. The success of such measures depends on the ability to predict floods and the effectiveness of the warning systems. A flood warning system, broadcast by radio or television, or from helicopters or vehicles, can be used to alert a community to the danger of flooding. Warning systems may use rain gauges and stream sensors equipped with self-activating radio transmitters to convey data to a central computer for analysis. Alternatively, a radar rainfall scan can be combined with a computer model of a flood hydrograph to produce real-time forecasts as a flood develops. However, widespread use of flood warning usually is only available in highly sensitive areas. The success of a flood warning system depends largely on the hydrological characteristics of a river. Warning systems often work well in large catchment areas that allow enough time between a rainfall or snowmelt event and the resultant flood peak to permit evacuation, and any other measures to be put into affect. By contrast, in small tributary areas, especially those with steep slopes or appreciable urban development, the lag-time may be so short that, although prompt action may save lives, it is seldom possible to remove or protect property.

Financial assistance in the form of government relief or insurance pay-outs do nothing to reduce flood hazard. Indeed, by attempting to reduce the economic and social impact of a flood, they encourage repair and rebuilding of damaged property that may lead to the next flood of similar size giving rise to more damage. What is more, the expectation that financial aid will be made available in such an emergency may result in further development of flood-prone areas.

River control refers to projects designed to hasten the run-off of flood waters or confine them within restricted limits, to improve drainage of adjacent lands, to check stream bank erosion or to provide deeper water for navigation. What has to be borne in mind, however, is that a river in an alluvial channel is continually changing its position due to the hydraulic forces acting on its banks and bed. As a consequence, any major modifications to the river regime that are imposed without consideration of the channel will give rise to a prolonged and costly struggle to maintain the change.

Artificial strengthening and heightening of levées, or constructing artificial levées, are frequent measures employed as protection against flooding. Because levées confine a river to its channel its efficiency is increased. The efficiency of a river also is increased by cutting through the constricted loops of meanders. River water can be run off into the cut-off channels when high floods occur. Canalization or straightening of a river can help to regulate flood flow, and improves the river for navigation. However, canalization in some parts of the world has only proved a temporary expedient. This is because canalization steepens the channel, as sinuosity is reduced. This, in turn, increases the velocity of flow and therefore the potential for erosion. Moreover, the base level for the upper reaches of the river has, in effect, been lowered, which means that channel incision will begin there. Although the flood and drainage problems are solved temporarily, the increase in erosion upstream increases sediment load. This then is deposited in the canalized sections that can lead to a return of the original flood and drainage problems.

Diversion is another method used to control flooding, this involves opening a new exit for part of the river water. However, any diversion must be designed in such a way that it does not cause excessive deposition in the main channel, otherwise it is defeating its purpose. If in some localities little damage will be done by flooding, then these areas can be inundated during times of high flood and so act as safety valves.

Channel regulation can be brought about by training walls or dams that are used to deflect channels into more desirable alignments or to confine them to lesser widths. Walls and dams can be used to close secondary channels and so divert or concentrate a river into a preferred course. In some cases ground sills or weirs need to be constructed to prevent undesirable deepening of the river bed by erosion. Bank revetment by pavement, rip-rap or protective mattresses to retard erosion usually is carried out along with channel regulation or it may be undertaken independently for the protection of the lands bordering a river.

The construction of a dam across a river in an attempt to control flood water can lead to other problems. It decreases the peak discharge and reduces the quantity of bedload through the river channel. The width of a river may be reduced downstream of a dam as a response to the decrease in flood peaks. Moreover, scour normally occurs immediately downstream of a dam. Removal of the finer fractions of the bed material by scouring action may cause armouring of a river channel.

River channels may be improved for navigation purposes by dredging. When a river is dredged its floor should not be lowered so much that the water level is appreciably lowered. In addition, the nature of the materials occupying the floor and their stability should be investigated. This provides data from which the stability of the slopes of the projected new channel can be estimated and indicates whether blasting is necessary. The rate at which sedimentation takes place provides some indication of the regularity with which dredging should be carried out.

3.5 Marine action

Waves, acting on beach material, are a varying force. They vary with time and place due, firstly, to changes in wind force and direction over a wide area of sea, and secondly, to changes in coastal aspects and offshore relief. This variability means that a beach is rarely in equilibrium with the waves, in spite of the fact that it may only take a few hours for equilibrium to be attained under new conditions. Such a more or less constant state of disequilibrium occurs most frequently where the tidal range is considerable, as waves are continually acting at a different level on the beach.

Coastal erosion is not necessarily a disastrous phenomenon, however, problems arise when erosion and human activity come into conflict. Increases in the scale and density of development along coastlines has led to increased vulnerability, which means that better coastal planning is necessary. Those waves with a period of approximately four seconds usually are destructive while those with a lower frequency, that is, a period of about seven seconds or over are constructive. The rate at which coastal erosion proceeds is influenced by the nature of the coast itself and is most rapid where the sea attacks soft unconsolidated sediments. When soft deposits are actively eroded, a cliff displays signs of landsliding, together with evidence of scouring at its base. For erosion to continue the debris produced must be removed by the sea. This normally is accomplished by longshore drift. If, on the other hand, material is deposited to form extensive beaches and the detritus is reduced to a minimum size, then the submarine slope becomes very wide. Wave energy is dissipated as the water moves over such beaches and cliff erosion ceases.

3.5.1 Beaches

The beach slope is produced by the interaction of swash and backwash. Beaches undergoing erosion tend to have steeper slopes than prograding beaches. The beach slope also is related to the grain size, in general the finer the material, the gentler the slope and lower the permeability of a beach. For example, the loss of swash due to percolation into beaches composed of grains of 4 mm in median diameter is some ten times greater than into those where the grains average 1 mm. As a result there is almost as much water in the backwash on fine beaches as there is in the swash. The grain size of beach material, however, is continually changing because of breakdown and removal or addition of material.

Storm waves produce the most conspicuous constructional features on shingle beaches, but they remove material from sandy beaches. A small foreshore ridge develops on a shingle beach at the limit of the swash when constructional waves are operative. Similar ridges or berms may form on a beach composed of coarse sand, these being built by long low swells. Berms represent a marked change in slope and usually occur at a small distance above high-water mark. They are not conspicuous features on beaches composed of fine sand.

Sandy beaches are supplied with sand that usually is derived primarily from the adjacent sea floor, although in some areas a larger proportion is produced by cliff erosion. During periods of low waves the differential velocity between onshore and offshore motion is sufficient to move sand onshore except where rip currents are operational. Onshore movement is notable, particularly when long period waves approach a coast whereas sand is removed from the foreshore during high waves of short period.

The amount of longshore drift along a coast is influenced by coastal outline and wave length. Short waves can approach the shore at a considerable angle, and generate consistent downdrift currents. This is particularly the case on straight or gently curving shores and can result in serious erosion where the supply of beach material reaching the coast from updrift is inadequate. Conversely, long waves suffer appreciable refraction before they reach the coast.

Dunes are formed by onshore winds carrying sand-sized material landward from the beach along low-lying stretches of coast where there is an abundance of sand on the foreshore. They act as barriers, energy dissipators and sand reservoirs during storm conditions. Broad sandy beaches and high dunes along a coast present a natural defence against inundation during storm surges. Because dunes provide a natural defence against erosion, once they are breached the ensuing coastal changes may be long-lasting. In spite of the fact that dunes inhibit erosion, without beach nourishment (i.e. artificial replacement of beach material) they cannot be relied upon to provide protection in the long-term along rapidly eroding shorelines. Beach nourishment widens the beach and maintains the proper functioning of the beach-dune system during normal and storm conditions.

3.5.2 Storm surges

Except where caused by failure of protective works, marine inundation is almost always attributable to severe meteorological conditions giving rise to abnormally high sea levels, referred to as storm surges. Low pressure and driving winds during a storm may lead to marine inundation of low-lying coastal areas, particularly if this coincides with high spring tides. This is especially the case when the coast is unprotected. Floods may be frequent, as well as extensive where flood-plains are wide and the coastal area is flat. Coastal areas that have been reclaimed from the sea and are below high-tide level are particularly suspect if coastal defences are breached (Fig. 3.19). A storm surge can be regarded as the magnitude of sea level along a shoreline that is above the normal seasonally adjusted high-tide level. Storm surge risk often is associated with a particular season. The height and location of storm damage along a coast over a period of time, when analyzed, provides some idea of the maximum likely elevation of surge effects. The seriousness of the damage caused by storm surge tends to be related to the height and velocity of water movement.

Factors that influence storm surges include the intensity in the fall of atmospheric pressure, the length of water over which the wind blows, the storm motion and offshore topography. Obviously, the principal factor influencing a storm surge is the intensity of the causative storm, since this affects the speed of the wind piling up the sea against a coastline. For instance, threshold windspeeds of approximately 120 km h^{-1} tend to be associated with central pressure drops of around 34 mbar. Normally, the level of the sea rises with reductions in atmospheric pressure associated with intense low pressure systems. In addition, the severity of a surge is influenced by the size and track of a storm and, especially in the case of open coastline surges, by the nearness of the storm track to a coastline. The wind direction and the length of fetch (i.e. the length of water over which wind blows) also are important, both determining the size and energy of waves. Because of the influence of the topography of the sea floor, wide shallow areas on the continental shelf are more susceptible to damaging surges than those where the shelf slopes

Fig. 3.19 left: Marine inundation flooding polderland near Philipsand, the Netherlands, January 1953. right: Inundation of New Orleans in 2005 caused by breaching of levées due to Hurricane Katrina (Courtesy Global Security.org).

steeply. Surges are intensified by converging coastlines that exert a funnel effect as the sea moves into such inlets.

In addition to the storm surge, enclosed bays may experience seiching. Seiches involve the oscillatory movement of waves generated by a hurricane. Seiching also may occur in open bays if the storm moves forward very quickly and in low-lying areas this can be highly destructive.

Obviously, prediction of the magnitude of a storm surge in advance of its arrival can help to save lives. Storm tide warning services have been developed in various parts of the world. Warnings usually are based upon comparisons between predicted and observed levels at a network of tidal gauges. To a large extent the adequacy of a coastal flood forecasting and warning system depends on the accuracy with which the path of the storm responsible can be determined. Storms can be tracked by satellite, and satellite and ground data used for forecasting. Protective measures are not likely to be totally effective against every storm surge. Nevertheless, where flood protection structures are used extensively they often prove substantially effective.

3.5.3   Coastal protection

Before any coastal protection project can be started a complete study of the beach concerned must be made. The preliminary investigation of the area should first consider the landforms and geological formations along the beach and adjacent rivers, giving particular attention to their stability. The stage of development of the beach area in terms of the cycle of shore erosion, the sources of supply of beach material, probable rates of growth, and probable future history merit considerable discussion. Estimates of the rates of erosion, or the proportion and size of material contributed to the beach must be made. Consideration must be given to the width and slope of the beach, the presence of cliffs, dunes, marshy areas or vegetation in the backshore area, rock formations that compose headlands and the presence of any beach structures such as groynes. Samples of the beach and underwater material have to be collected and analyzed for such factors as their particle size distribution. Particle size analyses may prove useful in helping to determine the amount of material that is likely to remain on the beach, for beach sand is seldom finer than 0.1 mm in diameter. The amount of material moving along the shore must be investigated since the effectiveness of any structures erected may depend upon the quantity of drift available.

Fig. 3.20 Permeable groynes running from the sea wall, north of Clynnogfawr, North Wales.

Topographic and hydrographic surveys of an area allow the compilation of maps and charts from which a study of the changes along a coast may be made. Observations should be taken of winds, waves and currents. The contributions made by rivers that enter the sea in or near the area concerned vary. For example, material brought down by large floods may cause a temporary, but nevertheless appreciable, increase in the beach width around the mouths of rivers.

Groynes are the most important structures used to stabilize or increase the width of the beach by arresting longshore drift (Fig. 3.20). Consequently, they are constructed at right angles to the shore. Groynes should be approximately 50% longer than the beach on which they are erected. Standard types usually slope at about the same angle as the beach. Permeable groynes have openings that increase in size seawards and thereby allow some drift material to pass through them. The selection of the type of groyne and its spacing depends upon the direction and strength of the prevailing or storm waves, the amount and direction of longshore drift, and the relative exposure of the shore. With abundant longshore drift and relatively mild storm conditions almost any type of groyne appears satisfactory whereas when the longshore drift is lean, the choice is much more difficult. Groynes, however, reduce the amount of material passing downdrift and therefore can prove detrimental to those areas of the coastline. Therefore, their effect on the whole coastal system should be considered.

Beach nourishment or artificial replenishment of the shore by building beach-fills is used either if it is economically preferable or if artificial barriers fail to defend the shore adequately from erosion. Sand dunes and high beaches offer natural protection against marine inundation and can be maintained by artificial nourishment with sand. Dunes also can be constructed by tipping and bulldozing sand or by aiding the deposition of sand by the use of screens and fences. In fact, beach nourishment represents the only form of coastal protection that does not adversely affect other sectors of the coast. Unfortunately, it is often difficult to predict how frequently a beach should be renourished. Ideally, the beach fill used for renourishment should have a similar particle size distribution to the natural beach material.

Sea walls may be constructed of concrete or sometimes of steel sheet piling. These protective waterfront structures may rise vertically or they may be curved or stepped in cross-section. They are designed to prevent wave erosion and overtopping by high seas by dissipating or absorbing destructive wave energy. Sea walls must be stable, and this is usually synonymous with weight, impermeability and resistance to marine abrasion. The toe of the wall should not be allowed to become exposed. If basal erosion does becomes serious, then a row of sheet piling can be driven along the seaward edge of the foot of the wall. Unless the eroding foreshore in front of a sea wall is stabilized, the apron of the wall must be repeatedly extended seawards as the foreshore falls. Permeable sea walls may be formed of rip-rap or concrete tripods.

Bulkheads are vertical walls formed of either timber or steel sheet piling. Careful attention must be paid to the ground conditions in which these retaining structures are founded and due consideration given to the likelihood of scour occurring at the foot of the wall, as well as to changes in beach conditions. Cut-off walls of steel sheet piling below reinforced concrete super-structures provide an effective method of construction, ensuring protection against scouring.

Embankments frequently are used as a means of coastal protection and should be composed of impervious material or provided with a clay core to ensure that seepage does not occur. Clay is the normal material used for embankment construction. However, clay embankments may fail due to direct frontal erosion by wave action; by flow through any fissured zones causing shallow slumps to occur on the landward face of the wall; by scour of the back of the wall by overtopping; or by rotational slip. A revetment of stone, rip-rap or concrete affords an embankment protection against wave erosion.

Offshore breakwaters and jetties are designed to protect inlets and harbours. Breakwaters disperse the waves of heavy seas while jetties impound longshore drift material up-beach of the inlet and thereby prevent sanding of the channel they protect. Offshore breakwaters commonly run parallel to the shore or at slight angles to it, chosen with respect to the direction that storm waves approach the coast. Jetties usually are built at right angles to the shore, although their outer segments may be set at an angle. Two parallel jetties may extend from each side of a river for some distance out to sea. Such structures inhibit the downdrift movement of material, the deprivation of which may result in serious erosion.

Barrages may be used to shorten the coastline along highly indented coastlines and at major estuaries. Although expensive, barrages can serve more than one function, not just flood protection. They may produce hydro-electricity, carry communications or be used to create freshwater lakes. Flood barriers may be constructed in estuaries to protect urbanized areas from storm surges.

Because of the large cost of engineering works and the expense of rehabilitation, coastal management represents an alternative way of tackling the problem of coastal erosion and involves prohibiting or even removing developments that are likely to destabilize the coastal regime. The location, type and intensity of development must be planned and controlled. Public safety is the primary consideration. Generally, the level at which risk becomes unacceptable is governed by the socio-economic cost. Conservation areas and open space zones may need to be established. However, in areas that have been developed, the existing properties along a coastline may be expensive, so that public purchase may be out of the question.

3.5.4 Tsunamis

One of the most terrifying phenomena that occur along coastal regions is inundation by large masses of water called tsunamis (Fig. 3.21). Most tsunamis originate as a result of earthquakes on the sea floor, though they also can be generated by submarine landslides or volcanic activity. Seismic tsunamis are most common in the Pacific Ocean and usually are generated when submarine faults undergo vertical movement. Resulting displacement of the water surface generates a series of tsunami waves, which travel in all directions across the ocean. As with other forms of waves, it is the energy of tsunamis that is transported not the mass. Oscillatory waves are developed with periods of 10 to 60 min that affect the whole column of water from the bottom of the ocean to the surface. The magnitude of an earthquake together with its depth of focus and the amount of vertical crustal displacement determine the size, orientation and destructiveness of tsunamis.

In the open ocean tsunamis normally have very long wavelengths and their amplitude is hardly

Fig. 3.21   The effects of the tsunami at Seward, Alaska, associated with the 1964 earthquake.

noticeable. Successive waves may be from five minutes to an hour apart and they travel rapidly across the ocean at speeds of around 650 km h^{-1}. Because their speed is proportional to the depth of water, this means that the wave fronts always move towards shallower water. It also means that in coastal areas the waves slow down and increase in height, rushing onshore as highly destructive breakers. The size of a wave as it arrives at the shore depends upon the magnitude of the original displacement of water at the source and the distance it has travelled, as well as underwater topography and coastal configuration. Large waves are most likely to occur when tsunamis move into narrowing inlets. Waves have been recorded up to almost 20 m in height above normal sea level. Such waves cause terrible devastation. For example, if the wave is breaking as it crosses the shore, it can destroy houses merely by the weight of water. The subsequent backwash may carry many buildings out to sea and may remove several metres depth of sand from dune coasts. A classification of tsunami intensity is given in Table 3.3.

Fortunately, the tsunami hazard is not frequent and when it does affect a coastal area its destructiveness varies with both location and time. Accordingly, analysis of the historical record of tsunamis is required in any risk assessment. This involves a study of the seismicity of a region to establish the potential threat from earthquakes of local origin. In addition, tsunamis generated by distant earthquakes must be evaluated. The data gathered may highlight spatial differences in the distribution of the destructiveness of tsunamis that may form the basis for zonation of the hazard. If the historic record provides sufficient data, then it may be possible to establish the frequency of recurrence of tsunami events, together with the area that would be inundated by a 50, 100 or even 500 year tsunami event. On the other hand, if insufficient information is available, then tsunamis modelling may be resorted to, either using physical or computer models. The latter are now much more commonly used and provide reasonably accurate predictions of potential tsunami inundation that can be used in the management of tsunamis hazard. Such models permit the extent of damage to be estimated and the limits for evacuation to be established. The ultimate aim is to produce maps that indicate the degree of tsunami risk that, in turn, aids the planning process thereby allowing high risk areas to be avoided or used for low intensity development. Models also facilitate the design of defence works.

Various instruments are used to detect and monitor the passage of tsunamis. These include sensitive seismographs that can record waves with long period oscillations, pressure recorders placed on the sea floor in shallow water, and buoys anchored to the sea floor and used to measure changes in the level of

Table 3.3   Scale of tsunami intensity.*

	Run-up height(m)	Description of tsunami	Frequency in Pacific Ocean
I	0.5	*Very slight.* Wave so weak as to be perceptible only on tide gauge records.	One per hour
II	1	*Slight.* Waves noticed by people living along the shore and familiar with the sea. On very flat shores waves generally noticed.	One per month
III	2	*Rather large.* Generally noticed. Flooding of gently sloping coasts. Light sailing vessels carried away on shore. Slight damage to light structures situated near the coast. In estuaries, reversal of river flow for some distance upstream.	One per eight months
IV	4	*Large.* Flooding of the shore to some depth. Light scouring on made ground. Embankments and dykes damaged. Light structures near the coast damaged. Solid structures on the coast lightly damaged. Large sailing vessels and small ships swept inland or carried out to sea. Coasts littered with floating debris.	One per year
V	8	*Very large.* General flooding of the shore to some depth. Quays and other heavy structures near the sea damaged. Light structures destroyed. Severe scouring of cultivated land and littering of the coast with floating objects, fish and other sea animals. With the exception of large ships, all vessels carried inland or out to sea. Large bores in estuaries. Harbour works damaged. People drowned, waves accompanied by a strong roar.	Once in three years
VI	≥16	*Disastrous.* Partial or complete destruction of man-made structures for some distance from the shore. Flooding of coasts to great depths. Large ships severely damaged. Trees uprooted or broken by the waves. Many casualties.	Once in ten years

*Soloviev, S.L. 1978. Tsunamis. In: *The Assessment and Mitigation of Earthquake Risk*. UNESCO, Paris, 91–143

the sea surface. The locations of places along a coast affected by tsunamis can be hazard mapped, for example, to show the predicted heights of tsunami at a certain location for given return intervals (e.g. 25, 50 or 100 years). Homes and other buildings can be removed to higher ground and new construction prohibited in the areas of highest risk. However, the resettlement of all coastal populations away from possible danger zones is not a feasible economic proposition. Hence, there are occasions when evacuation is necessary. This depends on estimating just how destructive any tsunami will be when it arrives at a particular coast. Furthermore, evacuation requires that the warning system is effective and that there is an adequate transport system to convey the public to safe areas.

Breakwaters, coastal embankments, and groves of trees tend to weaken a tsunami wave, reducing its height and the width of the inundation zone. Sea walls may offer protection against some tsunamis. Buildings that need to be located at the coast can be constructed with reinforced concrete frames and elevated on reinforced concrete piles with open spaces at ground level (e.g. for car parks). Consequently, the tsunami may flow through the ground floor without adversely affecting the building. Buildings usually are orientated at right angles to the direction of approach of the waves, that is, perpendicular to the shore. It is, however, impossible to protect a coastline fully from the destructive effects of tsunamis.

3.6 Wind action and arid regions

In arid regions, in particular, because there is little vegetation, wind action is much more significant than elsewhere. By itself, wind can only remove soil and uncemented rock debris but once armed with particles, the wind becomes a noteworthy agent of abrasion. Wind erosion takes place when air pressure overcomes the force of gravity on surface particles. At first particles are moved by saltation (i.e. in a series of jumps). The impact of saltating particles on others may cause them to move by traction, saltation or suspension. Saltation accounts for three-quarters of the grains transported by wind, most of the remainder being carried in suspension, the rest are moved by traction.

3.6.1 Wind erosion

One of the most important factors in wind erosion is its velocity. Its turbulence, frequency, duration and direction also are important. As far as the mobility of particles is concerned, the important factors are their size, shape and density. Particles less than 0.1 mm in diameter usually are transported in suspension, those between 0.1 and 0.5 mm normally are transported in saltation and those larger than 0.5 mm tend to be moved by traction. Grains with a specific gravity of 2.65, such as quartz sand, are most suspect to wind erosion in the size range 0.1 to 0.15 mm. A wind blowing at 12 km h^{-1} can initiate movement in grains 0.2 mm diameter, a lesser velocity can keep the grains moving.

Because wind can only remove particles of a limited size range, if erosion is to proceed beyond the removal of existing loose particles, then remaining material must be sufficiently broken down by other agents of erosion or weathering. Material that is not sufficiently reduced in size seriously inhibits further wind erosion. Obviously, removal of fine material leads to a proportionate increase in that of larger size that cannot be removed. The latter affords increasing protection against continuing erosion and eventually a wind-stable surface is created. Binding agents, such as silt, clay and organic matter, hold particles together and so make wind erosion more difficult, as does vegetation. Soil moisture also contributes to cohesion between particles. Generally, a rough surface reduces the velocity of the wind immediately above it. Consequently, particles of a certain size are not as likely to be blown away as they would on a smooth surface. In addition, the longer the surface distance over which a wind can blow without being interrupted, the more likely it is to attain optimum efficiency.

There are three types of wind erosion, namely, deflation, attrition and abrasion. Deflation by wind lowers land surfaces by blowing loose unconsolidated particles away. The effects of deflation are most acutely seen in arid and semi-arid regions where it can lead to scour and undermining of railway lines, roads and structures. However, deflation can give rise to a wind stable surface such as a stone pavement or surface crust. On the other hand, human interference can initiate deflation on a stable surface by causing its destruction, for example, by the construction of unpaved roads and other forms of urbanization. Basin-like depressions can be formed by deflation. The suspended load carried by the wind is further comminuted by attrition, turbulence causing the particles to collide vigorously with one another. When the wind is armed with grains of sand it possesses great erosive force, the effects of which are best displayed in rock deserts. As a consequence, any surface subjected to prolonged attack by wind-blown sand is polished, etched or fluted. Abrasion has a selective action, picking out the weaknesses in rocks. Hence, the differential effects of wind erosion are well illustrated in areas where alternating beds of hard and soft rock are exposed. Abrasion is at its maximum near to the ground since it is there where the heaviest rock particles are transported.

3.6.2 Dunes

About one-fifth of the land surface of the Earth is desert. Approximately four-fifths of this desert area consists of exposed bedrock or weathered rock waste. The rest is mainly covered with deposits of sand. Several factors control the form that an accumulation of sand adopts. Firstly, there is the rate at which sand is supplied; secondly, there is the wind speed, frequency and constancy of direction; thirdly, there is the size and shape of the sand grains; and fourthly, there is the nature of the surface across which the

Fig. 3.22   Barkhan dunes in Death Valley, California.

sand is moved. Sand drifts accumulate at the exits of the gaps through which wind is channelled and are extended down-wind, but are dispersed if they are moved further down-wind. Whalebacks are large mounds of comparatively coarse sand, which probably represent the relics of seif dunes. These features develop in regions devoid of vegetation. By contrast, undulating mounds are found in the peripheral areas of deserts where the patchy cover of vegetation slows the wind and creates sand traps. Sand sheets also are developed in the marginal areas of deserts.

Two types of true dunes have been recognized, namely, the barkhan and the seif. A barkhan is crescentic in outline and is orientated at right angles to the prevailing wind direction, while a seif is a long ridge-shaped dune running parallel to the direction of the wind (Fig. 3.22). Seif dunes are much larger than barkhans, they may extend lengthwise for up to 90 km and reach heights up to 100 m. Barkhans are rarely more than 30 m in height and their widths are usually about twelve times their height. Generally, seifs occur in great numbers running approximately equidistant from each other, the individual crests being separated from one another by anything from 30 to 500 m.

The problems of sand and dust movement are most acute in areas of active sand dunes or where dunes have been destabilized by human interference disturbing the surface covers and vegetation of fixed dunes. Dust storms and sand storms are common in arid and semi-arid regions, and dust from deserts often is transported hundreds of kilometres. The former, in particular, at times cover huge areas. Sand and dust storms can reduce visibility, bringing traffic and airports to a halt. During severe sand storms the visibility can be reduced to less than 10 m. In addition, dust storms may cause respiratory problems and even lead to suffocation of animals, disrupt communications, and spread disease by transport of pathogens. The deposition of dust may bury young plants, contaminate food and drinking water, can clog equipment so that it has to be maintained more frequently and can make roads impassable. As noted, the abrasive effect of moving sand is most notable near the ground (i.e. up to a height of 250 mm). However, over hard man-made surfaces, the abrasion height may be higher because the velocity of sand movement and saltation increase. Structures may be pitted, fluted or grooved depending on their orientation in relation to the prevailing wind direction.

The rate of movement of sand dunes depends on the velocity and persistence of the winds, the constancy of wind direction, the roughness of the surface over which they move, the presence and density of vegetation, the size of the dunes and the particle size distribution of the sand grains. Large dunes may move up to 6 m per year and small ones may exceed 23 m per year. Movement of dunes can bury obstacles in their path such as roads and railways or accumulate against large structures. Only a few centimetres of sand on a road surface or rail track can constitute a major hazard. Such moving sand necessitates continuous and often costly maintenance activities. Indeed, areas may be abandoned as a result of sand encroachment.

Aerial photographs and remote sensing imagery of the same area, and taken at successive time intervals, can be used to study dune movement, rate of land degradation and erodibility of surfaces. Field mapping involves the identification of erosional and depositional evidence of sediment movement, which can provide useful information on the direction of sand movement but less on dust transport. Ultimately, hazard maps of dune morphology and migrating dunes can be produced (see Fig. 2.1). The methods available for monitoring sand movement, in addition to sequential photography, include surveys, tracing with fluorescent sand grains, and the use of trenches and collectors to monitor movement. The variability of dust movement makes monitoring it difficult but settling jars and sediment collectors have been used.

Of particular importance in an assessment of the sand and dust hazard are the recognition of sources and stores of sediment. Sources are sediments that contribute to the contemporary aeolian transport system and include weathered rock outcrops, fans, dunes, playa/sabkha deposits and alluvium. Stores are unconsolidated or very weakly consolidated surface deposits that have been moved by contemporary aeolian processes and have accumulated temporarily in surface spreads, dunes, sabkha and stone pavements.

3.6.3 Dealing with problems associated with wind

The best way to deal with a hazard is to avoid it, this being both more effective and cheaper than resorting to control measures. This is particularly relevant in the case of aeolian problems, notably those associated with large active dune fields where even extensive control measures are likely to prove only temporary. Hence, sensible site selection, based on thorough site investigation, is necessary prior to any development.

It is extremely unlikely that all moving sand can be removed. This can only happen where the quantities of sand involved are small and so can only apply to small dunes (a dune 6 m in height may incorporate 20 000–26 000 m³ of sand with a mass of 30 000–45 000 tonnes). Even so, this is expensive and the excavated material has to be disposed of. Often the removal of dunes is only practical when the sand can be used as fill or ballast, however, the difficulty of compacting aeolian sand frequently precludes its use. In addition, flattening dunes does not represent a solution since the wind will develop a new system of dunes on the flattened surface within a short period of time. Accordingly, mobile sand must be stabilized. Plants used for the stabilization of dunes must be able to exist on moving sand and either survive temporary burial or keep pace with deposition. Mulches can be used to check erosion and to provide organic material. Brush matting has been used to check sand movement temporarily where wind velocities do not exceed 65 km h⁻¹. Natural geotextiles can be placed over sand surfaces after seeding, to protect the seeds, to help retain moisture and to provide organic matter eventually. These geotextiles can contain seeds within them. A layer of coarse gravel or aggregate, over 50 mm thick, can be placed over a sand surface to prevent its deflation. Artificial stabilization, which provides a protective coating or bonds grains together, may be necessary on loose sand. In such cases rapid-curing cutback asphalt and rapid-setting asphalt emulsion have been used. Many chemical stabilizers are only temporary, breaking down in a year or so and therefore they tend to be used together with other methods of stabilization, such as the use of vegetation.

Windbreaks frequently are used to control wind erosion and obviously are best developed where groundwater is near enough to the surface for trees and shrubs to have access to it. It is essential to select species of trees that have growth rates that exceed the rate of sand accumulation and that have a bushy shape. Windbreaks can act as effective dust and sand traps. Their shape, width, height and permeability influence their trap efficiency and the amount of turbulence generated as the wind blows over them. Fences can be used to impound or divert moving sand. They can be constructed of various materials from brushwood or palm fronds, wood or metal panels, stone walls or earthworks. The surfaces around fences should be stabilized in order to avoid erosion and undermining with consequent collapse. Individual fence designs have different sand trapping abilities at different wind velocities so that fence design must be tailored to local environmental conditions. They must be located in areas where the development of a large artificial dune will not pose problems. In fact, the use of fences over a long period of time represents a commitment to a sand control policy based on dune building. Alternatively, a number of fences can be positioned at an acute angle (usually around 45°) to the prevailing wind or a 'V' shaped barrier pointing into the sand stream can be used. Such fences not only trap sand but they also deflect the wind, thereby removing some of the accumulated sand.

3.6.4 Stream action in drylands

In arid regions rainfall is erratic and in semi-arid regions it is seasonal but in both rain often falls as heavy and sometimes violent showers. The result is that the river channels cannot cope with the amount of rainwater and so extensive flooding takes place. These floods develop with remarkable suddenness and either form raging torrents, which tear their way down slopes excavating gullies as they go, or they may assume the form of sheet floods. Dry wadis are rapidly filled with swirling water and are thereby enlarged. However, such floods are short-lived since the water soon becomes choked with sediment and the consistency of the resultant mudflow eventually reaches a point when further movement is checked. Much water is lost by percolation, and mudflows are checked where there is an appreciable slackening in gradient. Gully development frequently is much in evidence.

High run-off associated with torrential rainfall can flush sediment from gullies. Small fans may develop at the exits from gullies spreading into stream channels. This, in turn, may led to the formation of gravel sheets, lobes and bars on the one hand and/or bank erosion on the other. Major flows can transport massive quantities of sediment, which when deposited raises the floor levels of valleys. Terraces frequently flank channels. Sediment load increases downstream as slopes and tributaries contribute more load, and as water is lost by infiltration. Hence, the lower reaches of streams are dominated by deposition. Obviously, the annual sediment yield is highly variable. Extensive deep deposits of coarse-grained alluvium may occur in many ephemeral stream channels, and such material possesses a high infiltration capacity and a potential for significant storage of water. Because of high infiltration in the upper parts of stream systems, groundwater recharge downstream is affected adversely that, in turn, may mean that less water is available for any irrigation schemes.

Avoidance of flood-prone areas is much more important in arid regions than elsewhere because of the unpredictability of flash floods. Although flood warning systems do exist in certain arid regions, it cannot be expected that they could be established in most arid areas. Indeed, in many arid regions the lack of data mitigates against the establishment of flood warning systems. However, hazard maps of flooding indicating areas of low, medium and high risk can be used for planning purposes, including land zonation. Any development that occurs, should be restricted to those areas of low risk. Furthermore, conventional engineering flood control methods are not always appropriate in arid regions and design should be more in keeping with the landscape. Disturbance should be kept to a minimum.

3.6.5 Salt weathering

Because of the high rate of evaporation in hot arid regions, the capillary rise of near-surface groundwater normally is very pronounced. The type of soil governs the height of capillary rise but normally in desert conditions this does not exceed 2–3 m. Although the capillary fringe can be located at depth within the ground, obviously depending on the position of the water table, it also can be very near or at the ground surface. Hence, in the latter case, salts are precipitated in the upper layers of the soil. Among the most frequently occurring salts are calcium carbonate, gypsum, anhydrite and halite. Salt concentrations are highest in areas where saline water, either surface or subsurface, is evaporated, such as low-lying areas of internal drainage, especially with salinas and saline bodies of water, or along coasts particularly where sabkhas are present. New layers of minerals can be formed within months and thin layers can be dissolved just as quickly. The groundwater regime can be changed by construction operations and so may lead to changes in the positions at which mineral solution or precipitation occurs.

Salt weathering leading to rock disintegration is brought about as a result of stresses set up in the pores, joints and fissures in rock masses due to the growth of salts, the hydration of particular salts, and the volumetric expansion that occurs due to the high diurnal range of temperature. The hydration, dehydration, and rehydration of hydrous salts may occur several times throughout a year and depends upon the temperature and relative humidity conditions on the one hand and dissociation vapour pressures of the salts on the other. In fact, crystallization and hydration-dehydration thresholds of the more soluble salts such as sodium chloride, sodium carbonate, sodium sulphate and magnesium sulphate may be crossed at least once daily. The aggressiveness of the ground depends on the position of the water table and the capillary fringe above in relation to the ground surface, the chemical composition of the groundwater and the concentration of salts within it, the type of soil and the soil temperature. Also, when some salts such as $CaSO_4$ and NaCl are combined in solution their effectiveness as agents of breakdown is increased.

Salt weathering attacks structures and buildings, leading to cracking, spalling and disintegration of concrete, brick and stone. The extend to which damage due to salt weathering occurs depends upon climate, soil type and groundwater conditions since they determine the type of salt, its concentration and mobility in the ground; and the type of building materials used. One of the most notable forms of damage, especially to concrete, is brought about by sulphate attack. This can take the form of corrosion, and heaving and cracking of ground slabs due to the precipitation of salts beneath them. Salt attack can be prevented prior to construction by placing a vapour barrier (e.g. 6 mm polythene in a layer of sand) over a layer of gravel. Salt attack on brick foundations has sometimes led to their disintegration to a powdery material that, in turn, results in the building undergoing settlement. Obviously, any such damage can be avoided by not undertaking construction in low-lying areas with aggressive ground conditions. Alternatively, foundations can be provided with protective coverings, together with damp-proof courses to prevent capillary rise in buildings. In addition, it is advisable to use building materials that offer a high degree of resistance to salt weathering. Salt weathering of bituminous paved roads built over areas where saline groundwater is at or near the surface is likely to result in notable signs of damage such as heaving, cracking, blistering, stripping, pot-holing, doming and disintegration. Aggregates used for base and sub-base courses can be attacked and this can mean that roads undergo settlement and cracking. If the problems associated with aggressive salty ground are to be avoided, then the first objective must be to identify the limits of the hazard zone and the spacial variability of the hazard within it. The concentration of salts in the groundwater can allow hazard zones to be subdivided into those of high, moderate or low salinity.

3.7 Glacial hazards

Although the potential hazards of glaciers may be appreciable, their impact on people is not significant since less than 0.1% of the world's glaciers occur in inhabited areas. Glacial hazards can be divided into two groups, that is, those that are a direct action of ice or snow such as avalanches, and those that give rise to indirect hazards. The latter include glacier outbursts and flooding.

The rapid movement of masses of snow or ice down slopes as avalanches can pose a serious hazard in many mountain areas. For example, avalanches, particularly when they contain notable amounts of debris, can damage buildings and routeways, and may lead to loss of life. The path of an avalanche consists of three parts, namely, the starting area, the track and the run-out-deposition zone. Loose snow avalanches generally occur in the cohesionless surface layers of dry snow or wet snow containing water. In many instances a rotational failure occurs when the angle of repose is exceeded, that is, the weight of the snow mass exceeds the frictional resistance of the surface on which it lies. Wet or slush avalanches do not need steep slopes in the areas of initiation as do dry snow avalanches, the critical angle of repose can be as low as 15°. Dry avalanches commonly travel along a straight path as compared with wet-snow avalanches that tend to hug the contours and become channelled in small valleys. Slab avalanches take place when a slab of cohesive snow fails. These tend to be more dangerous. Ice avalanches are released by frontal block failure, ice slab failure and ice bedrock failure from a mass of ice. Avalanche location often can be predicted from historical evidence relating to previous avalanches combined with topographical data. As a consequence, hazard maps of avalanche prone areas can be produced. Avalanche forecasting is related to weather conditions and monitoring of snow.

Glacier floods result from the sudden release of water that is impounded in, on, under or adjacent to a glacier. Glacier outburst and jökulhlaup both refer to a rapid discharge of water, which is under pressure, from within a glacier. Water pressure may build up within a glacier to a point where it exceeds the strength of the ice, with the result that the ice is ruptured. In this way water drains from water-pockets so that subglacial drainage is increased and may be released as a frontal wave many metres in height. Progressive enlargement of internal drainage channels also leads to an increase in discharge. In both cases, the quantity of discharge declines after the initial surge. Water that is dammed by ice eventually may cause the ice to become detached from the rock mass on which it rests, the ice then is buoyed up allowing the water to drain from the ice-dammed lake. Discharges of several thousand cubic metres per second have occurred from such lakes and can cause severe damage to any settlements located downstream. Most of these outbursts occur during the summer months. Nevertheless, it can be difficult to predict the location and timing of the release of water so that the floods are unexpected. The resulting floods from these outbursts often carry huge quantities of debris. Débâcles are rapid discharges of water from proglacial moraine dammed lakes, while aluvions involve the rapid discharge of liquid mud that may contain large boulders. Such releases of large quantities of water on occasions have led to thousands of people being killed.

Glacier fluctuations in which glaciers either advance or retreat occur in response to climatic changes. Rapid changes in the position of the snout of a glacier are referred to as surges. The advance of a glacier into a valley can lead to the river that occupies the valley being dammed with land being inundated as a lake forms.

3.8 Dissolution of rocks

3.8.1 Carbonate rocks

Limestones and dolostones are commonly transected by joints. These frequently have been subjected to varying degrees of dissolution so that some may gape. Rainwater generally is weakly acidic and further acids may be taken into solution from carbon dioxide or organic or mineral matter in the soil. The degree of aggressiveness of water to limestone can be assessed on the basis of the relationship between the dissolved carbonate content, the pH value and the temperature of the water. If solution continues its rate slackens and it eventually ceases when saturation is reached. Hence, solution is greatest when the bicarbonate saturation is low. This occurs when water is circulating so that fresh supplies with low lime saturation are continually made available. However, the dissolution of limestone is a very slow process. For instance, mean rates of surface lowering of limestone areas in the British Isles range from 0.041 to 0.099 mm annually.

Joints are opened by dissolution of limestone to form grykes. This means that surface water begins to flow underground, which in thick massive limestones ultimately may lead to an integrated system of subterranean galleries and caverns. Such features are associated with karstic landscapes. Sinkholes are one of the commonest dissolution features of limestone terrains and can be classified on a basis of origin into dissolutional, collapse, caprock, dropout, suffusion and buried solutional sinkholes. Dissolutional sinkholes form when drainage takes place through caverns and micro-caverns in the underlying bedrock. Collapse sinkholes are formed by the collapse of roofs of caverns, which have become unstable due to removal of support. Rapid collapse of a roof is very unusual, progressive collapse of a heavily fissured zone being more common. The collapse process usually is associated with solution erosion. Collapse sinkholes are steep-sided. Caprock sinkholes involve undermining and collapse of an insoluble caprock

Fig. 3.23   Sinkhole collapse beneath a house south of Pretoria, South Africa.

over a cavity in underlying limestone. Dropout sinkholes are formed in cohesive cover soils. For example, clay soil may be capable of bridging a void developed in limestone beneath but as the cavity is enlarged, ravelling takes place in the clay cover until it eventually fails (Fig. 3.23). A number of conditions accelerate the development of cavities in the clay soil and initiate collapse. Rapid changes in moisture content lead to aggravated slabbing or roofing in clay soils. Lowering the water table increases the downward seepage gradient and accelerates downward erosion. It also reduces capillary attraction and increases instability of flow through narrow openings and gives rise to shrinkage cracks in highly plastic clays that weaken the mass in dry weather and produce concentrated seepage during rains. Increased infiltration often initiates failure, particularly when it follows a period when the water table has been lowered. Ravelling failures are the most widespread and probably the most dangerous of all the subsidence phenomena associated with limestones and dolostones. Suffusion sinkholes are developed in non-cohesive soils where percolating rain water washes the soil into cavities in the carbonate rock below. Buried sinkholes occur in limestone rockhead and are filled or buried with sediment. They may be developed by subsurface solution or as normal subaerial sinkholes that later are filled with sediment.

Solution cavities present numerous problems in the construction of large foundations such as for dams where bearing strength and watertightness are paramount. However, few sites are so bad that it is impossible to construct safe and successful structures upon them, but the cost of the necessary remedial treatment may be prohibitive.

The various features associated with karst terrains can be aggravated by environmental changes such as uncontrolled seepage of water into the ground and so one of the best ways to keep sinkhole failures in karst areas to a minimum is the proper control of water. In particular, groundwater lowering brought about by excessive drawdown from wells should be avoided. This entails monitoring groundwater abstraction in order to prevent the water table from being lowered. Sinkholes or voids require treatment, especially prior to any construction. Location of buildings and/or structures away from these features is the simplest course of action to take if the site conditions allow. Otherwise the throat areas of sinkholes can be sealed with concrete plugs or, if small enough, capped with reinforced concrete. The inverted filter method essentially consists of backfilling a sinkhole initially with boulders and soil-cement slurry to choke the throat, followed by further backfilling using progressively finer material. Anchored geogrid may be used to enhance the strength of the compacted backfill. Compaction grouting is the staged injection of low slump grout to improve soil properties. As such, it has been used to improve soft soils over areas of karstic limestone or as a remedial measure to underpin a building after ground movements have taken place. Compaction grouting uses highly viscous grout (mixtures of cement, sand, clay and pulverized fuel ash, PFA) to displace and compress loose soil as a grout bulb grows and expands. It sometimes has been used to form plugs in the throats of small sinkholes that are partially choked with clayey material that has been washed by any water running into the sinkhole concerned. Cap grouting has been used to reduce or prevent gaping grykes or voids in the bedrock immediately below the soil horizon from either being further enlarged by dissolution or of collapsing respectively. Jet grouting can be used when the soil infill in the throat of a sinkhole is too stiff to displace by compaction grouting. It also can be used to form a barrier around a sinkhole when the soil around its sides is potentially unstable and may collapse. In its simplest form, jet grouting involves inserting an injection pipe into the soil to the required depth. The soil then is subjected to a horizontally rotating jet of water and at the same time mixed with grout to form plastic soil-cement. The injection pipe is raised gradually. Jet grouting can be employed in all types of soils. Grout curtains are used extensively in karst terrain to reduce the flow of groundwater beneath dams. Ideally, a grout curtain is taken to a depth where the requisite degree of tightness is available naturally. In multiple row curtains, the outer rows are completed first thereby allowing the innermost row to effect closure on the outer rows. Bulk grouts are relatively cheap pumpable slurries consisting of mixtures of cement, sand, gravel and PFA, that can be pumped via drillholes into cavities in carbonate rocks. Foam grouts also are used to fill voids. Soils above karstic terrains may be subjected to dynamic compaction, vibrocompaction or cement/lime stabilization to improve their engineering performance (see Chapter 11).

Chalk, being a carbonate rock, is subject to dissolution along discontinuities and solution pipes may develop at the intersection of discontinuites. Sinkholes, however, tend to be small and of more limited distribution. Buried sinkholes and solution pipes frequently are found near the contact of the Chalk and the overlying Tertiary and superficial deposits. This is largely due to the concentration of run-off at such locations. Lowering of the Chalk surface beneath overlying deposits due to solution can occur, disturbing the latter deposits and lowering their degree of packing. Hence, the Chalk surface may be extremely irregular in places.

3.8.2   Evaporitic rocks

Gypsum is more readily soluble than limestone. Sinkholes and caverns therefore can develop in thick beds of gypsum more rapidly than they can in limestone. Indeed, in the United States such features have been known to form within a few years where beds of gypsum are located beneath dams. Extensive surface cracking and subsidence has occurred in certain areas of Oklahoma and New Mexico due to the collapse of cavernous gypsum. The problem is accentuated by the fact that gypsum is weaker than limestone and therefore collapses more readily. The presence of sinkholes and caves in gypsum has led to subsidence problems and in some instances sinkholes have collapsed suddenly. Many collapses occur after periods of prolonged rain, sometimes making their appearance within minutes. Where beds of gypsum approach the surface their presence sometimes can be traced by broad funnel-shaped sinkholes. Massive deposits of gypsum usually are less dangerous than those of anhydrite because gypsum tends to dissolve in a steady manner forming caverns or causing progressive settlements. Massive anhydrite can be dissolved to produce uncontrollable runaway situations in which seepage flow rates increase in a rapidly accelerating manner. Even small fissures in massive anhydrite can prove dangerous. By contrast, anhydrite on contact with water may become hydrated to form gypsum, which can lead to uplift of the ground surface.

The solution rate of gypsum or anhydrite is controlled principally by their surface area in contact with water and the flow velocity of water associated with a unit area of the material. Hence, the amount of fissuring in a rock mass, and whether it is enclosed by permeable or impermeable beds, is most important.

Solution also depends on the sub-saturation concentration of calcium sulphate in solution. The salinity of the water also is influential. For example, the rates of solution of gypsum and anhydrite are increased by the presence of sodium chloride, carbonate and carbon dioxide in solution.

Where soluble minerals occur in particulate form in the ground, then their removal by solution can give rise to significant settlements. In such situations the width of the solution zone and its rate of progress are obviously important as far as the location of hydraulic structures are concerned. Anhydrite is less likely to undergo catastrophic solution in a fragmented or particulate form than gypsum. Another point, and this particularly applies to conglomerates cemented with soluble material, is that when this is removed by solution, the rock mass is reduced greatly in strength.

Rock salt is yet more soluble than gypsum. Evidence of slumping, brecciation and collapse structures in rock masses that overlie saliferous strata bear witness to the fact that rock salt has gone into solution in past geological times. It generally is believed, however, that in areas underlain by saliferous beds, measurable surface subsidence is unlikely to occur except where salt is being extracted. Perhaps this is because equilibrium has been attained between the supply of unsaturated groundwater and the rock salt available for solution. Exceptionally, cases have been recorded of rapid subsidence, such as the 'Meade salt sink' in Kansas. In arid regions, however, salt karst may be developed. For example, more than 300 sinkholes, fissures and depressions are associated with the salt karst zone of north eastern Arizona. Jointing in the rock mass overlying the rock salt is a major influence on the karst features that form and the walls of many sinkholes represent expressions of the major joint systems in the area.

3.9 Gases

3.9.1 Radon

Radon (Rn) is a naturally occurring radioactive gas that is produced by the radioactive decay of uranium (U) and thorium (Th). It is the only radioactive gas, and is colourless, odourless and tasteless. Although Rn frequently is present in notable amounts in areas where there is no uranium mineralization, it normally is associated with rocks with high concentrations of U. Rocks such as sandstone generally contain less than 1 mg kg^{-1}, whilst some carbonaceous shales, some rocks rich in phosphates and some granites may contain more than 3 mg kg^{-1}. Uranium can be concentrated by weathering processes and re-sedimentation, and so redistributed in a form more amenable to Rn release. Such processes may give rise to local Rn anomalies in areas of otherwise low Rn exposure. Faults and shear zones often are enriched with U and so are associated with elevated concentrations of Rn gas in the overlying soils. In addition, the presence of particularly high levels of Rn along a fault may be due to the increased rate of migration of Rn, because the permeability is higher along the fault or to the presence of Rn or Rn-bearing groundwater within the fault. As Rn is moderately soluble in water it can be transported over considerable distances, and so anomalous concentrations can occur far from the original sources of U or Th. Transport by fluids is especially rapid in limestones and along faults.

Radon represents a health hazard since it emits alpha particles. Outside the body, these do not present a problem, because their large size and relatively large charge mean that they cannot pass through the skin. However, when alpha particles are inhaled or ingested, they can damage tissue. Indeed, inhalation of Rn and its daughter products accounts for about one half of the average annual exposure to ionizing radiation of people in the United States and Britain. Although the inhalation of Rn is the principal way alpha particles enter the human body, most is breathed out. The solid daughter products (e.g. ^{218}Po that adheres to dust), however, are also alpha particle emitters but are more dangerous because they often are retained in lung tissue where they increase the risk of lung cancer. Some Rn may be dissolved in body fats and its daughter products transferred to the bone marrow. The accumulated dose in older people can be high and may give rise to leukaemia. Radon also has been linked with melanoma, cancer of the kidney and some childhood cancers. The United States Environmental Protection Agency (USEPA) regards 4 pCi l^{-1} as the limit beyond which Rn is considered a hazard. The average concentration of Rn outdoors is around 0.2 pCi l^{-1} as compared with approximately 1.0 pCi l^{-1} for indoors.

Radon and its daughter products accumulate in confined spaces such as buildings. Soil gas is drawn into a building by the slight underpressure indoors, which is attributable to the warmer air rising. A relatively small contribution is made by building materials. Radon can seep through concrete floors and foundations, drains, small cracks or joints in walls below ground level, or cavities in walls. Emission of Rn from the ground can vary, for example, according to barometric pressure and the moisture content of the soil. Accumulation of Rn also is affected by how well a building is ventilated. It also can enter a building via the water supply, particularly if a building is supplied by a private well. Public supplies of water, however, usually have relatively low Rn contents as the time the water is stored helps Rn release. Reduction of the potential hazard from Rn in homes involves locating and sealing the points of entry of Rn, and improved ventilation by keeping more windows open or using extraction fans. Ventilation systems can be built into a house during its construction, for example, a system can be installed beneath the house.

Surveys of soil gas, based on geological information, can be used to estimate the Rn potential of an area. The resulting maps show the levels of Rn in the soil gas in relation to rock types (Fig. 3.24). Each rock formation should be tested several times and some traverses should be repeated to test the variability. The best time to undertake a survey in temperate climatic regions is during the spring to autumn months, when the soils are less wet and thus are more permeable.

Radon in soil gas is measured by means of a hollow spike hammered into the ground and linked to a gas pump and detection unit. Detection of Rn normally is by the zinc sulphide scintillation method. Alternative, passive methods employ detectors that are buried in the soil and recovered some time later, often up to a month. This procedure is used when long-term monitoring is required to overcome the problems of variation in concentration of Rn due to changing weather conditions.

3.9.2 Methane

Methane and carbon dioxide are generated by the breakdown of organic matter and frequently are associated with coal bearing strata. Methane is a by-product formed during coalification, that is, the process that changes peat into coal. Biodegradation takes place in the early stages of accumulation of

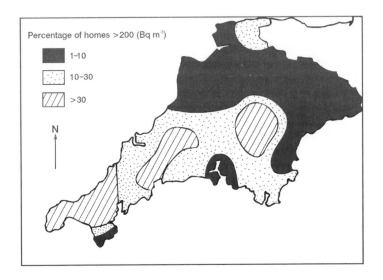

Fig. 3.24 Estimated percentage of houses above the radon action level in Devon and Cornwall, England.

plant detritus with the evolution of some methane and carbon dioxide. The major phase of methane production, however, probably takes place at a later stage in the coalification process after the deposits have been buried. During this process approximately 140 m^3 of methane is produced per tonne of coal. As a consequence, the quantity of methane produced during coalification exceeds the holding capacity of coal, resulting in excess methane migrating into reservoir rocks that surround or overlie a coal deposit. Dissolved gases may be advected by groundwater and only when the pressure is reduced and the solubility limit of the gas in water exceeded, do they come out of solution and form a separate gaseous phase. Such pressure release occurs when coal is removed during mining, tunnelling or shaft sinking operations in strata so affected. It is essential that such degassing is not allowed to occur in confined spaces where an explosive mixture could develop (5 –15% methane mixed with air is explosive). Methane can be oxidized during migration to form carbon dioxide. However, carbon dioxide can be generated both microbially and inorganically in a number of ways that do not involve methane.

Gas problems are not present in every coal mine. Nonetheless, gas may accumulate in abandoned mines and methane, in particular, since it is lighter than air may escape from old workings via shafts, and via crownholes where the workings are at shallow depth. Methane can move through coal by diffusion. This is a relatively slow process but as coal seams are likely to be fractured the diffusion rate is increased. Gas also migrates through rocks via intergranular pore space or, more particularly, along discontinuities. Where strata have been disturbed by mining subsidence gas permeability is enhanced, as is that of groundwater holding gas.

Methane can have more than one source, it need not be coal bearing strata, it could be from a landfill (see Chapter 7). In major construction operations on the one hand or in the case of domestic dwellings on the other, the sources of methane need to be considered and measures taken to minimize the risk posed. Cases are on record of explosions having occurred in buildings due to the presence of methane.

Generally, the connection between a source of methane and the location where it is detected can be verified by detecting a component of the gas that is specific to the source or by establishing the existence of a migration pathway from the source to the location where the gas is detected. An analysis of the gas obviously helps identify the source. For example, methane from landfill gas contains a larger proportion of carbon dioxide (16–57%) than does coal gas (up to 6%). Analysis also may involve trace components or isotopic characterization using stable isotope ratios ($^{13}C/^{12}C$; $^2H/^1H$) or the radioisotope ^{14}C. This again allows distinction between coal gas and landfill gas as in the former all the radiocarbon has decayed.

Determining the migration pathway of gas involves both geological and hydrogeological investigation. Groundwater flow needs to be determined if gases are dissolved in groundwater, particular note being taken of areas of discharge. Hydrogeological assessments require accurate measurement of piezometric pressure, and sampling and chemical analysis of water. If methane is dissolved in groundwater, then samples should be obtained at *in situ* pressure and they must not be allowed to de-gas as they equilibrate to atmospheric pressure.

4

Water Resources

4.1 Basic hydrology

The hydrological cycle can be visualized as starting with the evaporation of water from the oceans followed by the transport of the resultant water vapour by winds and moving air masses. Some water vapour condenses over land and falls back to the surface of the Earth as precipitation. This precipitation then makes its way back to the oceans via streams, rivers or underground flow, although some precipitation may be evapotranspired and go through several subcycles before completing its journey. Groundwater forms an integral part of the hydrological cycle. Although groundwater represents only 0.5% of the total water resources of the Earth, and not all this is available for exploitation, about 98% of the usable fresh water is stored underground.

Precipitation is dispersed as run-off, as infiltration/percolation and as evapotranspiration. Run-off consists of two basic components, that is, surface water run-off and groundwater discharge. The former is usually the more important and is responsible for the major variations in river flow. Run-off generally increases in magnitude as the time from the beginning of precipitation increases. Infiltration refers to the process whereby water penetrates the ground surface and starts moving down through the zone of aeration. The subsequent gravitational movement of the water down to the zone of saturation is termed percolation, although there is no clearly defined point where infiltration becomes percolation. Whether infiltration or run-off is the dominant process at a particular time depends upon several factors, such as the intensity of the rainfall, and the porosity and permeability of the surface. For example, if the rainfall intensity is much greater than the infiltration capacity of the soil, then run-off is high. The amount of water lost by evaporation is influenced by the climatic regime, especially the temperatures concerned.

If lower strata are less permeable than the surface layer, the infiltration capacity is reduced so that some of the water that has penetrated the surface moves parallel to the water table and is called interflow. Part of the water that becomes interflow is discharged to a river at some point to form the baseflow of the river. The remaining water may continue down through the zone of aeration until it reaches the water table and becomes groundwater recharge.

4.2 Reservoirs

Although most reservoirs are multipurpose, their principal function is to stabilize the flow of water in order to satisfy a varying demand from consumers or to regulate water supplied to a river course. Hence, the most important physical characteristic of a reservoir is its storage capacity. Probably the most important aspect of storage in reservoir design is the relationship between capacity and yield. The yield is the quantity of water that a reservoir can supply at any given time. The maximum possible yield equals the mean inflow less evaporation and seepage loss. In any consideration of yield the maximum quantity of water that can be supplied during a critical dry period (i.e. during the lowest natural flow on record) is of prime importance and is defined as the safe yield.

The maximum elevation to which the water in a reservoir basin rises during ordinary operating conditions is referred to as the top water or normal pool level (Fig. 4.1). For most reservoirs this level is fixed by the top of the spillway. Conversely, minimum pool level is the lowest elevation to which the water is drawn

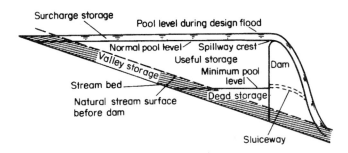

Fig. 4.1   Zones of storage in a reservoir.

under normal conditions, this being determined by the lowest outlet. Between these two levels the storage volume is termed the useful storage, while the water below the minimum pool level, because it cannot be drawn upon, is the dead storage.

Problems may emerge both upstream and downstream in any adjustment of a river regime to the new conditions imposed by a reservoir. Deposition around the head of a reservoir may cause serious aggradation upstream resulting in a reduced capacity of the stream channels to contain flow. Hence, flooding becomes more frequent and the water table rises. Removal of sediment from the outflow of a reservoir can lead to erosion in the river regime downstream of the dam, with consequent acceleration of headward erosion in tributaries and lowering of the water table.

In an investigation of a potential reservoir site, consideration must be given to the amount of rainfall, run-off, infiltration, and evapotranspiration that occurs in the catchment area. The climatic, topographical, and geological conditions therefore are important, as is the type of vegetative cover. Accordingly, the two essential types of basic data needed for reservoir design studies are adequate topographical maps and hydrological records. Indeed, the location of a large, impounding direct supply reservoir is very much influenced by topography since this governs its storage capacity. Initial estimates of storage capacity can be made from topographic maps or aerial photographs, more accurate information being obtained, where necessary, from subsequent surveying. Catchment areas and drainage densities also can be determined from maps and air photos.

Records of stream flow are required for determining the amount of water available for conservation purposes. Such records contain flood peaks and volumes, which are used to determine the amount of storage capacity needed to control floods, and to design spillways and other outlets. Records of rainfall are used to supplement stream flow records or as a basis for computing stream flow where there are no flow records obtainable. Losses due to seepage and evaporation also must be taken into account.

The most attractive site for a large impounding reservoir is a valley constricted by a gorge at its outfall with steep banks upstream so that a small dam can impound a large volume of water with a minimum extent of water spread. However, two other factors have to be taken into consideration, namely, the watertightness of the basin and bank stability. The question of whether or not significant water loss will take place is determined chiefly by the groundwater conditions, more specifically by the hydraulic gradient. Consequently, once the groundwater conditions have been investigated an assessment can be made of watertightness and possible groundwater control measures. In this context, seepage is a more discreet flow than leakage, which spreads out over a larger area, but may be no less in total amount.

Apart from the conditions in the immediate vicinity of a dam, the two factors that determine the retention of water in reservoir basins are the piezometric conditions in, and the permeability of, the floor and flanks of the basin. If the groundwater divide and piezometric level are at a higher elevation than that of the proposed top water level, then no significant water loss occurs. Seepage can take place through a separating ridge into an adjoining valley where the groundwater divide, but not the piezometric level, is above the top water level of a reservoir. The flow rate of the seepage is determined by the *in situ* permeability and the hydraulic head. When both the groundwater divide and piezometric level are at a lower elevation than the top water level but higher than that of the reservoir floor, then the increase in groundwater head is low and the flow from the reservoir may be initiated under conditions of low piezometric pressure in the reservoir flanks. A depressed water table does not necessarily mean that reservoir construction is out of the question but groundwater recharge will take place on filling, which will give rise to a changed hydrogeological environment as the water table rises. In such instances the impermeability of the reservoir floor is important. When impermeable beds are more or less saturated, particularly when they have no outlet, seepage is appreciably decreased. At the same time the accumulation of silt on the floor of the reservoir tends to reduce seepage. If, however, any permeable beds present contain large pore spaces or discontinuities and they drain from the reservoir, then seepage continues.

Although a highly leaky reservoir may be acceptable in an area where run-off is evenly distributed throughout the year, a reservoir basin with the same rate of water loss may be of little value in an area where run-off is seasonally deficient. Downstream of the dam site, leakage from a reservoir can take the form of sudden increases in stream flow with boils in the river and the appearance of springs on the valley sides. It may be associated with major defects in the geological structure. Serious leakage has taken place at reservoirs via karst conditions in limestone and sites are best abandoned where large numerous solution cavities extend to considerable depths. Where the problem is not so severe, solution cavities can be cleaned and grouted. Sinkholes and caverns can develop in thick beds of gypsum more rapidly than they can in limestone (see Chapter 3). Another point is that when soluble material that forms the cement within a rock mass is removed by solution, then the rock mass is reduced greatly in strength. A classic example of this is associated with the failure of the St Francis Dam in California in 1928. One of its abutments was founded in conglomerate cemented with gypsum that was gradually dissolved, the conglomerate losing strength, and ultimately the abutment failed.

Buried channels may be filled with coarse granular stream deposits or deposits of glacial origin and if they occur near the perimeter of a reservoir they almost invariably pose leakage problems. A thin layer of relatively impermeable superficial deposits does not necessarily provide an adequate seal against seepage. Where artesian conditions exist (see below), springs may break through the thinner parts of the superficial cover. If the water table below superficial deposits is depressed, then there is a risk that the weight of water in the reservoir may puncture the cover. In addition, on filling a reservoir, it is possible that the superficial material may be ruptured or partially removed to expose the underlying rocks. Leakage along faults, generally, is not a serious problem as far as reservoirs are concerned since the length of the flow path concerned is usually too long. However, fault zones occupied by permeable fault breccia running beneath the dam must be given special consideration. Open discontinuities also represent pathways for water leakage.

Some soil or rock masses, which are brought within the zone of saturation by the rising water table on reservoir filling, may become unstable and fail. This can lead to slumping and sliding on the flanks of a reservoir. Landslides that occur after a reservoir is filled reduce its capacity. Also, ancient landslipped areas that occur on the rims of a reservoir can be reactivated (as well as presenting a potential leakage problem). The most disastrous example of a landslide moving into and displacing water from a reservoir occurred in northern Italy in 1963. More than $300 \times 10^6 \, m^3$ of rock material slid into the Vajont reservoir with such momentum that it rode 135 m up the opposite flank of the reservoir, filled approximately one third of the reservoir, and formed a wave 100 m in height at the crest of the dam. Five villages downstream of the reservoir were destroyed and some 3000 people killed.

Sedimentation in a reservoir may lead to one or more of its major functions being seriously curtailed or even to it becoming inoperative. In a small reservoir, sedimentation may seriously affect the available carry-over water supply and ultimately necessitate abandonment. The size of a drainage basin is the most important consideration as far as sediment yield is concerned – the rock types, drainage density and gradient of slope also being important. The sediment yield also is influenced by the amount and seasonal distribution of precipitation, and the vegetative cover. In those areas where streams carry heavy sediment

(a)

(b)

(c)

Fig. 4.2 (a) Meldon Dam, Devon, England. A gravity dam. (b) Katse Dam, an arch dam nearing completion, Lesotho Highlands Water Project, Lesotho. (c) Wimbleball Dam, Somerset, UK, an example of a buttress dam (courtesy of South West Water).

loads the rates of sedimentation must be estimated accurately in order that the useful life of any proposed reservoir may be determined. A good cover of vegetation over the catchment area of a reservoir is one of the best ways to reduce sedimentation. Sediment traps at times have been used to reduce sedimentation and consist of basins constructed across the main sediment-contributing streams. Some accumulated sediment may be removed by flushing through sluices in the dam.

4.3 Dam sites

The type and size of dam constructed depend upon the need for and the amount of water available, the topography and geology of the site, and the construction materials that are readily obtainable. Dams can be divided into two major categories according to the type of material with which they are constructed, namely, concrete dams and earth dams. Concrete dams can be subdivided into gravity, arch, and buttress dams (Fig. 4.2). Earth dams comprise rolled-fill and rockfill embankments (Fig. 4.3). As the prime concern in dam construction is safety (this coming before cost), the foundations and abutments must be adequate for the type of dam selected. Some sites that are geologically unsuitable for a specific type of dam design may support one of composite design (Fig. 4.4).

Percolation of water through the foundations beneath concrete dams, even when the rock masses concerned are of good quality and of minimum permeability, is always a decisive factor in the safety and performance of dams. Such percolation can remove filler material that may be occupying joints, which in turn can lead to differential settlement of the foundations. It also may open joints, which decreases the strength of the rock mass. In highly permeable rock masses excessive seepage beneath a dam may damage the foundation. Seepage rates can be lowered by reducing the hydraulic gradient beneath a dam by incorporating a cut-off into the design. Uplift pressure acts against the base of a dam and is caused by water seeping beneath it that is under hydrostatic head from the reservoir. The hydrostatic pressure can be reduced by allowing water to be conducted downstream by drains incorporated into the foundation and base of the dam.

Fig. 4.3 Scammonden Dam, West Yorkshire, England, an example of an embankment dam.

Fig. 4.4 Inanda Dam, Natal, South Africa, a composite dam.

4.3.1 Geological conditions and dam sites

Geological conditions represent the most important natural factors that directly influence the design of dams. Not only do they control the character of the foundation but they govern the materials available for construction. In the case of rock masses, the main information required regarding the location of a dam includes the depth at which adequate foundations exist, the strengths of the rock masses involved, the likelihood of water loss, and any special features that have a bearing on excavation. The character of the foundations upon which dams are built and their reaction to the new conditions (of stress and strain, of hydrostatic pressure, and of exposure to weathering) must be ascertained so that appropriate safety factors may be adopted to ensure against subsequent failure.

In their unaltered state plutonic igneous rocks such as granite are essentially sound and durable with adequate strength for any engineering requirement. In some instances, however, major intrusions may be highly altered by weathering or hydrothermal alteration. As far as volcanic rocks are concerned, thick massive basalts make satisfactory dam sites but many basalts of comparatively young geological age are highly permeable, transmitting water via their open joints, pipes, cavities, tunnels, and contact zones. Foundation problems in young volcanic sequences also can be due to weak beds of ash and tuff that may occur between the basalt flows and give rise to differential settlement or sliding. In addition, weathering during periods of volcanic inactivity may have produced fossil soils that are of much lower strength. Pyroclasts usually give rise to extremely variable foundation conditions due to wide variations in strength, durability, and permeability. Ashes are invariably weak and often highly permeable; they also may undergo hydrocompaction on wetting.

Fresh (unweathered) metamorphosed rocks may be very strong and so afford excellent dam sites. Marble has the same advantages and disadvantages as other carbonate rocks. Cleavage and schistosity in regional metamorphic rock masses may adversely affect their strength and make them more susceptible to decay. For instance, certain schists and phyllites are of poor rock quality and so wholly undesirable in foundations and abutments.

Joints and shear zones may be responsible for unsound rock encountered at dam sites located on plutonic and metamorphic rock masses. Unless they are sealed, they may permit leakage through foundations and abutments. Slight opening of joints on excavation may lead to imperceptible rotations and sliding of rock blocks, large enough to reduce the strength and stiffness of a rock mass appreciably. Sheet or flat-lying joints tend to be approximately parallel to the topographic surface and introduce a dangerous element of weakness into valley slopes.

Sandstones have a wide range of strength depending largely upon the amount and type of cement-matrix material occupying the voids. With the exception of shaley sandstone, sandstone is not subject to rapid surface deterioration on exposure. As a foundation rock, even poorly cemented sandstone is not susceptible to plastic deformation. However, many sandstones are highly vulnerable to the scouring and plucking action at the overflow from a dam and have to be adequately protected by suitable hydraulic structures. Sandstones frequently are interbedded with beds of shale that may constitute potential sliding surfaces.

Limestone sites vary widely in their suitability for dam location. Thick-bedded horizontally lying limestones that are relatively free from solution cavities afford excellent dam sites. On the other hand, thin-bedded, highly folded or cavernous limestones are likely to present serious foundation or abutment problems involving bearing capacity, watertightness, or both. If the rock mass is thin-bedded, a possibility of sliding may exist. Similarly, beds separated by layers of clay or shale, especially those inclined downstream, may act as sliding planes and give rise to failure. Some solution features are always present in limestone. The size, form, abundance, and downward extent of these features depend upon the geological structure and the presence of interbedded impervious layers. Individual cavities may be open, they may be partially or completely filled with clay, silt, sand, or gravel mixtures, or they may be water-filled conduits. As can be inferred from above, solution cavities present numerous problems in the construction of large dams, among which bearing capacity and watertightness are paramount.

Well cemented shales, under structurally sound conditions, present few problems at most dam sites, although their strength limitations and elastic properties may be factors of importance in the design of concrete gravity dams of appreciable height. However, they have lower moduli of elasticity and lower shear strength values than concrete and therefore are unsatisfactory foundation materials for arch dams. Moreover, if the lamination is horizontal and well developed, then a foundation may offer little shear resistance to the horizontal forces exerted by a dam. A structure keying the dam into such a foundation is then required. Severe settlements may take place in low-grade compaction shales. Thus, such sites generally are developed with earth dams, but associated concrete structures such as spillways still involve these problems. The stability of slopes, both during and after construction, is one of the major problems of shale.

Earth dams usually are constructed on clay soils because clay lacks the load-bearing properties necessary to support concrete dams. Clay soils beneath valley floors often are contorted, fractured, and softened due to valley bulging so that the load of an earth dam may have to be spread over wider areas than is the case with shales and mudstones. Rigid ancillary structures necessitate spread footings or raft foundations. Slope stability problems also arise, with rotational slides representing a hazard.

Glacial deposits may be notoriously variable in composition, both laterally and vertically. As a result, dam sites in glaciated areas are often among the most difficult to appraise on the basis of surface evidence. A primary consideration in glacial terrains is the location of sites where rock foundations are available for spillway, outlet, and powerhouse structures. Normally, earth dams are constructed in areas of glacial deposits.

The major problems associated with foundations in alluvial deposits generally result from the fact that the deposits are poorly consolidated. Silts and clays are subject to plastic deformation or shear failure under relatively light loads and undergo consolidation for long periods of time when subjected to appreciable loads. Many large earth dams have been built upon such materials but this demands a thorough exploration and testing programme in order to design safe structures. The slopes of an embankment dam may be reduced in order to mobilize greater foundation shear strength, or berms may be introduced. Where soft alluvial clay soils are not more than 2.3 m thick they should consolidate during construction if covered with a drainage blanket (especially if resting on sand and gravel). With thicker deposits it may be necessary to incorporate vertical drains (e.g. band drains or sandwicks) within the clay soil. However, coarser sands and gravels undergo comparatively little consolidation under load and therefore afford excellent foundations for earth dams. Their primary problems result from their permeability. Alluvial sands and gravels form natural drainage blankets under the higher parts of an earth or rockfill dam, so that seepage through them beneath the dam must be cut off.

Landslides are a common feature of valleys in mountainous areas and a large slip often causes narrowing of a valley that therefore looks topographically suitable for a dam site. Unless landslides are shallow seated and can be removed or effectively drained, it is prudent to avoid landslipped areas in dam location, because their unstable nature may result in movement during construction or subsequently on inundation by reservoir water.

Fault zones may be occupied by shattered or crushed material and so represent zones of weakness that may give rise to sliding upon excavation for a dam. Movement along faults in active seismic regions occurs not only in association with large and infrequent earthquakes but also in association with small shocks and continuous slippage known as fault creep. Zoned embankment dams can be constructed at sites with active faults (Fig. 4.5). They should be designed with a higher freeboard and a wider crest than embankment dams in more stable regions, and with flatter slopes to provide against slumping and sliding. The cores of such zoned dams are wider and the filter zones larger, with suitable transition zones. In addition, filters are incorporated wherever different materials with different percolation characteristics are opposite each other. This ensures control of leakage if transverse cracks form in the core during an earthquake. Large outlets are desirable in zoned dams in order to provide a means of lowering the water in the reservoir basin quickly in an emergency.

4.3.2 Construction materials for earth dams

Wherever possible, construction materials for an earth dam should be obtained from within the future reservoir basin. Embankment soils need to develop high shear strength, low permeability and low water absorption, and undergo minimal settlement. This is achieved by compaction. Well-graded clayey sands and sand-gravel-clay mixtures develop the highest construction pore water pressures whereas uniform silts and fine silty sands are the least susceptible. In some cases only one type of soil is obtainable easily for an earth dam. If this is impervious, the design will consist of a homogeneous embankment, which incorporates a small amount of permeable material, in the form of filter drains, to control internal seepage. On the other hand, where sand and gravel are in plentiful supply, a very thin clay core may be built into an embankment dam if enough impervious soil is available, otherwise an impervious membrane may be

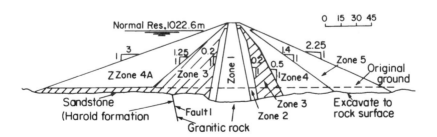

Fig. 4.5 Cross-section of Cedar Springs Dam, California, a zoned embankment dam. Zone 1 = clay core from lake bed deposit; zone 2 = silty sand from Harold Formation; zone 3 = processed sand-gravel transition from river alluvium or crushed rock; zones 4 and 4a = rolled processed rockfill (75 mm minimum, 750 mm maximum); zone 5 = dumped processed rockfill (450 mm minimum).

constructed of concrete or interlocking steel-sheet piles. However, since concrete can withstand very little settlement such core walls must be located on sound foundations. Sites that consist of a variety of soils lend themselves to the construction of zoned dams. The finer, more impervious materials are used to construct the core whilst the coarser materials provide strength and drainage in the upstream and downstream zones.

4.4 Basic hydrogeology

The principal source of groundwater is meteoric water, that is, precipitation (rain, sleet, snow, and hail). However, two other sources are very occasionally of some consequence, that is, juvenile water and connate water. The former is derived from magmatic sources whilst connate water was trapped in the pore spaces of sedimentary rocks as they were formed.

The amount of water that infiltrates into the ground depends upon how precipitation is dispersed, that is, on what proportions are assigned to immediate run-off and to evapotranspiration, the remainder constituting the proportion allotted to infiltration/percolation. The rate at which groundwater is replenished is dependent basically upon the quantity of precipitation falling on the recharge area of an aquifer, although rainfall intensity also is very important. Frequent rainfall of moderate intensity is more effective in recharging groundwater resources than short concentrated periods of high intensity, because the rate at which the ground can absorb water is limited, any surplus water tending to become run-off.

The retention of water in a soil depends upon the capillary force and the molecular attraction of the particles. As the pores in a soil become thoroughly wetted, the capillary force declines so that gravity becomes more effective. In this way downward percolation can continue after infiltration has ceased, but as the soil dries so capillarity increases in importance. No further percolation occurs after capillary and gravity forces are balanced. Thus, water percolates to the zone of saturation when the retention capacity is satisfied. This means that the rains that occur after the deficiency of soil moisture has been catered for are those that count in terms of groundwater recharge.

4.4.1 Water tables and aquifers

The pores within the zone of saturation are filled with water, generally referred to as phreatic water. The upper surface of this zone therefore is known as the phreatic surface but is more commonly termed the water table. Above the zone of saturation is the zone of aeration in which both air and water occupy the pores. The water in the zone of aeration is commonly referred to as vadose water.

The water table fluctuates in position, particularly in those climates where there are marked seasonal changes in rainfall. Hence, permanent and intermittent water tables can be distinguished, the former marking the level beneath which the water table does not sink, whilst the latter is an expression of the fluctuation. Usually, water tables fluctuate within the lower and upper limits rather than between them, especially in humid regions, since the periods between successive recharges are small. The position at which the water table intersects the surface is termed the spring line. Intermittent and permanent springs similarly can be distinguished. A perched water table is one that forms above a discontinuous impermeable layer such as a lens of clay soil in a formation of sand, the clay soil impounding a water mound.

An aquifer is the term given to a rock or soil mass that not only contains water but from which water can be abstracted readily in significant quantities. The ability of an aquifer to transmit water is governed by its permeability. Indeed, the permeability of an aquifer usually is in excess of 10^{-5} m s^{-1}. By contrast, a formation with a permeability of less than 10^{-9} m s^{-1} is one that, in engineering terms, is regarded as impermeable and is referred to as an aquiclude. An aquitard is a formation that transmits water at a very slow rate but that, over a large area of contact, may permit the passage of large amounts of water between adjacent aquifers that it separates. An aquifer is described as unconfined when the water table is open to the atmosphere, that is, the aquifer is not overlain by material of lower permeability (Fig. 4.6a). Conversely, a confined aquifer is one that is overlain by impermeable rocks. The water in a confined aquifer may be under piezometric pressure, that is, there is an excess of pressure sufficient to raise the water above the base of the overlying bed when the aquifer is penetrated by a well. Piezometric pressures are developed

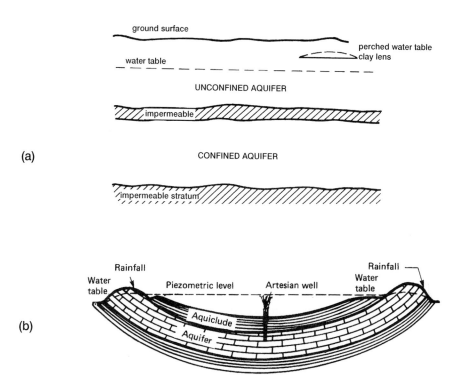

Fig. 4.6 (a) Diagram illustrating unconfined and confined aquifers, with a perched water table in the vadose zone. (b) An artesian basin.

when the buried upper surface of a confined aquifer is lower than the water table in the aquifer at its recharge area. Where the piezometric surface is above ground level, water will overflow from a well, the well and groundwater conditions being described as artesian (Fig. 4.6b).

4.4.2 Porosity, permeability and specific yield

Porosity and permeability are the two most important factors governing the accumulation, migration and distribution of groundwater. The former can be defined as the percentage pore space within a given volume of rock or soil. Total or absolute porosity is a measure of the total volume of the voids whereas the effective or net porosity excludes the volume of occluded pore space (occluded pores are not in connection with others). The effective porosity may be regarded as the pore space from which water can be removed. The porosity of a deposit does not necessarily provide an indication of the amount of water that can be obtained from a rock or soil mass.

The capacity of a material to yield water is of greater importance than is its capacity to hold water, as far as supply is concerned. Even though a rock or soil mass may be saturated, only a certain proportion of groundwater can be removed by drainage under gravity or by pumping, the remainder being held in place by capillary or molecular forces. The ratio of the volume of water retained to that of the total volume of rock or soil mass expressed as a percentage is referred to as the specific retention. The amount of water retained varies directly in accordance with the surface area of the pores and indirectly with regard to the pore space. The specific yield of a rock or soil mass refers to its water yielding capacity and is the ratio of the volume of water, after saturation, that can be drained by gravity to the total volume of the aquifer,

Table 4.1   Some examples of specific yield.

Materials	Specific yield (%)
Gravel	15–30
Sand	10–30
Dune sand	25–35
Sand and gravel	15–25
Loess	15–20
Silt	5–10
Clay	1–5
Till (silty)	4–7
Till (sandy)	12–18
Sandstone	5–25
Limestone	0.5–10
Shale	0.5–5

expressed as a percentage. The specific yield plus the specific retention is equal to the porosity of the material when all the pores are interconnected. Examples of the specific yield of certain common types of soil and rock are given in Table 4.1 (it must be appreciated that individual values of specific yield can vary from those quoted).

The storage coefficient or storativity of an aquifer is defined as the volume of water released from or taken into storage per unit surface area of the aquifer, per unit change in head normal to that surface. It is a dimensionless quantity. The storage coefficient of an unconfined aquifer virtually corresponds to the specific yield as more or less all the water is released from storage by gravity drainage. In confined aquifers water is not yielded simply by gravity drainage from the pore space because there is no falling water table and the material remains saturated. Therefore, much less water is yielded by confined than unconfined aquifers.

A material is termed permeable when it allows the passage of a measurable quantity of fluid in a finite period of time and impermeable when the rate at which it transmits that fluid is slow enough to be negligible under existing temperature–pressure conditions. The permeability of a particular soil or rock mass is defined by its coefficient of permeability, it being the flow in a given time through a unit cross-sectional area of a material (Table 4.2). As far as the permeability of intact rock (primary permeability) is concerned, it usually is several orders less than the *in situ* permeability (secondary permeability) of a rock mass. In other words, the permeability of rock masses is governed by fissure flow, that is, it normally is the interconnected system of discontinuities that determine rock mass permeability. The transmissivity or flow in cubic metres per day through a section of aquifer 1 m wide under a hydraulic gradient of unity is sometimes used in the calculation of groundwater flow.

4.5   Wells

The commonest way of recovering groundwater is to sink a well and lift water from it. The most efficient well is developed so as to yield the greatest quantity of water with the least drawdown and the lowest velocity of flow in the vicinity of the well. The specific capacity of a well is expressed in litres of yield per metre of drawdown when the well is being pumped. It is indicative of the relative permeability of the aquifer. Location of a well obviously is important if an optimum supply is to be obtained and a well site should always be selected after a careful study of the hydrogeological conditions. Completion of a well in an unconsolidated formation requires that it is cased so that the surrounding deposits are supported. Sections of the casing must be perforated to allow the penetration of water from the aquifer into the well, or screens can be used. The casing should be as permeable as, or more permeable than, the deposits it confines.

Table 4.2 Relative values of permeabilities.

Rock types	Porosity — Primary (pore) %	Porosity — Secondary (fracture)*	Permeability range (m s⁻¹)	Well yields	Type of water bearing unit
Sediments, unconsolidated					
Gravel	30–40		Very high–High	High	Aquifer
Coarse sand	30–40		Very high–High	High	Aquifer
Medium to fine sand	25–35		High–Med	Med	Aquifer
Silt	40–50	Occasional, often fissured	Med–Low	Med	Aquiclude
Clay, till	45–55	Occasional, often fissured	Low–Very low	Low	Aquiclude
Sediments, consolidated					
Limestone, dolostone	1–50	Solutions joints, bedding planes	Very high–Low	High	Aquifer or aquiclude
Coarse, medium sandstone	<20	Joints and bedding planes	High–Low	Med	Aquifer or aquiclude
Fine sandstone	<10	Joints and bedding planes	Med–Very low	Med	Aquifer or aquiclude
Shale, siltstone	–	Joints and bedding planes	Very low–Impermeable	Low	Aquiclude or aquifer
Volcanic rocks, e.g. basalt	–	Joints and bedding planes	Very high–Low	High	Aquifer or aquiclude
Plutonic and metamorphic rocks	–	Weathering and joints decreasing as depth increases	Very low–Impermeable	Low	Aquiclude or aquifer

Permeability range scale (m s⁻¹): 10^0 / 10^{-2} / 10^{-4} / 10^{-6} / 10^{-8} / 10^{-10} corresponding to Very high, High, Med, Low, Very low, Impermeable.

*Rarely exceeds 10%.
Lines represent approximate range of permeability and well yield.

119

The long-term yield of a well depends upon a number of factors. Ideally, a well should be located where the saturated thickness is greatest. If the aquifer material has only a small variation in permeability, then transmissivity increases with increasing saturated thickness. This facilitates flow to a well. There may be some advantage in siting a well near to a recharge area, so that a surface water resource is diverted underground to augment aquifer flow by induced infiltration, providing that the surface water will not affect the well adversely (see below). This can increase well yields. Alternatively, a well may be sited near the discharge area of an aquifer with the objective of diverting as much of the natural discharge as possible to the well. Neither of these options should be undertaken without a careful evaluation of the possible consequences. An alternative strategy may be to exploit the storage of an aquifer without interfering with the flow of groundwater through the aquifer. In this case, the well should be sited some distance from the areas of natural recharge and discharge, so that it takes some significant time for the pumping effects to reach them. The annual rate of groundwater recharge influences the rate of flow in an aquifer and therefore the amount of water available for abstraction. Obviously, the higher the permeability, the easier it is for water to flow to a well during periods of abstraction. Faults or notable discontinuities may act as preferred flow channels and greatly increase the flow to a well. Many wells in relatively impermeable material have been successful as a result of flow via faults or discontinuities. However, a fault may also act as a barrier to flow if it is filled with impermeable gouge material or if the throw of the fault places the aquifer against an impermeable stratum. Finally, the location of wells must be made with respect to any features that may jeopardize the quality and quantity of the discharge, or the groundwater resource as a whole. In addition, wells that supply drinking water must be properly sealed at the surface in order to avoid being contaminated. It follows from this that a well should be sealed after it has been abandoned, otherwise the aquifer can be contaminated.

4.6 Safe yield

The yield of a surface water resource generally is related to a return interval, it being defined as the steady supply that can be maintained through a drought of specified severity. For example, the yield that can be maintained through a 1 in 50 year (2%) drought is greater than that which could be maintained through a 1 in 100 year (1%) drought. Hence, yield is not an absolute quantity, but a variable that depends upon the specified frequency of occurrence of the limiting drought conditions. Yield also may be defined as the steady supply that could be maintained through the worst drought on record. In this case the severity of the limiting drought conditions depends upon the rainfall recorded in the years preceding the drought period, the length of the record available for analysis, and chance – for instance, a short record may contain a particularly severe drought, while a much longer record elsewhere may not. For this reason the first definition is to be preferred.

The concept of safe yield has been used to express the quantity of groundwater that can be withdrawn without impairing an aquifer as a water source. The abstraction of groundwater at a greater rate than it is being recharged leads to the water table being lowered, and upsets the equilibrium between discharge and recharge. Draft in excess of safe yield is overdraft. Estimation of the safe yield is a complex problem that must take into account the climatic, geological, and hydrogeological conditions. As such, the safe yield is likely to vary appreciably with time. Nonetheless, the recharge–discharge equation, the transmissivity of the aquifer, the potential sources of contamination, and the number of wells in operation must all be given consideration if an answer is to be found. The transmissivity of an aquifer may place a limit on the safe yield, in that a large yield can only be realized if the aquifer is capable of transmitting water from the source area to wells at a rate high enough to sustain the draft. Where contamination of the groundwater is possible, then the location of wells, their type, and the rate of abstraction must be planned in such a way that (conditions permitting) contamination cannot be developed.

Once an aquifer is developed as a source of water supply, then effective management becomes increasingly necessary if it is not to suffer deterioration. Management should not merely be concerned with the abstraction of water but should also consider its utilization, since different qualities of water can be put to different uses. A progressive decline in groundwater level is often the consequence of exceeding the safe yield of an aquifer, leading to an unacceptable pumping lift, to a reduced yield due to the restricted drawdown available, and possibly to a deterioration in water quality. The latter often occurs as a result of old highly mineralized water being drawn from deep in the aquifer into the wells, or through induced infiltration or saline intrusion. These problems can necessitate a reduction in the output of a wellfield, or even its abandonment. Falling groundwater levels also may result in the loss of marshes and wetlands, with potentially serious agricultural and ecological implications.

4.7 Artificial recharge

Artificial recharge may be defined as an augmentation of the natural replenishment of groundwater storage by artificial means. Its main purpose is water conservation, often with improved quality as a second aim. It therefore is used for reducing overdraft, for conserving and improving surface run-off, and for increasing available groundwater supply. An aquifer subjected to artificial recharge must have adequate storage and the bulk of the water recharged should not be lost rapidly by discharge into a nearby river. The source of water for artificial recharge may be storm run-off, river or lake water, water used for industrial cooling purposes, industrial waste water, or sewage water. Many of these sources require some kind of pre-treatment. Interaction between artificial recharge and groundwater may lead to precipitation, for example, of calcium carbonate, and iron and magnesium salts, resulting in a reduction of permeability, which ideally should be avoided.

Artificial recharge may be brought about by various surface-spreading methods utilizing basins, ditches, or flooded areas; by spray irrigation; or by pumping water into the ground via vertical shafts, horizontal collector wells, pits or trenches. The most widely practised methods are those of water spreading, which allow increased infiltration to occur over a wide area when the aquifer outcrops at or is near the surface. These methods require that the ground has a high infiltration capacity. Care has to be taken in surface-spreading methods not to flush salts and nutrients from the soil into the groundwater. In regions with hot, dry climates there may be a danger of excessive evaporation of recharge water leading to salinization. Recharge wells are employed most frequently when the aquifer to be recharged is deep or confined, or there is insufficient space for recharge basins.

4.8 Conjunctive use

The combined use of surface and groundwater is referred to as conjunctive use. By adopting a conjunctive-use approach the differing characteristics of surface and groundwater can be used to optimize the yield of the total water resource. For instance, surface waters may be available seasonally, but usually some uncertainty surrounds the time and amount available. Additionally, surface systems are characterized by flooding, not all the water of which can be captured by impounding reservoirs or used for water supply. While surface reservoirs can be filled rapidly, they are subject to losses by evaporation and seepage. On the other hand, groundwater often is available in large aquifers in large quantities, with relatively little variation over time. Groundwater reservoirs tend to react comparatively slowly to changes in inflow or outflow. Hence, less uncertainty is involved in predicting future groundwater availability than in predicting surface stream flow. A conjunctive use approach to water supply aims to manage the surface and groundwater resources of an area in order to obtain a net gain in yield.

Groundwater can be used to augment river flow during a dry part of the year or during droughts in a conjunctive use scheme. The quantity of groundwater required depends on the variability of river flow and the level of river regulation adopted. The drawdown experienced by an aquifer, and consequently the time required for groundwater levels to recover, depends on the properties of the aquifer and the level of river regulation adopted. Groundwater levels can be increased during periods of surplus river flow by using artificial recharge if natural recharge is insufficient or too slow. Additionally, wells should be concentrated in restricted areas, to limit the area affected by pumping and so reduce the length of river over which decreased flow can be expected. With unconfined aquifers, particularly when they have a fast response time, it may be desirable to site abstraction wells some distance away from a river. If the wells are too near to the river, induced infiltration may cause very rapid circulation of water around the aquifer–river system with a negligible net gain. However, siting wells remote from a river unfortunately increases pumping and pipeline costs. Confined aquifers, because of their small coefficient of storage and fast response time, are not always suitable for conjunctive use, although the apparent isolation of the surface and groundwater resources initially may make them appear attractive propositions.

Conjunctive use schemes not only provide optimization of water use but also offer greater flexibility when responding to an increase in demand, since more than one source is available. Moreover, since water can be transferred from impounding reservoirs to underground storage, the level in surface reservoirs can be dropped to allow for increased flood storage. If water is derived from different sources at different times, the water supplied to the consumer may change from soft to hard. Hence, some blending of the water from the different sources may be required.

4.9 Water quality

In an evaluation of water resources, the quality of the water is of almost equal importance to the quantity available. In other words, the chemical, physical and biological characteristics of water are of major importance in determining whether or not it is suitable for domestic, industrial or agricultural use. However, with regard to groundwater, the number of major dissolved constituents is quite limited and natural variations usually are not large (Table 4.3). The quality of water in the zone of saturation reflects that of the water that has percolated to the water table and the subsequent reactions between water and soil/rock that occur. The quality of phreatic water also is affected by fluctuations of the water table. In particular, if the water table occurs at shallow depth, then losses by transpiration, and possibly evaporation, increase when it rises. This means that the salt content increases. Conversely, when the water table is lowered this may cause lateral inflow from surrounding areas with a consequent change in salinity. The factors that influence the solute content include the original chemical quality of water entering the zone of saturation; the distribution, solubility, exchange capacity and exchange selectivity of the minerals involved in reaction; the porosity and permeability of the rock masses; and the length of flow path of the water. Of critical importance in this context is the residence time of the water since this determines whether there is sufficient time for dissolution of minerals to proceed to the point where the solution is in equilibrium with the reaction. Residence time depends on the rate of groundwater movement and this usually is very slow beneath the water table.

Table 4.3 Examples of groundwater quality from different types of rock masses (in mg l⁻¹).

Rock type	TDS	Ca	Mg	HCO₃	Na	K	Cl	SO₄	Fe	SiO₄	Location
Granite	223	27	6.2	93	9.5	1.4	5.2	32	1.6	39	Maryland
Basalt	505	62	28	294	24	–	37	30	–	30	Hyderabad
Diorite	346	72	4.1	114	10	2.8	6.5	115	0.04	22	N. Carolina
Andesite	70	12	0.5	38	1.8	2.6	–	6.3	–	8.9	Idaho
Marble	236	39	10	162	2.7	0.3	3.8	2.4	0.03	9.9	Alabama
Gneiss	135	19	5.1	39	4.4	3.2	5.8	30	0.09	13	Connecticut
Sandstone	210	40	12	67	7.6	0.4	19	26	–	12	Worcestershire
Limestone	247	48	5.8	168	4	0.7	0.1	4.8	0.05	8.9	Florida
Chalk	384	115	5	152	10.2	1.2	20	39	–	1.1	Hertfordshire
Dolostone	546	67	39	390	7.6	0.4	–	17	–	24	South Africa
Gypsum	2480	636	43	143	16.1	0.9	24	1570	–	29	New Mexico
Shale	260	29	16	126	12	1.1	12	22	0.02	16	New Jersey

The uppermost layers of soil and rock act as purifying agents. In the soil, organisms such as fungi and bacteria attack pathogenic bacteria, as well as reacting with certain other harmful substances. The other important factor in purifying groundwater is the filtering action of soil and rock. This depends on the size of the pores, the proportion of argillaceous and organic matter present, and the distance travelled by the groundwater involved. Unconfined groundwater at shallow depth is highly susceptible to pollution but at greater depth recharge is by water that has been partially or wholly purified. Nonetheless, there is a tendency for the salt content of groundwater to increase with depth. The reasons for this are, firstly, that the greater the depth at which groundwater occurs, then the slower it moves and so it is less likely to be replaced by other water, especially that infiltrating from the surface. Secondly, the longer residence time of the groundwater provides more time for reaction with the host rock and so more material goes into solution until an equilibrium condition is attained. Thirdly, connate or fossil water may occur at greater depth. As the character of groundwater in an aquifer frequently changes with depth, it is possible at times to recognize zones of different quality of groundwater. With increasing depth, cation exchange reaction increases in importance, and there is a gradual replacement of calcium and magnesium in the groundwater by sodium. Any nitrates present near the surface of an aquifer invariably decrease with depth. On the other hand, sulphates tend to increase with depth. However, at appreciable depth, sulphates are reduced, which produces low sulphate-high bicarbonate water. The chloride content also tends to increase with depth. In fact, with increasing depth groundwater may become non-potable due to its high chloride content. Most highly saline groundwater (not associated with evaporites) occurring at depth in sedimentary rocks, where groundwater circulation is restricted, is connate water.

As noted, ion exchange affects the chemical nature of groundwater. It involves the replacement of ions adsorbed on the surfaces of clay minerals, organic compounds and zeolites by ions in solution. Glauconite also is an important ion exchange mineral in some sedimentary rocks. The process usually is associated with cations and therefore is referred to as cation or base exchange. For instance, the replacement of Ca^{2+} and Mg^{2+} by Na^+ may occur when groundwater moves beneath argillaceous rocks into a zone of more restricted circulation. This produces soft water. The cation exchange process means that fine-grained formations, notably clay soils and shales, can behave like semi-permeable membranes, retarding the movement of ions. Such behaviour can produce osmotic pressure differences, salt sieving or ultrafiltration, and electro-potential differences. For example, the salt content of certain groundwaters may be changed by salt sieving, that is, as water flows out of a fine-grained formation that behaves as a semi-permeable membrane, then the salt is retained. Changes in temperature–pressure conditions may result in precipitation (e.g. a decrease in pressure may liberate CO_2 causing the precipitation of calcium carbonate).

The total dissolved solids (TDS) in a sample of water include all solid material in solution, whether ionized or not. Water for most domestic and industrial uses should contain less than 1000 mg l^{-1} and the TDS content of water for most agricultural purposes should not be above 3000 mg l^{-1}. Groundwater has been classified according to its TDS content as follows:

Fresh	less than 1000 mg l^{-1}
Slightly saline	1000 to 3000 mg l^{-1}
Moderately saline	3000 to 10 000 mg l^{-1}
Very saline	10 000 to 35 000 mg l^{-1}
Briny	over 35 000 mg l^{-1}

Several minor elements in drinking water are a matter of concern because of their toxic effects. These include arsenic, barium, copper, chromium, cadmium, lead and mercury (Table 4.4).

The hardness of water relates to its reaction with soap, and to the scale and encrustations that form in boilers and pipes where water is heated and transported. It is attributable to the relative abundance of calcium and magnesium in groundwater. Groundwater derived from limestone or dolostone aquifers containing gypsum or anhydrite may contain 200 to 300 mg l^{-1} hardness or more. Water for domestic use should not contain more than 80 mg l^{-1} total hardness. Hardness, H_T, generally is expressed in terms of the

Table 4.4 Minor and trace elements, and compounds in drinking water.

Element	WHO (1993) guideline maxima (mg l^{-1})	CEC (1980) maximum admissible concentrations (mg l^{-1})
Aluminium	0.2	0.2
Ammonium	1.5	0.5
Antimony	0.005 (P)	0.01
Arsenic	0.01	0.05
Barium	0.7	
Boron	0.3	
Cadmium	0.003	0.005
Chromium	0.05 (P)	0.05
Copper	2 (P)	
Fluoride	1.5	1.5
Lead	0.01	0.05
Manganese	0.5 (P)	0.05
Mercury	0.001	0.001
Molybdenum	0.07	
Nickel	0.02	0.05
Nitrate	50	50
Selenium	0.01	0.01
Zinc	3	5

WHO – World Health Organization
CEC – Commission of European Community
P – Provisional value

equivalent of calcium carbonate, hence:

$$H_T = Ca \times (Ca \times O_3/Ca) + Mg \times (CaCO_3/Mg) \qquad (4.1)$$

where H_T, Ca and Mg are measured in mg l^{-1} and the ratios in equivalent weights. This equation reduces to

$$H_T = 2.5Ca + 4.1Mg \qquad (4.2)$$

The degree of hardness is described in Table 4.5.

Temperature, colour, turbidity, odour and taste are the most important physical properties of water in

Table 4.5 Hardness of water.

Description	Hardness (mg l^{-1} as CaCO$_3$)	
	(After Sawyer and McCarthy, 1967)*	(After Hem, 1985)*
Soft	below 75	below 60
Moderately hard	75–150	61–120
Hard	150–300	121–180
Very hard	over 300	over 180

*See Suggestions for Further Reading

relation to water supply, and drinking water especially, should be aesthetically acceptable. Groundwater only undergoes appreciable fluctuations in temperature at shallow depth, beneath which temperatures remain relatively constant. Below the zone of surface influence, groundwater temperatures increase by approximately 1°C for every 30 m of depth, that is, in accordance with the geothermal gradient. The colour of groundwater may be attributable to organic or mineral matter carried in solution. For example, light to dark brown discolorations can occur in groundwater that has been in contact with peat or other organic deposits. Brownish discoloration also can result from groundwater that contains dissolved ferrous iron that has been exposed to the atmosphere. This leads to insoluble ferric hydroxides being formed. Oxidation of manganese in groundwater can produce black stains and therefore the maximum permitted concentration of manganese for domestic water is 0.05 mg l⁻¹. Turbidity of groundwater is mainly caused by the presence of clay and silt particles derived from the aquifer. Oxidation of dissolved ferrous iron to form insoluble ferric hydroxides also contributes towards turbidity. The natural filtration that occurs when groundwater flows through unconsolidated deposits largely removes such material from groundwater. Tastes and odours may be derived from the presence of mineral matter, organic matter, bacteria or dissolved gases. For example, the maximum concentration of chloride in drinking water is 250 mg l⁻¹, primarily for reasons of taste. Again for reasons of taste, and also to avoid staining, the recommended maximum concentration of iron in drinking water is 0.3 mg l⁻¹. The pH value of drinking water should be close to 7 but treatment can cope with a range of 5 to 9. Beyond this range treatment to adjust the pH to 7 becomes less economical.

Certain dissolved gases such as oxygen and carbon dioxide affect groundwater chemistry. Others affect the use of water, for example, hydrogen sulphide in concentrations greater than 1 mg l⁻¹ render water unfit for consumption because of the objectionable odour. Methane coming out of solution may accumulate and present a fire or explosion hazard. The minimum concentration of methane in water sufficient to produce an explosive methane–air mixture above the water from which it has escaped depends on the volume of air into which the gas evolves. Theoretically, water containing as little as 1 to 2 mg l⁻¹ of methane can produce an explosion in poorly ventilated air space.

Obviously, water for human consumption must be free from organisms, as well as chemical substances, in concentrations large enough to affect health adversely. Standard tests used to determine the safety of water for drinking purposes involve identifying whether or not bacteria belonging to the coliform group are present. One of the reasons for this is that this group of bacteria (*Escherichia coli* in particular) are relatively easy to recognize. Since the whole coliform group is foreign to water, a positive *E. coli* test indicates the possibility of bacteriological contamination. If *E. coli* are present, then it is possible that the less numerous or harmful pathogenic bacteria, which are much more difficult to detect, also are present. On the other hand, if there are no *E. coli* in a sample of water, then the chances of faecal contamination and of pathogens being present generally are regarded as negligible. Viruses in groundwater are more critical than bacteria in that they tend to survive longer and some viruses are more resistant to disinfectant. In addition, one virus unit (one plague-forming unit in a cell culture) may cause an infection when ingested. By contrast, ingestion of thousands of pathogenic bacteria may be required before clinical symptoms are developed. Groundwater generally is free from pathogenic bacteria and viruses, except perhaps that from very shallow aquifers. Micro-organisms can be carried by groundwater but tend to attach themselves by adsorption to the surfaces of clay particles. In fine-grained soils bacteria generally move less than a few metres, but they can migrate much larger distances in coarse-grained soils or discontinuous rock masses. The maximum rate of travel of bacterial pollution appears to be about two-thirds that at which the groundwater is moving. Viruses that retain all their characteristics for more than 50 days may migrate 250 m or more in soils where organic matter is present to supply a food source.

The quality of water required for different industrial processes varies appreciably, indeed it can differ within the same industry. Nonetheless, salinity, hardness and silica content are three parameters that usually are important in terms of industrial water. Water used in the textile industries should contain a low amount of iron, manganese and other heavy metals likely to cause staining. Hardness, total dissolved solids, colour and turbidity also must be low. The quality of water required by the chemical industry varies widely depending on the processes involved. Similarly, water required in the pulp and paper industry is governed by the type of products manufactured. Groundwater generally is preferable to surface water since it shows less variation in chemical and physical quality.

4.10   Pollution

Pollution can be regarded as impairment of water quality by chemicals, heat, or bacteria to an extent that does not necessarily create an actual public health hazard, but that it does adversely affect such waters for domestic, agricultural or industrial use. The greatest danger of groundwater pollution is from surface sources such as excessive application of fertilizers, leaking sewers, polluted streams, mining and mineral wastes, domestic and industrial waste, and so on. Usually, the most dangerous forms of groundwater pollution are those that are miscible with the groundwater in an aquifer. Any possible source of pollution should be carefully evaluated, both before and after well construction, and the viability of groundwater protection measures considered. However, the slow rate of travel of pollutants through most subsurface strata means that a case of pollution may go undetected for a number of years. During this period a large part of an aquifer may become polluted and cease to have any potential as a source of water. By contrast in areas of karstic limestone, pollutants may be able to travel quickly over large distances (e.g. over 30 km in approximately 3 months). Deeply buried aquifers overlain by relatively impermeable beds of shale or clay generally have a low pollution potential. Areas with thin soil cover, particularly when the overlying material consists of sand or gravel, or where an aquifer is exposed, such as the recharge area, are the most critical from the point of view of pollution potential. In general, the concentration of a pollutant decreases as the distance it travels through the ground increases.

Perhaps the greatest risk of groundwater pollution arises when pollutants can be transferred directly from the ground surface to an aquifer. In this case the purifying processes that take place within the soil are bypassed and attenuation of the pollutant is reduced. A common cause of such groundwater pollution is poor well design, construction and maintenance. When a drill hole penetrates an aquifer it can act as a conduit for the transfer of pollution from the ground surface. With multiple aquifers there is the additional possibility of inter-aquifer flow. In such cases, each aquifer must be isolated using cement-grout seals in the intervening strata.

Attenuation of a pollutant occurs as it moves through the ground as a result of four major processes. First, the soil has an enormous purifying power due to the communities of bacteria and fungi that live in the soil. These organisms are capable of attacking pathogenic bacteria and also can react with certain other harmful substances. Secondly, as water passes through fine-grained porous media suspended impurities are removed by filtration. Thirdly, some substances react with minerals in the soil/rock, and some are oxidized and precipitated from solution. Adsorption also may occur in argillaceous or organic material. Fourthly, dilution and dispersion of a pollutant may lead to its concentration being reduced until it eventually becomes negligible at some distance from the source.

In many instances of groundwater pollution, the contaminating fluid is discharged into the aquifer as a continuous or nearly continuous flow. Hence, the polluted groundwater often takes the form of a plume or plumes. An important part of an investigation of groundwater pollution is to locate and define the extent of the contaminated body of groundwater in order to establish the magnitude of the problem and, if necessary, to design an efficient abatement system.

Frequently, it is not practical to initiate counter measures to combat groundwater pollution once it has occurred, the eventual attenuation of the pollutant being the result of time, degradation, dilution and dispersion. In the case of a recent accidental spill, however, it may be possible to excavate affected soil, and use scavenger wells to intercept and recover some of the pollutant before it has dispersed significantly. Alternatively, in a few situations, artificial recharge may be undertaken in an attempt to dilute the pollutant, or to form a hydraulic pressure barrier that will divert the pollutant away from abstraction wells. Unfortunately, none of these options provides a reliable method of dealing with groundwater pollution and they should only be considered as a last resort.

4.10.1 Induced infiltration

Induced infiltration occurs where a stream is hydraulically connected to an aquifer and lies within the area of influence of a well. When water is abstracted from a well the water table in the immediate vicinity is lowered and assumes the shape of an inverted cone that is referred to as a cone of depression (Fig. 4.7). The steepness of the cone is governed by the soil/rock types concerned, it being flatter in highly permeable materials. The size of the cone depends on the rate of pumping, equilibrium being achieved when the rate of abstraction balances the rate of recharge. However, if abstraction exceeds recharge, then the cone of depression increases in size. As the cone of depression spreads, groundwater is withdrawn from storage over a progressively increasing area of influence and groundwater levels about the well continue to be lowered. Cones associated with some larger wells may have radii of up to 1.5 km. With time the aquifer is recharged by influent seepage from the surface stream. As pumping continues the proportion of water derived from the stream that enters the cone of depression increases progressively. Whether or not induced infiltration gives rise to pollution depends upon the quality of the surface water source, the nature of the aquifer, the quantity of infiltration involved and the intended use of the abstracted groundwater. Induced infiltration at one extreme can have potentially disastrous consequences whilst at the other can provide a valuable addition to the overall water resources of an area.

4.10.2 Landfill and groundwater pollution (see also Chapter 7)

The disposal of wastes in landfill sites leads to the production of leachate and gases, which may present a health hazard as a consequence of pollution of groundwater supply. Leachate often contains high concentrations of dissolved organic substances resulting from the decomposition of organic material such

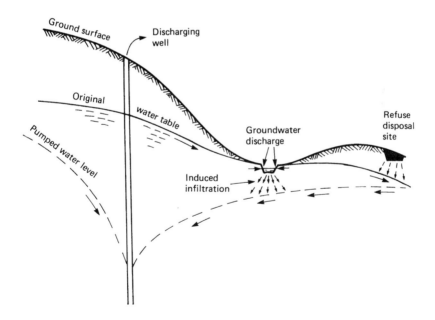

Fig. 4.7 An example of induced infiltration caused by overpumping. The original hydraulic gradient over much of the area has been reversed so that pollutants can travel in the opposite direction, that is, towards the well. Additionally, the aquifer has become influent (i.e. water drains from the river into the aquifer) instead of effluent, as it was originally.

as vegetable matter and paper. Recently emplaced wastes may have a chemical oxygen demand (COD) of around 11 600 mg l^{-1} and a biochemical oxygen demand (BOD) in the region of 7250 mg l^{-1}. The concentration of chemically reduced inorganic substances varies according to the hydrology, and the chemical and physical conditions at the site. For example, 1000 m^3 of waste can yield 1.25 tonnes of potassium and sodium, 0.8 tonnes of calcium and magnesium, 0.7 tonnes of chloride, 0.19 tonnes of sulphate and 3.2 tonnes of bicarbonate.

Site selection for waste disposal must take into account the character of the material that is likely to be tipped. For instance, toxic or oily liquid waste represent a serious risk, although sites on impermeable substrata often merit a lower assessment of risk. Consequently, selection of a landfill site for a particular waste or a mixture of wastes involves a consideration of the geological and hydrogeological conditions. Argillaceous sedimentary, massive igneous and metamorphic rocks have low permeabilities and therefore afford the most protection to water supply. By contrast, the least protection is offered by rock masses intersected by open discontinuities or in which solution features are developed, or by open-work gravel deposits.

Leachate pollution can be tackled by either concentrating and containing, or by diluting and dispersing. Infiltration through sandy ground of liquids from a landfill may lead to their decontamination and dilution. Hence, sites for disposal of domestic refuse can be chosen where decontamination has the maximum chance of reaching completion and where groundwater sources are located far enough away to enable dilution to be effective. Consequently, domestic waste can be tipped at dry sites on sandy material, which has a thickness of at least 15 m. Water supply sources should be located at least 0.8 km away from the landfill site. They should not be located on discontinuous rocks unless overlain by 15 m of clay deposits. Potentially toxic waste should be contained. Such sites should be underlain and confined by at least 15 m of impermeable strata and any source abstracting groundwater for domestic use should be at least 2 km away. Furthermore, the topography of the site should be such that run-off can be diverted from the landfill so that it can be disposed of without causing pollution of surface waters. Containment can be achieved by an artificial impermeable lining being placed over the base of a site. However, there is no guarantee that this will remain impermeable. Drains can be installed beneath a landfill to convey leachate to a sump, which then can be either pumped to a sewer, transported away by tanker, or treated on site.

4.10.3   Saline intrusion

Although saline groundwater, originating as connate water or from evaporitic deposits, may be encountered, the problem of saline intrusion is specifically related to coastal aquifers. Near to the coast there is an interface between the overlying fresh groundwater and the underlying saline groundwater (Fig. 4.8). Excessive lowering of the water table along a coast leads to saline intrusion, the salt water entering the aquifer via submarine outcrops thereby displacing fresh water. However, the fresh water still overlies the saline water and continues to flow from the aquifer to the sea. Overpumping – whereby the natural flow of freshwater to the sea is interrupted or significantly reduced by abstraction – gives rise to salt water encroachment. Continuous pumping or inappropriate location and design of wells may also contribute to salt encroachment. Generally, saline water is drawn up towards a well, this being referred to as 'upconing'. This is a dangerous condition and a well may be ruined by an increase in salt content even before the actual 'cone' reaches the bottom of the well due to 'leaching' of the interface by freshwater.

The encroachment of salt water may extend for several kilometres inland leading to the abandonment of wells. The first sign of saline intrusion normally is a progressively upward trend in the chloride concentration of groundwater obtained from the affected wells. Chloride levels may increase from a normal value of around 25 mg l^{-1} to something approaching 19 000 mg l^{-1}, which is the concentration in sea water. Once saline intrusion develops in a coastal aquifer it is not easy to control. The slow rates of groundwater flow, the density differences between fresh and salt waters, and the amount of remedial flushing required, usually mean that pollution, once established, may take years to remove under natural conditions. The encroachment of salt water, however, can be checked by maintaining a fresh water hydraulic gradient towards the sea. This gradient can be maintained naturally, or by artificial recharge or by an extraction

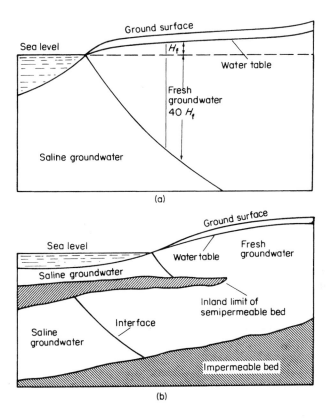

Ground surface

(a)

Ground surface

(b)

Fig. 4.8 Diagrams illustrating the Ghyben-Herzberg hydrostatic relationship in (a) a homogeneous coastal aquifer (b) a layered coastal aquifer.

barrier. Artificial recharge involves injecting water into the aquifer via wells so as to form a groundwater mound between the coast and the area where abstraction is taking place. An extraction barrier, consisting of a line of wells parallel to the coast, abstracts salt water before it reaches an inland wellfield. Neither of these two methods of controlling saline intrusion offer a foolproof solution. Consequently, when the possibility of saline intrusion exists, then the best policy is to locate wells as far from the coast as possible, select the design discharge with care and use an intermittent seasonal pumping regime. Ideally, at the first sign of progressively increasing salinity, pumping should be stopped.

4.10.4 Nitrate pollution

There are at least two ways in which nitrate pollution of groundwater is suspected of being a threat to health. Firstly, the build-up of stable nitrate compounds in the bloodstream reduces its oxygen-carrying capacity. Infants under one year old are most at risk and excessive amounts of nitrate can cause infantile methaemoglobinaemia, commonly called 'blue-baby'. Consequently, if the limit of 50 mg l^{-1} of NO_3 recommended by the World Health Organisation is exceeded, then bottled low-nitrate water should be provided for infants. Secondly, there is a possibility that a combination of nitrates and amines through the action of bacteria in the digestive tract can result in the formation of nitrosamines, which are potentially carcinogenic.

Nitrate pollution is basically the result of intensive use of synthetic nitrogenous fertilizer, although over-manuring with natural organic fertilizer can have the same result. Regardless of the form in which the nitrogen fertilizer is applied, within a few weeks it will have been transformed to NO_3^-. This ion is neither adsorbed nor precipitated in the soil and therefore is easily leached by heavy rainfall and infiltrating water. However, the nitrate does not have an immediate affect on groundwater quality, possibly because most of the leachate that percolates through the unsaturated zone has a typical velocity of about 1 m year^{-1}. Thus, there may be a considerable delay between the application of the fertilizer and the subsequent increase in the concentration of nitrate in the groundwater. The effect of the time lag, which is frequently 10 years or more, makes it very difficult to correlate fertilizer application with nitrate concentration in groundwater. Measures that can be taken to alleviate nitrate pollution include better land-use management, mixing water from various sources, or treatment of high nitrate water before it is put into supply.

4.10.5 Other causes of groundwater pollution

The list of potential groundwater pollutants is almost endless. For instance, volatile organic chemicals (VOCs) are the most frequently detected organic contaminants in water supply wells in the United States. Of the VOCs, by far the most commonly found are chlorinated hydrocarbon compounds. Conversely, petroleum hydrocarbons are rarely present in supply wells. This may be due to their *in situ* biodegradation. Many of the VOCs are non-aqueous phase liquids (NAPLs), which are sparingly soluble in water. Those that are lighter than water, such as the petroleum hydrocarbons are termed LNAPLs whereas those that are denser than water, that is, the chlorinated solvents are called DNAPLs. Of the VOCs, the DNAPLs are the least amenable to remediation. If DNAPLs penetrate the ground in a large enough quantity, depending on the hydrogeological conditions, they may percolate downwards into the saturated zone. This can occur in granular soils or discontinuous rock masses. Plumes of dissolved VOCs develop from the source of pollution. Because VOCs are sparingly soluble in water the time taken for complete dissolution (especially of DNAPLs) by groundwater flow in granular soils is estimated to be decades or even centuries.

The run-off from roads can contain chemicals from many sources, including those that have been dropped, spilled, or deliberately spread on the road. For instance, hydrocarbons from petroleum products, and urea and chlorides from de-icing agents are potential pollutants and have caused groundwater contamination. There also is the possibility of accidents involving vehicles carrying large quantities of chemicals.

Sewage sludge arises from the separation and concentration of most of the waste materials found in sewage. Since the sludge contains nitrogen and phosphorous it has a value as a fertilizer. While this does not necessarily lead to groundwater pollution, the presence in the sludge of contaminants such as metals, nitrates, persistent organic compounds and pathogens, does mean that the practice must be carefully controlled. The widespread use of chemical and organic pesticides and herbicides is another possible source of groundwater pollution.

Cemeteries and graveyards form another possible health hazard. Decomposing bodies produce fluids that can leak to the water table if a non-leakproof coffin is used. Typically the leachate produced from a single grave is of the order of 0.4 m^3 year^{-1} and this may constitute a threat for about 10 years. The water table in cemeteries should be at least 2.5 m deep and an unsaturated depth of 0.7 m should exist below the bottom of the grave.

The problem of pollution due to acid mine drainage is dealt with in Chapter 6.

4.11 Groundwater monitoring

A groundwater monitoring programme should determine the extent, nature and degree of pollution, as well the propagation mechanism and hydrogeological parameters so that the appropriate counter measures can be initiated. In almost all situations where groundwater monitoring is undertaken, it is important that adequate background samples are obtained before a groundwater abstraction scheme is inaugurated. Without this background data it is impossible to assess the effects of any new developments. It also should be borne in mind that groundwater can undergo cyclic changes in quality, so any apparent changes must be interpreted with caution. Hence, routine all-year-round monitoring is essential.

The design of a water quality monitoring well and its method of construction must be related to the geology of the site. After a well has been completed it should be developed, by pumping or bailing, until the water becomes clear. Background samples then should be collected over a lengthy period prior to the commencement of monitoring.

When designing a monitoring network to detect pollution, the problem is to ensure that there are sufficient wells to allow the extent, configuration and concentration of the pollution plume to be determined, without incurring the unnecessary expense of constructing more wells than are actually required. The network of monitoring wells must be designed to suit a particular location and modified, as necessary, as new information is obtained or in response to changing conditions. At least three monitoring wells should be used to observe the effects of a new development such as a landfill site. However, as a result of aquifer heterogeneity, non-uniform flow, variations in quality with depth and so on, well clusters (rather than individual wells) may be required. Each well cluster should contain two or more wells located at different depths within the aquifer, or within different aquifers in the case of a multiple aquifer system. One cluster should be located close to the source of the pollution, for early warning purposes, with another installation some suitable distance down gradient to assess the propagation of the plume. A third installation should be located up gradient of the monitored site to detect changes in background quality attributable to other causes.

Under favourable conditions the resistivity method can be used to determine the boundaries of a plume of polluted groundwater. Vertical electrical soundings are made in the area of known pollution in an attempt to define the top and bottom of the plume. Drill hole logs are used to establish geological control. Next, resistivity profiling is carried out to determine the lateral extent of the polluted groundwater. In this way a quantitative assessment can be made of the extent of groundwater pollution. The method is based on the fact that formation resistivity depends on the conductivity of the pore fluid, as well as the properties of the porous medium. Generally, the resistivity of a rock mass is proportional to the water with which it is saturated. The groundwater resistivity decreases if its salinity increases and hence the resistivity of the rock mass concerned decreases. Obviously, there must be a contrast in resistivity between the contaminant and groundwater in order to obtain useful results. Contrasts in resistivity may be attributed to mineralized groundwater with a higher than normal specific conductance due to pollution. In simple situations ground conductivity profiling can be used to detect plumes of polluted water. For example, a ground conductivity survey can be carried out if saline groundwater is near the surface. If the depth to groundwater is too great, then the thickness of overlying unsaturated sediments can mask any contrasts between polluted and natural groundwater. In addition, the geology of the area has to be relatively uniform so that the conductivity values and profiles can be compared with others. The ionic content of groundwater in a drill hole can be monitored by measuring the resistivity of the fluid, with a probe that has a short electrode spacing. The quality of groundwater also can be estimated from the spontaneous potential deflection on an electric log. The method tends to give better results as the salinity of the pore water increases.

5

Soil and the Environment

5.1 Origin of soil

Soil is one of the world's most important resources. It normally consists of an unconsolidated assemblage of particles between which there are voids that may contain air or water or both. Although soils are of most value in terms of agriculture, they also play an important role as construction materials. Soil is derived from the breakdown of rock material by weathering and/or erosion. It may accumulate in place or have undergone some amount of transportation prior to deposition. Soil also may contain organic matter or, in the case of peat, consist mainly of organic material. The type of breakdown process(es) and the amount of transport undergone by sediments influence the nature of the macro- and microstructure of the soil. A given type of rock can give rise to different types of soils, depending on the climatic regime and vegetative cover under which they have been developed. Time also is an important factor in the development of a mature soil, especially from the agricultural point of view.

Probably the most important methods of soil formation are mechanical and chemical weathering. The agents of weathering, however, are not capable of transporting material. Transport is brought about by gravity, water, wind or moving ice. If sedimentary particles are transported, then this affects their character, particularly their grain size distribution, sorting and shape (Table 5.1).

Plants affect the soil in several ways. For instance, when their roots die and decay, they leave behind a network of passages that allow air and water to move through the soil more readily. Roots, especially those of grasses, help bind soil together and so reduce erosion. Perhaps the major contribution to soil made by plants is the addition of organic matter. The total organic content of soils can vary from less

Table 5.1 Effects of transportation on sediments.

	Gravity	Ice	Water	Air
Size	Various	Varies from clay to boulders	Various sizes from boulder gravel to muds	Sand size and less
Sorting	Unsorted	Generally unsorted	Sorting takes place both laterally and vertically Marine deposits often uniformly sorted. River deposits may be well sorted	Uniformly sorted
Shape	Angular	Angular	From angular to well rounded	Well rounded
Surface texture	Striated surfaces	Striated surfaces	Gravel: rugose surfaces Sand: smooth, polished surfaces Silt: little effect	Impact produces frosted surfaces

Table 5.2 Capillary rise, capillary pressures and suction pressure in soil.

a) Capillary rise and corresponding pressure.

Soil	Capillary rise (mm)	Capillary pressure (kPa)
Fine gravel	up to 100	up to 1.0
Coarse sand	100–150	1.0–1.5
Medium sand	150–300	1.5–3.0
Fine sand	300–1000	3.0–10.0
Silt	1000–10 000	10.0–100.0
Clay	over 10 000	over 100.0

b) Suction pressure and pF values

pF value	Equivalent suction (mm water)	(kPa)
0	10	0.1
1	100	1.0
2	1000	10.0
3	10 000	100.0
4	100 000	1000.0
5	1 000 000	10 000.0

than 1% in the case of some immature or desert soils to over 90% in the case of peat. Microorganisms occur in greatest numbers in the surface horizons of the soil where there are the largest concentrations of food supply. They are particularly important in terms of the decomposition of organic matter, which leads to the formation of humus.

Changes occur in soils after they have accumulated. In particular, seasonal changes take place in the moisture content of sediments above the water table. Volume changes associated with alternate wetting and drying occur in fine-grained soils of high plasticity. Exposure of a soil to dry conditions means that its surface dries out and that water is drawn from deeper zones by capillary action. At each point where moisture menisci are in contact with soil particles the forces of surface tension are responsible for the development of capillary or suction pressure (Table 5.2). The air and water interfaces move into the smaller pores. In so doing the radii of curvature of the interfaces decrease and soil suction increases. Hence, the drier the soil, the higher is the soil suction. The capillary movement or rise therefore refers to the movement of moisture through minute pores between soil particles, which act as capillaries. Hence, groundwater can rise from the water table so that the capillary moisture is in hydraulic continuity with the water table and is raised against the force of gravity, with the degree of saturation decreasing from the water table upwards. The boundary separating capillary moisture from gravitational groundwater in the zone of saturation below the water table is ill-defined and cannot be accurately determined. The zone above the water table may be saturated with capillary moisture and is referred to as the capillary fringe. This supplementary pressure or capillary pressure has the same mechanical effect as a heavy surcharge. Therefore, surface evaporation from very compressible soils produces a conspicuous decrease in the pore space of the layer undergoing desiccation. If the moisture content in this layer reaches the shrinkage limit, then air begins to invade the pores and the soil structure begins to break down. The decrease in pore space consequent upon desiccation of fine-grained soil leads to an increase in its bearing strength. Thus, if a dry crust is located at or near the surface above softer material it acts as a raft. The thickness of dry crusts often varies erratically.

Chemical changes that take place in the soil, for example, due to the action of weathering, may bring about an increase in its clay mineral content, the latter developing from the breakdown of less stable minerals. In such instances, the plasticity of the soil increases while its permeability decreases. Leaching, whereby soluble constituents are removed from the upper horizons, to be precipitated in the lower horizons, occurs where rainfall exceeds evaporation. The porosity may be increased in the zone undergoing leaching.

Soil particles may form aggregates that are termed peds. Angular, blocky peds have sharp angular corners with flat, convex and/or concave faces. Subangular, blocky peds have convex and/or concave faces and rounded corners. Large peds that are elongated vertically are described as columnar. These peds commonly have domed upper surfaces and grade into the underlying material. Irregularly shaped peds with rough surfaces characterize soils with a crumb structure. Such soils normally have a high porosity and are well drained. Soils with granular structures contain subspherical peds. They also have an open texture. Horizontally elongate peds, with flat, parallel surfaces are described as laminar. Lenticular peds are lens shaped with convex surfaces and tend to overlap one another. Vertically elongate peds possess three or more vertically flat faces and, where the peds have somewhat indeterminate upper and lower boundaries, they are referred to as prismatic peds.

5.2 Soil fertility

Soil fertility refers to the capacity of a soil to supply nutrients needed for plant growth. More or less all soils have a certain inherent fertility. For example, soils that are acidic, alkaline, waterlogged or deficient in particular elements can support specific plant communities and, as such, can be regarded as fertile in relation to those communities. However, when people attempt to cultivate such soils, the inherent fertility may be unsuitable for particular crop(s). It then becomes necessary to alter the fertility by, for example, the addition of fertilizer or, installation of drainage measures, to suit the needs of these crops.

Be that as it may, there are a number of factors that influence plant growth. These include soil aeration and moisture content, soil temperature, pH value and essential elements. Good aeration usually is facilitated by a granular or crumb structure and free drainage. An adequate and balanced supply of moisture is essential for plant growth. The particle size distribution, texture, organic content and structure of a soil affect its retention of moisture. Generally, clays and organic soils have the highest moisture-retaining capacity, while organic soils have the highest available moisture. The moisture contained in the soil is lost mainly by evapotranspiration so that the rate of loss depends upon temperature and plant cover, as they increase so moisture losses increase. On the other hand, wet or waterlogged soils need drainage for satisfactory cultivation of most crops. Drainage is beneficial to agriculture as it improves the soil structure, facilitates aeration, increases the rate of decomposition of organic matter, leads to an increase in soil temperature and increases the bacterial population in the soil. In addition, the thickness of soil available for root growth and penetration is very important.

Two important moisture characteristics of soil are the field capacity and the permanent wilting point. When soil is saturated and the excess water drains away, the soil is described as being at its field capacity. Plants extract moisture from the soil until, as the soil dries out, it becomes impossible for them to continue doing so; they then wilt and die if the soil is not rewetted. The point at which permanent wilting starts is known as the permanent wilting point. The field capacity and permanent wilting point have been defined in terms of soil suction (the pF values being 2.0 and 4.2 respectively). Soil suction is a negative pressure and indicates the height to which a column of water could rise due to such suction (Table 5.2). Since this height or pressure may be very large, a logarithmic scale has been adopted to express the relationship between soil suction and moisture content, the latter is referred to as the pF value. The permanent wilting point varies from soil to soil and plant to plant. Water held between the field capacity and the permanent wilting point is the water available to plants, and similarly the amount varies considerably from soil to soil.

The pH value for soils ranges between 3 and 9, although the range for cultivated soils is more limited, tending to vary from 5.5 to 7.5. Very low values are sometimes found in swamps where the breakdown of organic material and sulphides gives rise to humic acid and sulphuric acid respectively. At the other extreme, very high pH values may be attributable to the presence of sodium carbonate. This may be the case in some salinas or sabkhas (see Chapter 9). Generally, pH values about neutral are associated with large amounts of exchangeable calcium and some magnesium. High acidity may lead to increasing amounts of manganese and aluminium in the pore water, with little formation of ammonia or nitrate, low availability of phosphorus and molybdenum deficiency.

A number of elements are essential for healthy plant growth. However, they must be present in the soil in the correct proportions. If there is an excess or deficiency in any one element then plant growth can be affected seriously, with plants developing symptoms of toxicity or nutrient starvation. Of the essential elements, calcium, carbon, hydrogen, magnesium, nitrogen, oxygen, phosphorous, potassium, and sulphur, should be present in soil in relatively large amounts whereas boron, chlorine, cobalt, copper, iron, manganese, molybdenum, and zinc are necessary in small amounts. Most of these elements (with the exception of those obtained from the air) are derived initially from the breakdown of minerals within parent rock, and are taken up by the plant roots. They subsequently are returned to the soil in the plant litter when it decomposes, thereby releasing the elements into the soil for the cycle to begin again.

5.3 Soil horizons

With continuing exposure, soil develops a characteristic profile from the surface downwards. A soil profile is divided into zones or horizons, which differ in character from the parent material to the soil surface. The master horizons are denoted by a capital letter as follows:

H: Formed by the accumulation of organic material deposited on the soil surface. Generally, contains 20% or more organic matter. May be saturated with water for prolonged periods.

O: Formed by the accumulations of organic material (plant litter) deposited on the surface. Contains 35% or more of organic matter. Is not saturated with water for more than a few days of a year.

A: A mineral horizon occurring at or adjacent to the surface that has a morphology acquired by soil formation. Generally, possesses an accumulation of humified organic matter intimately associated with the mineral fraction.

E: A pale-coloured mineral horizon with a concentration of sand and silt fractions high in resistant minerals from which clay particles, iron or aluminium or some combinations thereof have been leached. E horizons, if present, are eluvial horizons, which generally underlie H, O or A horizons and overlie B horizons.

B: A mineral horizon in which parent material is absent or faintly evident. It is characterized by one or more of the following features:
(a) an illuvial concentration of clay minerals, iron, aluminium, or humus, alone or in combinations;
(b) a residual concentration of sesquioxides relative to the source materials;
(c) an alteration of material from its original condition to the extent that clay minerals are formed, oxides are liberated, or both – or a granular, blocky, or prismatic structure is formed.

C: A mineral horizon of unconsolidated material similar to the material from which the solum is presumed to have formed and that does not show properties diagnostic of any other master horizons. Accumulations of carbonates, gypsum or other more soluble salts may occur in C horizon.

R: Rock that is sufficiently coherent when moist to make hand digging impracticable.

Not all the horizons mentioned above necessarily are present in a given soil profile. The H, O and A horizons tend to be dark in colour because of the presence of plentiful organic matter. The E horizon is light in colour due to leaching of iron oxides from it. Sometimes notable changes in colour may be displayed by the B horizon, varying from yellowish-brown to light reddish-brown to dark red. This depends primarily on the presence of iron oxides and clay minerals. The presence of carbonates may lighten the colours. Well drained soils are well aerated so that iron is oxidized to give a red colour. On the other hand, wet poorly drained soils generally mean that iron is reduced, which may give rise to a yellow colour.

As it takes time to develop a mature soil, different grades of soil profiles may be recognized from immature to mature. A poorly developed immature soil profile is one in which the A horizon may overlie the C horizon directly, and the C horizon may show signs of oxidization. In a moderately developed soil profile the A horizon may rest upon an argillic B horizon (i.e. one that contains illuvial clay particles and in which the peds may be coated with clay) that, in turn, is underlain by the C horizon. At times a carbonate B horizon also may be present. A mature well developed soil profile possesses a good soil structure in which the B horizon may contain a notable amount of clay material.

5.4 Pedological soil types

There are several pedological classifications of soil. That given below was developed by the Food and Agriculture Organization (FAO) of the United Nations in 1980.

The gleysols of tundra regions are characterized by a thin accumulation of organic matter at the surface, beneath which there is a dark greyish brown mixture of mineral and organic material lying on top of a wet mottled horizon about 500 mm thick. Then follows a sharp change to permafrost (i.e. permanently frozen ground). The upper horizons freeze in winter and thaw in summer.

South of the tundra there are extensive areas of histosols with marshy vegetation dominated by peaty soils. Peats tend to be acidic and peaty soils need to be drained and limed if they are to be cultivated.

The podsols tend to coincide with the coniferous forest regions. They are characterized by an accumulation of plant litter at the surface, below which there is a layer of dark brown, partially decomposed plant material (Fig. 5.1a). This grades into very dark brown amorphous organic matter. A dark grey

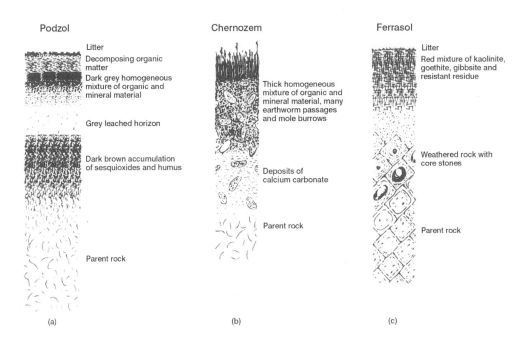

Fig. 5.1   Soil profiles (a) a podsol (b) a chernozem (c) a ferrasol.

layer of mixed organic and mineral material occurs beneath the organic matter. This is followed at increasing depth by a pale grey horizon, which has been leached of iron compounds and organic matter by downward percolating water. The leached material is precipitated beneath this horizon to form sesquioxides (e.g. Fe_2O_3, Al_2O_3). A layer of hard pan may be present at this level. Finally, the unaltered parent material occurs below this horizon. Podsols may be up to 2 m in thickness. These soils are usually acid, with pH values of less than 4.5 at the surface, increasing to about 5.5 in the lowest horizons. They also are very deficient in plant nutrients except in the surface organic matter, which releases them upon decomposition.

The cambisols, commonly known as brown earths, occur beneath some deciduous woodlands in the cool temperate regions. There is a continuous gradation from podsols to cambisols. At the surface of a cambisol there is a loose layer of plant litter resting on a brown or greyish brown horizon of mixed organic and mineral material. This overlies a brown, friable, loamy B horizon, which grades into unaltered material. This parent material at times is basic or calcareous and gives the soil a high base saturation. These soils generally are slightly acid to mildly alkaline. Organic matter is broken down rapidly and incorporated into the soil, giving it a high natural fertility.

Luvisols also are found in cool temperate regions beneath deciduous forests or mixed deciduous and coniferous forests. A thin layer of loose plant litter rests on a greyish brown mixture of organic and mineral material. This changes into a grey sandy horizon, beneath which there is a sharp change to a brown blocky or prismatic horizon that has a high content of clay particles. Then there is a gradation to the unaltered material. These soils have a middle horizon that contains more clay particles than those above or below (argillic B horizon), due to clay particles being washed from the upper horizons. Luvisols are weakly acid at the surface with medium base saturation, which means that they have a moderate natural fertility.

Precipitation declines moving towards the equator, and at less than 400 mm the deciduous forests are replaced by grasslands. Here are found the black earths or chernozems. These soils have a root mat at the surface that overlies a thick black horizon with a high humus content that may be up to 2 m in thickness (Fig. 5.1b). Thin thread-like deposits of calcium carbonate occur in the lower part of this horizon, as well as below. Concretions of calcium carbonate also may be present. The black horizon grades into yellowish brown material. Chernozems have neutral pH values, very high base saturation, and a very high inherent fertility. These soils have a well developed crumb structure and high water holding capacity due to their high organic content that develops from annual increments of dead grass.

With further decrease in precipitation, the vegetation changes from tall grass, to short grass, to bunchy grass, and then to species that can withstand long dry periods. Accordingly, there are similar changes in the soils with the thick black horizon characteristic of chernozems becoming lighter in colour and shallower, and the calcium carbonate horizon approaching the surface. Generally, three soil types can be recognized, namely, the kastanozems, the xerosols and the yermosols. The kastanozems are covered with short grass. They have a dark brown horizon, which may be up to about 300 mm thick, below which the calcium carbonate horizon is present, followed by the unaltered material. Xerosols are developed beneath bunchy grass and possess a thinner brown upper horizon than the lower calcium carbonate horizon. Yermosols are covered with sparse xerophytic desert type vegetation. They have a thin brownish grey upper horizon that rests on a carbonate or sometimes a gypsiferous layer. These three soils frequently have a high inherent fertility but, unless irrigated, crop production usually is restricted by the lack of moisture.

Soils with high salinity or alkalinity are found in some semi-arid or arid regions where evapotranspiration greatly exceeds precipitation. Ions such as calcium, chloride, magnesium, potassium, sodium and sulphate accumulate in the soils. Solonchaks are saline soils, which often have salt efflorescences on the ped surfaces and on the ground surface. The soil profile often has an upper grey organic mineral mixture that rests on a mottled horizon, beneath which is a grey or olive completely reduced horizon. Many solonchaks are potentially useful for agriculture if the excess ions can be dissolved and removed by large amounts of water provided by irrigation. One of the problems in such regions is salinization. This occurs when irrigated soils are not well drained.

Solonetz soils also occur in arid and some semi-arid regions. A thin layer of litter may occur at the surface, to be quickly followed by a thin very dark mixture of organic and mineral matter, and then a dark grey, frequently sandy horizon. Below this is a distinctive middle horizon with a marked increase in clay material. The latter is characterized by a prismatic or columnar structure. These soils have pH values that are often above 8.5 due to high levels of exchangeable sodium and magnesium but they generally have a low salt content. The high pH is harmful to plant growth and must be reduced if arable farming is to be practised. This usually is brought about by adding calcium sulphate.

Vertisols including black cotton soils occur in the savannah grassland regions of the tropics and subtropics, which are characterized by alternating wet and dry seasons. They are dark coloured, clayey soils in which montmorillonite is an important constituent. The surface horizon usually is granular but can be massive and rests upon a dense horizon with a prismatic or angular blocky structure that grades into similar material with a marked wedge structure. This is caused by the montmorillonite shrinking and expanding in response to seasonal drying and wetting. The pH value of vertisols is about neutral. They have a high base saturation and high cation exchange capacity. Vertisols are fertile soils with a high potential for agriculture. Irrigation frequently proves necessary in such regions as they tend to be affected by droughts.

The ferrasols (laterites) occur in the humid tropics, being highly weathered soils that are often red in colour. These soils possess a thin layer of plant litter at the surface, followed by a greyish red mixture of organic and mineral matter that is not more than 100 mm in thickness. This grades quickly into a bright red clayey horizon that may be several metres thick, which overlies the underlying rock (Fig. 5.1c). The red horizon consists largely of kaolinite with hydrated oxides of iron and aluminium. The change from lateritic soils to the parent rock can be gradual or sharp. There often is a red and cream mottled horizon beneath the red topsoil or a thick white horizon, known as the pallid zone, may be present. Because of the effectiveness of chemical weathering these soils are deficient in essential plant elements and so can be exhausted rapidly unless fertilizers are applied.

5.5 Soil surveys

Soil surveys are undertaken to assess the properties of soils and are of value in relation to rural and urban planning (especially with respect to agriculture and forestry); for feasibility and design studies in land development projects and for many engineering works. One of the principal objects of a soil survey is to produce a soil map and a report, which involves the delineation of soil types in the field in the case of a general purpose soil survey. The mapping unit for soil mapping is based primarily on soil profiles, although features of landforms may be included in the descriptions of mapping units. The report not only describes the soils and their properties but also should provide background information on environmental factors, on land potential and likely responses to various alternative forms of land management, and sometimes land use and management recommendations. However, soil surveys also can be carried out for a special purpose such as the development of a scheme for irrigation where the drainage properties and degree of salinity may be of particular importance; to demarcate areas prone to flooding; or for land reclamation.

Soil surveys can prove difficult in that the boundaries between different soils usually are not readily defined but grade into one another. The changes that do occur are not necessarily reflected in the surface expression. This means that soils have to be examined via holes sunk by augers or from small pits. Soils generally are mapped in a general-purpose survey according to the properties that are observed in the field whereas in a special-purpose survey the relevant properties to be mapped are chosen before the survey commences. General-purpose surveys involve the production of a pedological map that may be used in land evaluation. Such surveys prove useful in underdeveloped regions where little is known about either potential land use or appreciation of the physical environment. Hence, basic information on soils must be obtained in such areas prior to decisions on land use being made. In developed countries farmers tend only to be interested in soil surveys when significant changes in land use are contemplated. Nonetheless, soil surveys allow the results of experience of soil use to be communicated to those concerned within agriculture in order to bring about the optimum cropping system and management of soils. Soil surveys may investigate soil erodibility, depth of soil, steepness of slope, shrink–swell potential, frequency of rock outcrops, possibility of salinization, fertilizer–plant response and crop variety trials, especially of new crops. The most suitable soil conservation practices may be introduced with the aid of soil surveys. In cases where major changes in the type of land use are contemplated, an assessment must be made of the response of the soil to any proposed changes. Land suitability evaluation, giving the suitability of areas for each of several uses, may be undertaken in such cases. Accordingly, soil surveys often now form part of an environmental impact assessment.

Soil surveys may be used to assist in avoiding hazards such as flooding, salinization due to irrigation, or crop failure due to drought. The data from such surveys is used to adopt appropriate measures to counteract the hazard concerned. Alternatively, the data may form the basis of a decision not to use the land for the proposed purpose.

Data from soil surveys can be of value in the initial planning stages of construction projects since they may indicate where certain problem soils occur such as expansive clays, collapsible soils and peaty soils. They also may be of value in relation to the location of construction materials such as gravels and sands, and suitable supplies of soils for embankments. Such data may be particularly useful in terms of trench construction, low-cost housing, light industrial units, road construction and recreational areas.

5.6 Land capability studies

An important part of land-use planning is involved with matching an area of land with a land use to which it is most appropriate, taking into consideration the physical characteristics of the land, which includes any geological constraints. Land capability analysis involves evaluating these physical characteristics in relation to different types of land use. For example, areas threatened by landsliding can be allocated to recreation, forestry or grazing land. Hence, land capability analysis seeks to make geological data more understandable and so more useful to planners and politicians and, hopefully, makes planning more responsive to this information. This information can be used to relate geological factors to other environmental, social, political and economic factors, which can be expressed in monetary form, and so can contribute towards a land development decision. The following five steps usually are involved in a land capability study:

1. Identify the types of land use for which the land capability is to be determined.
2. Determine the natural factors that have a significant effect on the capability of the land to accommodate each use.
3. Develop a scale of values for rating each natural factor in relation to its effect on land capability.
4. Assign a weight to each natural factor indicting its importance relative to the other factors as a determinant of land capability.
5. Establish land units, rate each land unit for each factor, calculate the weighted ratings for each factor and aggregate the weighted ratings for each land unit.

Land capability classification was originally developed by the United States Soil Conservation Service for farm planning. In other words, it provides a means of assessing the extent to which limitations such as erosion risk, soil depth, wetness and climate hinder the agricultural use of land (Table 5.3). Three categories are recognized, capability classes, subclasses, and units. A capability class is a group of capability subclasses that have the same relative degree of limitation or hazard. Classes I to IV can be used for cultivation whereas classes V to VIII cannot because of permanent limitations such as wetness or steeply sloping land. The risk of erosion increases through the first four classes progressively, reducing the choice of crops that can be grown and requiring more expensive conservation practices. A capability subclass is a group of capability units that possess the same major conservation problem or limitation (e.g. erosion hazard; excess water; limited depth of soil or excessive stoniness). Lastly, a capability unit is a group of soil mapping units that have the same limitations and management responses. Soils within a capability unit can be used for the same crops and require similar conservation treatment.

Basic ENVIRONMENTAL AND ENGINEERING GEOLOGY

Table 5.3 Land use alternatives of capability classes.

Capability class	Limitations	Choice of crops	Conservation practices	Cultivation	Pasture (improved)	Range (unimproved pasture)	Woodland	Wildlife food and cover	Recreation water supply aesthetic
I	few	any	none	✓	✓	✓	✓	✓	✓
II	some	reduced or	moderate	✓	✓	✓	✓	✓	✓
III	severe	reduced and/or special		✓	✓	✓	✓	✓	✓
IV	very severe	restricted and/or very careful management		✓	✓	✓	✓	✓	✓
V	other than erosion				✓	✓	✓	✓	✓
VI	severe				✓	✓	✓	✓	✓
VII	very severe				✓	✓	✓	✓	✓
VIII	very severe							✓	✓

Classes I–IV denote soils suitable for cultivation
Classes V–VIII denote soils unsuitable for cultivation.

Characteristics and
recommended land use

Deep, productive soils easily worked, on nearly level land;
not subject to overland flow; no or slight risk of damage
when cultivated; use of fertilizers and lime, cover crops,
crop rotations required to maintain soil fertility and
soil structure.

Productive soils on gentle slopes; moderate depth; subject to
occasional overland flow; may require drainage; moderate risk
of damage when cultivated; use of crop rotations, water-control
systems or special tillage practices to control erosion.

Soils of moderate fertility on moderately steep slopes, subject
to more severe erosion; subject to severe risk of damage but can
be used for crops provided plant cover is maintained; hay or other
sod crops should be grown instead of row crops.

Good soils on steep slopes, subject to severe erosion; very severe
risk of damage but may be cultivated if handled with great care;
keep in hay or pasture but a grain crop may be grown
once in 5 or 6 years.

Land is too wet or stony for cultivation but of nearly level slope;
subject to only slight erosion if properly managed; should be
used for pasture or forestry but grazing should be regulated to
prevent plant cover from being destroyed.

Shallow soils on steep slopes; use for grazing and forestry; grazing
should be regulated to preserve plant cover, if plant cover is
destroyed, use should be restricted until cover is re-established.

Steep, rough, eroded land with shallow soils; also includes droughty
or swampy land; severe risk of damage even when used for pasture
or forestry; strict grazing or forest management must be applied.

Very rough land; not suitable even for woodland or grazing; reserve
for wildlife, recreation or watershed conservation.

5.7   Soil erosion

Loss of soil due to erosion by water or wind is a natural process. Soil erosion removes the topsoil, which usually contains a high proportion of the organic matter and finer mineral fractions in soil that provide nutrient supplies for plant growth. Unfortunately, soil erosion can be accelerated by human activity. It is difficult, however, to separate natural from human-induced changes in erosion rates. Ideally, as soil is being removed it should be replaced by newly formed soil. If this does not occur, then soil cover is removed gradually. This is accompanied by an increase in run-off by overland flow that, in turn, reduces the amount of water that infiltrates into the ground. Accordingly, chemical breakdown of rock material becomes less effective, so adversely affecting soil formation.

5.7.1   Soil erosion by water

Basically, soil erosion by water involves the separation of soil particles from the soil mass and their subsequent transportation. There are two agents responsible for this, namely, raindrops and flowing water. Erosion by raindrop action involves the detachment of particles from the soil by impact and their movement by splashing. Run-off removes soil by the action of water flowing either as sheets or in rills or gullies. Hence, the nature of soil erosion depends on the relationship between the erosivity of rain drops and run-off, and the erodibility of the soil. Splash and sheet erosion combined are referred to as interrill erosion. Such conditions are met most frequently in semi-arid and arid regions. Nevertheless, soil erosion occurs in many different climatic regions where vegetation has been removed, it being at a maximum when intense rainfall and the vegetation cover are out of phase, that is, when the surface is bare. Soil erosion by water is most active where the majority of the rainfall finds it difficult to infiltrate into the ground so that most flows as run-off over the surface and in so doing removes soil. The intensity of rainfall is at least as important in terms of soil erosion as the total amount of rainfall and as rainfall becomes more seasonal, the total amount of erosion tends to increase.

Sheet flow may cover up to 50% of the surface of a slope during a heavy rainfall but erosion does not take place uniformly across a slope. The velocity of sheet flow ranges between 15 and 300 mm s^{-1}. Velocities of 160 mm s^{-1} are required to erode soil particles of 0.3 mm diameter but velocities as low as 20 mm s^{-1} will keep these particles in suspension. Rills and gullies begin to form when the velocity of flow increases to rates in excess of 300 mm s^{-1} and flow is turbulent. Whether they form depends on soil factors, as well as velocity and depth of water flow. Rills and gullies remove much larger volumes of soil per unit area than sheet flow.

5.7.2   Soil erosion by wind

Surface winds that are capable of initiating soil particle movement are said to be turbulent. As far as wind erosion is concerned, the most important characteristics of soil particles are their size and density. In other words, there is a maximum particle size for a wind of a given velocity. The smaller grains of soil (e.g. silt-sized particles) and organic matter are readily subject to wind erosion. This action leads to a reduction in the fertility of the soil, as well as lowering its water-retention capacity. In terms of the wind itself, the force of the wind at the ground surface is the main factor affecting erosivity. Once particles have started moving, they exert a drag effect on the air flow, which then alters the velocity profile of the wind. The velocity of wind blowing over a surface increases with height above the surface, the roughness of the surface retarding the speed of the wind immediately above. Where the surface is rough, for instance, due to the presence of boulders, plants or other obstacles, this reduces the wind speed immediately above the ground. As more material is picked up, the surface wind speed is reduced. Hence, the more erodible a soil, the greater the proportion of grains involved in saltation and the greater the reduction in wind speed. Soil particles also are moved in suspension and by traction.

Although wind can only remove particles of a certain size, wind erosion is effective. For example, it is estimated that during the period when parts of Kansas formed the 'dust-bowl', wind erosion accounted for soil losses of up to 10 mm year^{-1} at certain locations. By comparison, in some semi-arid regions, water may remove around 1 mm year^{-1}. Wind erosion is most effective in arid and semi-arid regions where the ground surface is relatively dry and vegetation is absent or sparse. The problem is most acute in those regions where land-use practices are inappropriate and rainfall is unreliable. This means that the ground surface may be left exposed. Nonetheless, soil erosion by wind is not restricted to dry lands (Fig. 5.2). It also occurs in humid areas, though on a smaller scale.

In most soils the nature and stability of the textural aggregations govern erodibility. Hence, in a well-textured soil the number of soil particles that are small enough to be moved may be low. The stability of the textural aggregations depends upon the soil moisture content, organic content, cation exchange capacity, soil suction and cement bonds on the one hand and those processes responsible for breakdown on the other.

5.7.3 Importance of vegetation

Vegetation cover is of paramount importance as far as soil erosion is concerned. Generally, an increase in erosion occurs with increasing rainfall, and erosion decreases with increasing vegetation cover. However, the growth of vegetation depends on rainfall, so producing a rather complex variation of erosion with rainfall. Although vegetation cover depends primarily upon rainfall, agriculture – especially where irrigation occurs – can mean that vegetation becomes more or less independent of rainfall. Hence, farming practices have had, and do have, an influence on soil erosion since they alter the nature of the vegetation cover, the total rainfall and its intensity during times of low cover being most important. Consequently, land is most vulnerable to erosion during periods when ploughing and harvesting take place. In addition, overgrazing reduces the amount of vegetation cover and so can lead to an increase in the rate of soil erosion. As far as the rate of soil erosion is concerned, semi-arid regions appear to be the most sensitive to changes in the amount of rainfall and therefore vegetation cover.

Vegetation can increase the infiltration rate of the soil by maintaining a continuous pore system due

Fig. 5.2 Wind erosion of soil, north east of George, Western Cape Province, South Africa.

to root growth, by the presence of organic matter and by enhancing biological activity. This also leads to an increase in the moisture-storage capacity of the soil and a decrease in run-off. Vegetation also increases the time taken for run-off to occur and so heavier rainfall is required to produce a critical amount of run-off. Not only does vegetation affect the volume of rain reaching the ground, it also affects its local intensity and drop size distribution. Hence, the energy of rainfall that is available for splash erosion under a vegetation cover is governed by the proportions of rain falling as direct throughfall and as leaf drainage, as well as the height of the canopy (which determines the energy of leaf drainage).

The exhaustion of the organic matter and soil nutrients in soil is allied closely to soil erosion. Organic matter in the soil fulfils a similar role to clay minerals in holding water, as well as providing organic and inorganic nutrients. The organic matter in the soil is important in terms of aggregation of soil particles, and a good vegetation cover tends to increase biological activity and the rate of aggregate formation. Furthermore, increasing aggregate stability leads to increasing permeability and infiltration, as well as a moister soil. High permeability and aggregate strength minimize the risk posed by overland flow. Loss of organic matter depends largely on the vegetation cover and its management. Partial removal of vegetation or wholesale clearance prevents the addition of plant debris as a source of new organic material for the soil. Over a period of years this results in a loss of plant nutrients and in a dry climate this leads to a significant reduction in soil moisture. The process can turn a semi-arid area into a desert in less than a decade. This organic depletion leads to lower infiltration capacity and increased overland flow with consequent erosion on slopes of more than a few degrees.

The strength of the soil is increased by the presence of plant roots. Grasses and small shrubs can have a notable strengthening effect to depths of up to 1.5 m below the surface whereas trees can enhance soil strength to depths of 3 m or more. Roots have their greatest effects in terms of soil reinforcement near to the soil surface where the root density is highest. The reinforcing effect is limited with shallow rooting vegetation, where roots fail by pullout before their peak tensile strength is attained. In the case of trees, because their roots penetrate soil to greater depths, roots may be ruptured when the forces placed upon them exceed their tensile strength.

5.8 Erosion control and conservation practices

The aim of soil erosion control is to reduce soil loss in order that soil productivity can be maintained economically. In this context, the soil loss tolerance is defined as the maximum rate of soil erosion that allows a high level of productivity to be sustained. Tolerable rates of soil loss obviously vary from region to region depending upon type of climate, vegetation cover, soil type and depth, slope, and farming practices. A range of values has been proposed for tolerable mean yearly soil loss, from 0.2 kg m^{-2} year^{-1} for soils where the rooting depth is up to 250 mm, to 1.1 kg m^{-2} year^{-1} when the rooting depth exceeds 1.5 m. Nevertheless, where erosion rates are very high, higher threshold values may have to be adopted. Erosion control implies maintaining equilibrium in the soil system so that the rate of loss does not exceed the rate at which new soil is formed. However, this state of equilibrium is difficult to predict and therefore to attain in practice.

Conservation measures must protect the soil from raindrop impact; must increase the infiltration capacity of the soil to reduce the volume of run-off; must improve the aggregate stability of the soil to increase its resistance to erosion; and lastly, must increase the roughness of the ground surface to reduce run-off and wind velocities. In fact, soil conservation practices can be grouped into three categories, namely, agronomic measures, mechanical measures, and soil management practices. Sound soil management is vital to the success of any soil conservation scheme since it helps to maintain the texture and fertility of the soil.

5.8.1 Control of water erosion

As far as water erosion is concerned, conservation is directed mainly at the control of overland flow, since rill and gully erosion are reduced effectively if overland flow is prevented. There is a critical slope length at which erosion by overland flow is initiated. Hence, if slope length is reduced, for example, by terracing, then overland flow should be controlled. Terraces break the original slope into shorter units and may be regarded as small detention reservoirs that contain water long enough to help it infiltrate into the ground. The terraces follow the contours and are raised on the outer side so as to form a channel. Drainage ditches must be incorporated into a terrace system to remove excess water.

Contour farming involves following the contours to plant rows of crops. As with terraces, this interrupts overland flow and thereby reduces its velocity, and again can help conserve soil moisture. Contour farming probably is most successful in deep, well-drained soils, where the slopes are not too steep. Alternating strips of crops and grass, planted around the contours of slopes, form the basis of strip cropping. Not only do strips reduce the velocity of overland flow, they protect the soil from raindrop impact and aid infiltration. Strip cropping may be used on slopes that are too steep for terracing. The width of the strip is reduced with increasing slope, for example, on a slope of 3° the width could be 30 m, reducing to half that on a 10° slope. The widths also vary with the erodibility of the soil. The crop in each strip normally is rotated so that fertility is preserved. Grass strips may remain in place where the risk of erosion is high.

Soil or crop management practices include crop rotation where a different crop is grown on the same area of land in successive years in a four or five year cycle. This avoids exhausting the soil and can improve its texture. The use of mulches, that is, covering the ground with plant residues, affords protection to the soil against raindrop impact and reduces the effectiveness of overland flow. When mulches are ploughed into the soil they increase the organic content and thereby improve the texture and fertility of the soil. Reafforestation of slopes, where possible, also helps to control the flow of water on slopes and reduce the impact of rainfall. The tree roots also help improve the infiltration capacity of the soil.

Gullies can be dealt with in a number of ways. Most frequently some attempt is made at revegetating them by grassing or planting trees (Fig. 5.3). Small gullies can be filled or partially filled, or traps or small dams can be constructed to collect and control flow of sediment. A gully can be converted into an artificial channel with appropriate dimensions to convey water away.

Geotextiles can be used to simulate a vegetation cover so as to inhibit erosion processes and thereby reduce soil loss. They also can help maintain soil moisture and promote the growth of seed. Geotextiles used for erosion prevention can either be made from natural or synthetic materials to form an erosion blanket or mat. In the former case, the geotextile is biodegradable and so its role is temporary. Synthetic geotextiles remain in place to provide reinforcement for a soil. Both types of geotextile are used for slope protection. They can simply be rolled downslope and pegged into place. If, when vegetation is fully established it will control erosion, then temporary geotextiles can be used. However, the establishment

(a)   (b)

Fig. 5.3   (a) Serious gully erosion in Tennessee (b) after reclamation.

145

of vegetation may not coincide with the time at which the geotextile biodegrades. Seeding takes place prior to the positioning of temporary geotextiles, whereas synthetic geotextiles tend to be placed first, then seeded and covered with topsoil. In fact, some geotextile mats contain seeds.

5.8.2   Control of wind erosion

The problem of wind erosion can be mitigated against by reducing the velocity of the wind and/or altering the ground conditions so that particle movement becomes more difficult or particles are trapped. Windbreaks or shelter belts normally are placed at right angles to the prevailing wind. On the other hand, if winds blow from several directions, then a herringbone or grid pattern of windbreaks may be more effective in reducing wind speed. Windbreaks interrupt the length over which wind blows unimpeded and thereby reduce the velocity of the wind at the ground surface. The spacing, height, width, length and shape of windbreaks influence their effectiveness. Trees and bushes are fairly permanent features so their use should be planned with care when they are being planted to act as windbreaks. For example, their spacing should be related to the amount of shelter that each windbreak offers. As wind velocities may increase at the ends of windbreaks as a result of funnelling, they should be longer rather than shorter. The recommended length of a windbreak is about twelve times its height for prevailing winds that are at right angles to it. However, in order to allow for deviations in the direction of the wind this figure may be doubled. Furthermore, windbreaks should be permeable to avoid creating erosive vortices on their leeside. The shape of a vegetative windbreak can be influenced by careful selection of the bushes or trees involved. They preferably should have a triangular shape in cross-section. A windbreak consisting of trees and shrubs may need to be up to 9 m in width, comprising two rows of trees and three of bushes.

Geotextiles can interrupt the flow of wind over the ground surface and so reduce its velocity, particularly if coarse-fibred products are used. By reducing the initial erosion of particles, further erosion is reduced since there are fewer particles in saltation, suspension and traction. In addition, because of the moisture retentivity characteristics, especially of natural geotextiles, the soil remains relatively moist so that it is less susceptible to wind erosion.

Field cropping practices may be simpler and cheaper than windbreaks, and sometimes can be more effective. The effectiveness of the protection offered by plant cover against wind erosion is affected by the degree of ground cover provided, and therefore by the height and density of the plant cover. Plants should cover over 70% of the ground surface to afford adequate protection against erosion. Vegetative cover helps trap soil particles and is particularly important in terms of reducing the amount of saltating particles. Although the choice of crop may be limited by the availability of water, it would appear that legumes, small grain crops and grasses are reasonably effective in reducing the effects of wind erosion. Mulches may be used to cover the soil and should be applied at a minimum rate of 0.5 kg m^{-2}. As well as providing organic matter for the soil, mulches also trap soil particles and conserve soil moisture. Stubble may be left between strips of ploughed soil, the strip being wide enough to prevent saltating particles from leaping across it. Because wind erosion is affected adversely by surface roughness, the furrows produced by ploughing ideally should be orientated at right angles to the prevailing wind.

5.8.3   Hazard assessment and erosion surveys

The assessment of soil erosion hazard is a special form of land resource evaluation, the purpose of which is to identify areas of land where the maximum sustained productivity from a particular use of land is threatened by excessive loss of soil. Such an assessment subdivides a region into zones in which the type and degree of erosion hazard are similar. This can be represented in map form and used as the basis for planning and conservation within a region.

The first soil erosion surveys consisted of mapping sheet wash, rills and gullies within an area, frequently from aerial photographs. Indices such as gully density were used to estimate erosion hazard. Subsequently, sequential surveys of the same area were undertaken in which mapping, again generally from aerial photographs, was done at regular time intervals. Hence, changes in those factors responsible for erosion could be evaluated. They also could be evaluated in relation to other factors such as changing agricultural practices or land use. Any data obtained from imagery should be checked and supplemented with extra data from the field. It may be possible to rate the degree of severity of soil erosion as seen in the field.

5.9 Desertification

Desertification is a process of environmental degradation that occurs mainly in arid and semi-arid regions, and causes a reduction in the productive capacity of soil. Although deserts expand or contract naturally as precipitation varies over time, the prime cause of desertification is excessive human activity and demand in those regions with fragile ecosystems. The net result is that the productivity of agricultural land declines significantly, grasslands no longer produce sufficient pasture, dry farming fails and irrigated areas may be abandoned.

Deserts are encroaching into semi-arid regions largely as a consequence of poor farming practices such as overstocking. The problem is aggravated further by the improper use of water resources leading to inefficient use and to streams drying up. Excessive abstraction of water from wells lowers the water table, which adversely affects plant growth. Desertification can occur within a short time, that is, in 5 to 10 years. Whereas droughts come and go, desertification can be permanent if in order to reverse the situation substantial capital and resources are not available. When prolonged periods of drought are coupled with environmental mismanagement they may lead to the permanent degradation of the land.

The loss of vegetation at the margins of deserts leads to diminishing rainfall, increasing dust content in the air and accelerates the rate of desertification. An increased amount of dust in the atmosphere may adversely affect the radiation balance and deplete food production even further. It also may affect people with respiratory problems adversely. In the marginal and degraded areas there is a tendency for the number of plant and animal species to decline. For the people involved, desertification can mean loss of income and eventual starvation that, in turn, may lead to population migrations with severe social and economic repercussions.

One of the common causes of desertification is overgrazing of animals on a limited supply of forage. In many cases, especially in the developing world, there is a complete absence of range management. Even in the western United States overgrazing in 70% of the rangelands has led to a decline in the original forage potential by about 50%. Drought puts a severe strain on such regions. Recovery from drought may be a slow process and in some cases irreversible damage may be done to an ecosystem that already is undergoing serious degradation. It therefore is necessary to attempt to determine the carrying capacity of such grasslands, that is, the number of animals they can support without the vegetation being adversely affected.

Desertification brings with it associated problems such as removal of soil, as well as reduction in its fertility; deposition of windblown sand and silt (which can bury young plants and block irrigation canals and rivers; and moving sand dunes (Fig. 5.4). Also when rain does fall, a greater proportion of it contributes towards run-off so that erosion becomes more aggressive. This, in turn, means that the amount of sediment carried by streams and rivers increases. It also means that there is less water infiltrating into the ground for plant growth.

Initially, desertification may go unrecognized and it may not be until significant changes have occurred in the fertility of the soil that it is identified, hence it may be looked upon as a creeping disaster. Some indicators of desertification are given in Table 5.4. It has been suggested that the world loses some 20 million hectares of land to desertification each year with surveys classifying 18% of desertification as slight, 4% as moderate and 28% as severe. As remarked, however, desertification is not always readily identified with accuracy. There are a number of reasons for this. For instance, deterioration of land can take numerous forms and may be irregular in pattern rather than advancing as a recognizable front. In addition, the deterioration of soil affects areas of greater size than those that can be regarded as turned to desert. Furthermore, several years may need to elapse before expansion of desertification can be distinguished from areas that have been subjected to prolonged droughts. It has been estimated that 14% of the world's population live in drylands that are under threat, and that over 60 million people are

Table 5.4 Forms and severities of desertification.

Form	Severity: Slight	Moderate	Severe	Very severe
Water erosion	rills, shallow runnels	soil hummocks, silt accumulations	piping, coarse washout deposits, gullying	rapid reservoir siltation, landslides, extensive gullying
Wind erosion	rippled surfaces, fluting and small-scale erosion	wind mounds, wind sheeting	pavements	extensive active dunes
Water and **wind erosion**			scalding	extensive scalding
Irrigated land	crop yield reduced less than 10 per cent	minor white saline patches, crop yield reduced 10-50 per cent	extensive white saline patches, crop yield reduced more than 50 per cent	land unusable through excessive salinization, soils nearly impermeable, encrusted with salt
Plant cover	excellent-good range condition	fair range condition	poor range condition	virtually no vegetation

Fig. 5.4 Desertification. Trees buried by wind blown sand in Gu'an County, China.

affected by desertification. More or less one third of the land surface and one seventh of the population of the world are affected directly.

As desertification is a process of gradual degradation, it is important that it is monitored effectively so that the changes it brings about are detected as they occur. In this context, remote sensing imagery or aerial photographs taken at successive time intervals can prove useful in that they can record changes in the vegetation cover. Agricultural production also might be of some value in assessing degradation.

The cost of reversing desertification is very high so that it obviously is better to prevent it in the first place. In very degraded drylands, retrenchment may be the only solution and recovery may be out of the question in the short term. In areas that have suffered less deterioration, rotational cropping, reduction in numbers of livestock, use of special equipment for ploughing and sowing, use of specially adapted crops or the establishment of irrigation may be appropriate strategies.

5.10 Irrigation

Large irrigation schemes depend upon plentiful sources of water that can be supplied by large reservoirs or abstracted from deep wells. However, increasing demands for water, limited availability and concerns about water quality mean that water has to be used effectively, so that an irrigation system must be properly planned and designed, and operated efficiently. In order to make the best use of the water available, the seasonal water requirements of the crops grown must be known since plant requirements vary with the season. Furthermore, weather conditions affect water demand (i.e. temperature, wind and humidity). In particular, estimates of evapotranspiration must be determined when planning an irrigation system.

Easily the most common method of applying irrigation water, particularly in arid regions, is by flooding the ground surface. For example, wild flooding is where water is allowed to flow uncontrolled over the soil. Water also may be conveyed to the soil in a more regular manner via canals, ditches and furrows, or basins. Efficient surface irrigation requires that the ground surface is graded in order to control the flow of water. Obviously, the extent of grading is governed by the nature of the topography. Spray or sprinkler

Fig. 5.5 Spray irrigation in southern California.

Fig. 5.6   Salinization in south east South Australia.

irrigation systems provide a fairly uniform method of applying water and are more efficient than surface irrigation (Fig. 5.5). For example, conveying water via canals, ditches and furrows is regarded, at most, as being only 60% efficient, whereas sprinkler systems are about 75% efficient. Sprinkler irrigation does not require the land surface to be graded and it allows the rate of water application to be easily controlled. The infiltration capacity and the permeability of the soil determine how fast water can be applied in either ditch or sprinkler irrigation. Micro-irrigation or drip systems deliver water in small amounts (1–10 litre h^{-1}) via special porous tubes to the plant roots and are around 90% efficient. However, because of the high cost of drip irrigation it tends to be restricted to high value crops. Alternatively, perforated or porous tubes often are installed 0.1–0.3 m below the surface of the soil, supplying 1–5 litre min^{-1} per 100 m of tube.

The use of irrigation in semi-arid and arid regions to increase crop production can lead to the deterioration of water quality and salinity problems. As far as the quality of water for irrigation is concerned, the most important factors are the total concentration of salts; the concentration of potentially toxic elements; the bicarbonate concentration as related to the concentration of calcium and magnesium; and the proportion of sodium to other ions. Most water used for irrigation contains dissolved salts, some of which remain in the soil moisture as a result of evapotranspiration. Also, in hot dry climates plants abstract more moisture from the soil and so tend to concentrate dissolved solids in the soil moisture quickly. In addition to the potential dangers due to high salinity, a sodium hazard sometimes exists in that sodium in irrigation water can bring about a reduction in soil permeability and cause the soil to harden. Both effects are attributable to cation exchange of calcium and magnesium ions by sodium ions on clay minerals and colloids. Capillary action also brings salts to the surface. Inefficient irrigation leads to salinization of the soil, that is, the accumulation of salts near the ground surface (Fig. 5.6). Dissolved nitrates from the application of fertilizers are another source of contamination. In addition, the presence of trace elements derived from the rocks in the area, in amounts greater than their threshold values can prove toxic to plants. Such trace elements include arsenic, boron, cadmium, chromium, lead, molybdenum and selenium. High concentrations of bicarbonate ions can lead to the precipitation of calcium and magnesium bicarbonate from the soil water content, thereby increasing the relative proportions of sodium.

Generally, soil with more than 0.1% soluble salts within 0.2 m of the surface is regarded as salinized.

Heavily salinized land has been abandoned to agriculture in many parts of the world. If crop production levels are to be maintained and salinization avoided, then drainage systems are required to remove excess water and associated salts from the plant root zone. They should be installed prior to the commencement of irrigation. Drainage lowers the water table as well as conveying water away more quickly. Subsurface drainage can be used together with controlled surface drainage. Excess water can be applied to fields during the non-growing season to flush salts from the soil. Usually, the quantity of salt removed by crops is so small that it does not make a significant contribution to salt removal. The other problem that can result from inadequate drainage is waterlogging, which generally is brought about by a rising water table caused by irrigation.

The only way by which salts that have accumulated in the soil due to irrigation can be removed satisfactorily is by leaching. Hence, sufficient water must be applied to the soil so as to dissolve and flush out excess salts. Not only is adequate drainage required for conveying away excess water but it is also needed for water moving through the root zone. The traditional concept of leaching involved ponding of water so that uniform salt removal was achieved from the root zone. However, high frequency watering should be effective for salinity control. The salt content of the soil should be monitored to ensure that the correct amount of water is being used for irrigation.

6

Mining and the Environment

6.1 Introduction

Mining and associated mineral processing have an adverse impact on the environment. Unfortunately, this has frequently led to serious consequences. For example, the disposal of waste has led to unsightly spoils being left that disfigure the landscape, and to surface streams and groundwater being polluted. In addition, some urban areas have suffered serious subsidence damage due to undermining. The impact of mining depends on many factors, especially the type of mining and the size of the operation. However, greater awareness of the importance of the environment has led to tighter regulations being imposed by many countries to lessen the impact of mining. The concept of rehabilitation of a site after mining operations have ceased has become entrenched in law. An environmental impact assessment is necessary prior to the development of any new mine, and an environmental management programme has to be produced to show how the mine will operate and the site be restored after mining operations have ended.

Although the adverse impacts on the environment must be minimized some environmental degradation due to mining is inescapable. Mines, however, are local phenomena, although they may impact beyond mine boundaries. They also account for only a small part of the land area of a nation (e.g. the mining industry accounts for less than 1% of the total area of South Africa). Land that has become derelict by past mining activity can be restored, at a cost. Rehabilitated spoil heaps frequently become centres of social amenity such as parklands, golf courses and even artificial ski slopes. Open pits, when they fill with water, can be used as marinas, for fishing or as wildlife reserves. Even some underground mines can be used, such as those in limestone at Kansas City, Missouri, which are used as warehouses, cold storage facilities and offices. Mining therefore can be regarded as one of the stages in the sequential use of land.

6.2 Surface mining

Most surface mining methods are large scale, involving removal of massive volumes of material. Huge amounts of waste can be produced in the process. Obviously, one of the factors that has to be taken into account in the development of a surface mine is the stability of its slopes. This is influenced by the nature of the rock mass(es), the geological structure and the hydrogeological conditions. The latter determine whether dewatering is necessary. Another important factor is the stripping ratio, that is, the ratio of overburden to ore. This can determine the economic operation of a surface mine. A mineral occurring at a depth beyond the maximum stripping ratio will either have to remain unworked or be mined by underground methods.

Stratified mineral deposits that occur at or near the surface (such as coal or sedimentary iron ore) in relatively flat terrain generally are extracted either by strip mining or opencast mining. The deposit concerned is either horizontal or gently dipping and normally is within 60 m of the surface. All the strata overlying the mineral deposit are removed and placed in stockpiles. When necessary, blasting is used to break rock above the mineral deposit. The mineral deposit itself may have to be drilled and blasted, once exposed, prior to its removal by conventional loading and haulage equipment. When the overburden is removed from the working face in long parallel strips, and placed in long spoil heaps in the worked-out area, this is referred to as strip mining (Fig.6.1a). Parallel rows of spoil are produced as the face moves on. Soil is removed before the overburden and is placed in separate piles so that it can be used in the rehabilitation process. Rehabilitation of the spoil heaps can proceed before the mining operation ceases, the spoil heaps being regraded to fit into the surrounding landscape. Once the regrading of spoil has been completed, it is covered with soil, fertilized if necessary, and seeded.

Opencast coal mining involves the exploitation of shallow seams from relatively small sites (normally from 10 to 800 ha; Fig. 6.1b). This usually involves creating a box-cut to reveal the coal. Once the coal has been extracted the box-cut is filled with overburden from the next cut. Maximum depths of working are about 100 m, with stripping ratios up to 25:1. Progressive restoration of sites is undertaken so that handling of waste is limited. The vast majority of opencast sites are restored to agriculture. Nevertheless, due to increasing pressures on land use and the proximity of some sites to urban areas, backfilled opencast sites have been used for industrial and housing developments. Unfortunately, differential settlement can occur across a site, especially at the fill-solid interface, and can induce cracking in buildings, and adversely affect roads and services. The groundwater level in these pits may have to be lowered to permit extraction but after mining has ceased and the pit has been filled, the groundwater level begins to rise. This may cause settlement of the fill and so adversely affect any buildings that have been constructed on the site.

Open-pit mining is used to work ore bodies that occur at or near the surface where the ore body is steeply dipping or occurs in the form of a pipe. Initially, the overburden is stripped to expose the ore. The ore body is blasted in a series of benches, the walls of which are steeply dipping (Fig. 6.1c). Rock surrounding the ore body may have to be removed in order to maintain stable slopes in the pit. This represents waste and, together with the waste produced by processing the ore, is disposed of in spoil heaps or tailings lagoons. Because of the size of many open-pits, normally there is not enough waste material available to backfill them. Accordingly, the principal objectives after cessation of mining operations generally are to ensure that the walls of the pit are stable, and that the waste dumps and tailings dams are rehabilitated. Large-scale mining of coal is worked in similar fashion (Fig. 6.1d).

Dredge mining is used to recover minerals from alluvial, marine or aeolian (dune) deposits. It involves the underwater excavation of a deposit, usually carried out from a floating vessel that may incorporate processing and waste disposal facilities. The deposits that contain the minerals must be diggable. The dredge excavates the deposit and pumps the material to the separation plant, which may be separate from the dredge. If heavy minerals (e.g. ilmenite $FeTiO_3$, rutile TiO_2, cassiterite SnO_2, or chromite $(Fe,Mg)(Cr,Al,Fe)_2O_4$) are being worked, then they are concentrated and pumped ashore for further processing. Other materials such as gold or diamonds are collected. The tailings from the separation plant may be placed in the mined-out area for subsequent rehabilitation.

6.3 Subsidence

Surface subsidence occurs as a consequence of the extraction of mineral deposits or the abstraction of water, brine, oil or natural gas from the ground. The subsidence effects of mineral extraction depend on the type of deposit, the geological conditions, in particular the nature and structure of the overlying rock or soil masses, the mining methods and any mitigative action. In addition to subsidence due to the removal of support, mining often entails lowering the groundwater level that may lead to the consolidation of overburden (since the vertical effective stress is raised).

6.3.1 Metalliferous deposits

There are several underground mining methods used for the extraction of metalliferous deposits. In partial extraction methods, solid pillars are left unmined to provide support to the underground workings. Pillars may vary in shape and size, and in many cases, since pillars often represent reserves of ore, they are extracted or reduced in size towards the end of the life of the mine. Pillar collapse in former workings can result in localized or substantial surface subsidence. The resulting subsidence profile is likely to be

Fig. 6.1   (a) Jayant strip mine India (courtesy Dr L.J.Donnelly, Halcrow Group Ltd.) (b) St Aidens opencast site, Yorkshire (courtesy Dr L.J.Donnelly, Halcrow Group Ltd.) (c) Tiered working in Palabora open pit copper mine, South Africa. (d) Massive open pit working of deposits coal, near Lethbridge, southern British Columbia.

Fig. 6.2 Collapse of the surface outcrop of a reef into a stope, Johannesburg area, South Africa.

Fig. 6.3 Cratering due to block caving in a copper mine, Chile. (Courtesy of Prof. Dick Stacey.)

irregular, with large differential subsidence, tilts and horizontal strains at the perimeter of the subsidence area. The presence of faults or dykes can give rise to steps and where inclined ore bodies have been mined, the steeper the dip, the more localized the subsidence is likely to be. Fortunately, surface subsidence is preventable with modern mining practice.

Mining of tabular stopes with substantial spans is practised in the gold and platinum mines of southern Africa. Extraction of one, two, or more reefs has taken place, each typically with stoping widths of 1 to 2 m. Where this type of mining occurred at shallow depth detrimental subsidence frequently has taken place (Fig. 6.2). The surface profile is commonly very irregular and tension cracks often have been

155

observed. The effects are dependent on the dip of the reefs, which vary from as little as 10° to vertical. Closure of stopes is more likely for flatter dips and hence surface movements occur more often under such conditions. Steeply dipping stopes tend to remain open and present a longer-term hazard. Adverse subsidence is not usually observed when the mining depth exceeds 250 m. Most of the detrimental subsidence has occurred at the contacts with dyke and faults.

Many metalliferous deposits occur in large disseminated ore bodies and often the only way by which such ore bodies can be mined economically is by means of very high-volume production. When the rock mass is sufficiently competent, large open stopes can be excavated. The sizes of open stopes are the subject of design, and some form of pillar usually separates adjacent open stopes. The method is therefore a partial extraction approach, but open stopes may have spans of between 50 to 100 m. In caving, material from above is allowed to cave into the opening created by the mined extraction. Since collapse occurs above the ore body, it can progress through to the surface to form a subsidence crater (Fig. 6.3). Owing to the large volumes that are extracted during mining, the extent and depth of the subsidence crater can be very large.

The aim of any shallow ground stabilization methods that are required, is to develop a competent surface to promote arching across stopes and prevent collapses. The construction of flexible structures then can accommodate any subsequent minor surface deformation that occurs. Dynamic compaction (dropping a large weight suspended by a crane) has been used as a method of shallow stabilization where ground consisting of very loose sandy gravel and fill, occurs above old mine workings. The dynamic compaction improves the characteristics of the soil down to bedrock and closes any cavities due to shallow mine workings. Another, more common, form of shallow stabilization is to plug the stopes with concrete, to provide the necessary local support, and then to place backfill above the plug (Fig. 6.4). Deep stabilization of abandoned workings has to provide rigid support between roof and floor so that if collapse or instability occurs at greater depth, then the near-surface rigid zone forms an arch.

Fig. 6.4 Stabilization of a shallow outcrop of three adjacent stopes by a concrete plug, Johannesburg area, South Africa.

Fig. 6.5 (a) Abandoned pillar workings in the Bethaney Fall Limestone, Kansas City, United States. (b) Pillar collapse above an abandoned coal mine in the Witbank Coalfield, South Africa.

6.3.2 Subsidence due to abandoned mine workings

The mining method also influences the types of subsidence associated with stratified deposits. For instance, the mining of coal prior to the sixteenth century in Britain usually took the form of outcrop workings or bell-pits. The latter were shaft-like excavations up to about 12 m deep, the extent of the workings around the shaft being limited by their stability. Bell-pitting leaves an area of highly disturbed ground that generally requires treatment or complete excavation before it is suitable for surface development.

Later coal workings usually were undertaken by the pillar and stall method. This method entails leaving pillars of coal in place to support the roof of the workings. The method also is used at the present day to work other stratiform deposits, including limestone, gypsum, anhydrite, salt and sedimentary iron ores (Fig. 6.5a). However, in many old coal workings, pillars were robbed prior to abandonment of the mine – which increases the possibility of pillar collapse. Slow deterioration and failure of pillars may take place many years after mining operations have ceased. The yielding of a large number of pillars can bring about a shallow broad trough-like subsidence over a large area (Fig. 6.5b).

Fig. 6.6 Void migrations in a coal seam in an opencast working in Staffordshire, England.

157

Void migration develops if roof rock falls into the worked-out zones and can present a problem in areas of shallow abandoned mines (Fig. 6.6). It can occur within a few months, or a very long period of years after mining has ceased. The material involved in the fall bulks, so that migration eventually is arrested. Nevertheless, the process can, at shallow depth, continue upwards to the surface leading to the sudden appearance of a crown hole. The maximum height of migration in exceptional cases might extend to 10 times the height of the original stall, however, it generally is 3 to 5 times the stall height.

A site investigation is required in an area that is underlain by abandoned workings, especially if an important structure(s) is to be constructed. All strata likely to be significantly affected by the structural loading need to be investigated and sampled. The location of subsurface voids due to mineral extraction is of prime importance in this context. In other words, an attempt should be made to determine the number and depth of mined horizons, the proportion of mineral extracted (i.e. the extraction ratio), the layout of pillars, and the condition of the old workings. Careful note should be taken of whether the old workings are open, partially collapsed or collapsed, and the degree of fracturing and bed separation in the roof rocks should be recorded, if possible. This helps to provide an assessment of past and future collapse.

Assessments of mining hazards have usually been on a site basis, regional assessments being much less common. Nonetheless, regional assessments can offer planners an overview of the problems involved and can help them avoid imposing unnecessarily rigorous conditions on developers in areas where they really are not warranted. Hazard maps of areas where old mine workings are present ideally should represent a source of clear and useful information for planners. In this respect, they should realistically present the degree of risk (i.e. the probability of the occurrence of a hazard event). Descriptive terms such as high, medium and low risk must be defined, and the social and economic dangers in overstating the degree of risk considered (e.g. urban blight). Ideally, numerical values should be assigned to the degree of risk. However, this is no easy matter for numerical values can only be derived from a comprehensive record of events, which unfortunately in the case of old abandoned mine workings is available only infrequently. Since many events in the past may not have been recorded, this throws into question the reliability of any statistical analysis of data. The matter of risk assessment is complicated further in terms of tolerance, for example, people are less tolerant in relation to loss of life than to loss or damage of property. Hence, the likelihood of an event causing loss of life is assigned a higher value than one causing loss of property.

Hazard maps of areas of old mine workings frequently recognize safe and unsafe zones. The zones commonly are defined in terms of the thickness of cover rock, unsafe zones being regarded as not thick enough and safe zones as thick enough to preclude subsidence hazards. Intermediate zones may be recognized between the safe and unsafe zones, where buildings are constructed with reinforced foundations or rafts, as an added precaution against unforeseen problems. Development is prohibited in zones designated unsafe. Alternatively, building restrictions may be imposed on surface development related to the zones. Again development is prohibited where an area is deemed unsafe, while in other zones the permissible heights of proposed buildings are restricted in accordance with depth of mining below the site. These restrictions are progressively relaxed up to the point where they no longer apply. However, it must be borne in mind that thematic maps that attempt to portray the degree of risk of a hazard event represent generalized interpretations of the data available at the time of compilation. Therefore, they cannot be interpreted too literally and areas outlined as 'undermined' should not automatically be subject to planning blight.

Where a site that is proposed for development is underlain by shallow old mine workings there are a number of ways in which the problem can be dealt with. The first and most obvious method is to locate the proposed structure on sound ground away from old workings or over workings proved to be stable. Such location, of course, is not always possible. If old coal mine workings are at very shallow depth, then it might be feasible, by means of bulk excavation, to found on the strata beneath. This can be an economic solution, at depths of up to 7 m or on sloping sites. Where the allowable bearing capacity of the foundation materials has been reduced by mining, it may be possible to use a raft. Reinforced bored pile foundations also have been resorted to in areas of abandoned mine workings. In such instances, the piles bear on a competent stratum beneath the workings. Where old mine workings are believed to pose an unacceptable hazard to development and it is not practicable to use adequate measures in design or found below their level, then the ground itself can be treated. Such treatment involves filling the voids in order to prevent void migration and pillar collapse. For example, grouting can be undertaken by drilling holes from the surface into the mine workings and filling the remnant voids with an appropriate grout mix. Steel mesh reinforcement or geonets have been used in road construction over areas of potential mining subsidence. If a void should develop under the road, the reinforcing layer is meant to prevent the road from collapsing into the void.

6.3.3 Old mine shafts

Centuries of mining in many countries have left behind a legacy of old shafts. Unfortunately, many, if not most, are unrecorded or are recorded inaccurately. In addition, there can be no guarantee of the effectiveness of any shaft treatment unless it has been carried out in recent years. The location of a shaft is of great importance as far as the safety of a potential structure is concerned, for although shaft collapse is an infrequent event, its occurrence can prove disastrous (Fig. 6.7). The ground about a shaft may subside or collapse suddenly and the resulting crown hole may have a much larger diameter than the shaft itself. Consequently, the crown hole may affect adjacent shafts if they are interconnected. From the economic point of view the sterilization of land due to the suspected presence of a mine shaft is unrealistic.

Fig. 6.7 Collapse of an old coal mine shaft in Cardiff, South Wales.

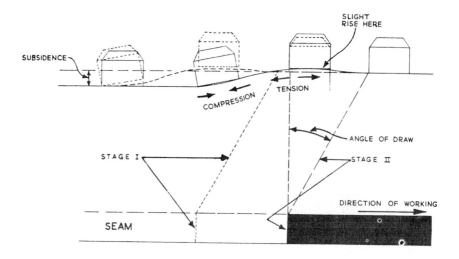

Fig. 6.8 Curve of subsidence developed above longwall working in a coal seam. Note that the area affected extends beyond the workings and is defined by the angle of draw. Note also the zones of tensile and compressive strain, and tilt that occur above the workings.

An investigation should include a survey of plans of abandoned mines, records of shaft sinking, geological maps, all available editions of topographic maps, aerial photographs, archival and other official records. The success of geophysical methods in locating old shafts depends on the existence of a sufficient contrast between the physical properties of the shaft and its surroundings to produce an anomaly. A shaft frequently will remain undetected when the top of the shaft is covered by more than 5 m of fill. The size, especially the diameter, of a shaft also influences whether or not it is likely to be detected. The most effective geochemical method yet used in the location of abandoned coal mine shafts is that of methane detection. Confirmation of the existence of a shaft, is accomplished by excavation, for example, with a mechanical boom-type digger that is anchored outside the search area. Alternatively a rig, placed on a platform, can be used to drill exploratory holes.

The search for old shafts on land that is about to be developed must extend outside the site in question for a sufficient distance to find any shafts that could affect the site itself if a collapse occurred. If the depth is not excessive and the shaft is open, it can be filled with suitable granular material, the top of the fill being compacted. If, as is more usual, the shaft is filled with debris in which there are voids, these should be filled with gravel and grouted. Concrete cappings are needed to seal mine shafts.

6.3.4 Longwall mining of coal

Mining coal by longwall methods involves total extraction of a panel of coal, 200 or more metres in width, within a seam. The working face is temporarily supported, the support being moved on as the face advances, leaving the roof from which support has been withdrawn to collapse. The resulting subsidence is largely contemporaneous with the mining. The surface effects include not only lowering of the ground surface but as a working face advances the ground above is subject to tilting accompanied by tension and then compression (Fig. 6.8). Once the subsidence front has passed by, the ground attains its previous slope and the ground strain returns to zero. However, permanent ground strains affect the ground above the edges of the extracted panel. Under certain conditions the maximum amount of subsidence can be up to around 90% of thickness of coal extracted. Normally, however, the amount of surface subsidence is significantly less than this. In many instances, subsidence effects are influenced by the geological

Fig. 6.9 Damage caused to houses by longwall working of coal at relatively shallow depth, Elsecar, South Yorkshire. The houses were evacuated, awaiting repair. Many houses in the village had to be demolished.

structure, notably the presence of faults, which can concentrate subsidence, and the character of the near-surface or surface rock masses, especially the nature of their discontinuities, and the overlying types of soil.

Structural damage is not simply a function of ground strain but the shape, size and form of construction are also important controls. Three classes of subsidence damage to buildings have been recognized, namely, architectural (characterized by small scale cracking of plaster and doors and windows sticking); functional damage (characterized by instability of some structural elements, jammed doors and windows, broken window panes, and restricted building services); and structural damage in which primary structural members are impaired, there is a possibility of structural collapse, and complete or large-scale rebuilding is necessary (Fig. 6.9). Fortunately, an important feature of subsidence due to longwall mining is its high degree of predictability so that buildings can be designed in relation to the effects of subsidence. In addition, the contemporaneous nature of subsidence associated with longwall mining sometimes affords the opportunity for planners to phase long-term surface development in relation to the cessation of subsidence.

The most common method of mitigating subsidence damage is by the introduction of flexibility into a structure. In flexible design, structural elements deflect according to the subsidence profile. Flexibility can be achieved by using specially designed rafts. The use of piled foundations in areas of mining subsidence presents its own problems, in that pile caps tend to move in a spiral fashion, and each cap moves at a different rate and in a different direction according to its position relative to the mining subsidence. Hence, it is often necessary to allow the structure to move independently of the piles by the provision of a pin joint or roller bearing at the top of each pile cap. Preventative techniques frequently can be used to reduce the effects of movements on existing structures. In the case of long buildings, damage can be reduced by cutting them into smaller, structurally independent units. Pillars of coal can be left in place to protect surface structures above them. Maximum subsidence also can be reduced by packing or stowing the mined out area.

Fig. 6.10 This sinkhole occurred at a miner's recreation centre at Venterspoort, South Africa, and swallowed part of the clubhouse and one person.

6.3.5   Subsidence and fluid abstraction

Subsidence of the ground surface occurs in areas where there is intensive abstraction of fluids, notably groundwater, oil or brine. In the case of groundwater, subsidence occurs where abstraction exceeds natural recharge and the water table is lowered, the subsidence being attributed to the consolidation of the sedimentary deposits as a result of increasing effective stress. For instance, if the groundwater level is lowered by 1 m, then this gives rise to a corresponding increase in average effective overburden pressure of 10 kPa. The amount of subsidence that occurs is governed by the increase in effective pressure (i.e. the magnitude of the decline in the water table); the thickness and compressibility of the deposits involved; the depth at which they occur; the length of time over which the increased loading is applied; and possibly the rate and type of stress applied. The rate at which consolidation occurs depends on the thickness of the beds concerned as well as the rate at which pore water can drain from the system, which is governed by permeability. Thick slow-draining fine-grained beds may take years or decades to adjust to an increase in applied stress, whereas coarse-grained deposits adjust rapidly. Methods that can be used to arrest or control subsidence caused by groundwater abstraction include reduction of pumping draft, artificial recharge of aquifers from the ground surface, repressurizing the aquifer(s) involved via wells, or any combination of these options. The aim is to manage the rate and quantity of water withdrawal so that its level in wells is either stabilized or raised somewhat so that the effective stress is not further increased. A reduction in the rate of abstraction can lead to a rise in groundwater levels and controlled withdrawal of groundwater permits the re-establishment of the natural hydraulic balance. Subsidence due to the abstraction of oil occurs for the same reasons as does subsidence associated with the abstraction of groundwater. It can be controlled by the repressurization of the oil reservoir by the injection of water.

Dewatering associated with mining in the gold bearing reefs of South Africa, which underlie dolostone and unconsolidated deposits, notably a residual clay deposit termed wad, has led to the formation of sinkholes and produced differential subsidence over large areas (Fig. 6.10). Consequently, it became a matter of urgency that the areas that were subject to the formation of sinkholes or subsidence depressions were delineated. Sinkholes formed concurrently with the lowering of the water table. Subsidence depressions also occur as a consequence of consolidation taking place in wad as the water table is lowered. The degree of subsidence that occurs reflects the thickness and proportion of wad that has consolidated,

Fig. 6.11 Subsidence caused by wild brine pumping of salt in Cheshire, England. Note the tension scars especially that immediately to the right of the fence. The subsidence trough is occupied in part by a flash (i.e. a small lake) in the top right-hand corner.

and can continue for a few years. The thickness of these deposits varies laterally, thereby giving rise to differential subsidence that, in turn, causes large fissures to occur at the surface. The total subsidence has varied from several centimetres to over 9 m.

Deposits that readily go into solution, particularly salt, can be extracted by solution mining. For example, salt has been obtained by brine pumping in a number of areas in Britain. Previously, wild brine pumping was carried out on the major natural brine runs. Active subsidence normally was concentrated at the head and sides of a brine run where fresh water first entered the system. Hence, serious subsidence

Fig. 6.12 Fractured brickwork due to the development of fissures within the ground consequent on groundwater lowering.

could occur at considerable distances, up to 8 km from pumping centres, and was unpredictable. In addition, tension cracks and small fault scars formed on the convex flanks of subsidence hollows (Fig. 6.11). Today wild brine pumping has been replaced almost totally by controlled solution mining with which no subsidence of any consequence has been associated. Sulphur is mined in Texas and Louisiana by the Frasch process, which involves pumping hot water into the beds of sulphur and then pumping the brine to the surface. Subsidence troughs with associated small faults at the periphery of the basins have formed.

Ground failure, that is, fissuring or faulting, often is associated with subsidence due to the abstraction of fluids (Fig. 6.12). Fissures may appear suddenly at the surface and the appearance of some may be preceded by the occurrence of minor depressions at the surface. Within a matter of a year or so most fissures become inactive. In a few instances new fissures have formed in close proximity to older fissures. Those faults that are suspected of being related to groundwater withdrawal are much less common than fissures. They frequently have scarps more than one kilometre in length and more than 0.2 m high.

A number of steps can be taken in order to evaluate the subsidence likely to occur due to the abstraction of fluids from the ground. These include defining the in situ hydraulic conditions; computing the reduction in pore pressure due to removal of a given quantity of fluid; conversion of the reduction in pore pressure to an equivalent increase in effective stress; and estimating the amount of consolidation likely to take place in the formation affected from consolidation data and the increased effective load. In addition to depth of burial, the ratio between maximum subsidence and reservoir consolidation also should take account of the lateral extent of the reservoir in that small reservoirs that are deeply buried do not give rise to noticeable subsidence, even if undergoing considerable consolidation, whereas extremely large reservoirs may develop significant subsidence.

6.4 Waste materials from mining

Mine wastes result from the extraction of metals and non-metals. In the case of metalliferous mining, high volumes of waste are produced because of the low or very low concentrations of metal in the ore. In fact, mine wastes represent the highest proportion of waste produced by industrial activity, billions of tonnes being produced annually. Wastes can be inert or contain hazardous constituents but generally are of low toxicity. The chemical characteristics of mine waste and associated waste water depend upon the type of mineral being mined, as well as the chemicals that are used in the extraction or beneficiation processes. Because of its high volume, mine waste historically has been disposed of at the lowest cost, often without regard for safety and often with considerable environmental impacts. The character of waste rock from a mine reflects that of the rock hosting and/or surrounding the mineral. The type of waste rock disposal facility depends on the topography and drainage at the site, and the volume of waste.

6.4.1 Spoil heaps

Generally, coarse mine waste is disposed of in spoil heaps. The configuration of a spoil heap depends upon the type of equipment used in its construction and the sequence of tipping the waste. Obviously, an important factor in the construction of a spoil heap is its stability, which includes its long-term stability. The shear strength of waste within a spoil heap, and therefore its stability, is influenced by the pore water pressures developed within it. Excess pore water pressures in spoil heaps may be developed as a result of the increasing weight of material added during their construction or by seepage though the spoil of natural drainage. High pore water pressures usually are associated with fine-grained materials that have a low permeability and high moisture content. A disastrous failure of a colliery spoil heap occurred at Aberfan, South Wales, in 1966, which resulted in 144 deaths, mainly of young children. A landslide took place on the spoil heap after heavy rain and, as it moved down-slope, it turned into a mudflow that flowed at approximately 32 km h^{-1}, engulfing a school and several houses.

Pyrite may be present in spoil heaps. When it weathers it may give rise to the formation of sulphuric acid, along with ferrous and ferric sulphates and ferric hydroxide that lead to acidic conditions in the weathered material. Significant concentrations of sulphates sometimes occur in low volume seepages from older, more permeable, spoil heaps. Acidic drainages from such sources, may have a pH of less than 5, and occasionally contain low concentrations (a few mg l^{-1}) of copper, nickel and zinc. Other heavy metals are rarely found in concentrations greater than 0.1 mg l^{-1}. Surface water run-off can leach out soluble salts from spoil heaps, especially chloride.

Spontaneous combustion of carbonaceous material is the most common cause of burning colliery spoil. It can be regarded as an atmospheric oxidation (exothermic) process in which self-heating occurs. Hot spots may develop under such conditions where temperatures up to 900°C may be recorded. In addition, oxidation of pyrite at ambient temperature in moist air is an exothermic reaction. When present in sufficient amounts, and especially when in finely divided form, pyrite associated with coaly material, increases the likelihood of spontaneous combustion. Compaction, digging out, trenching, injection with non-combustible material and water, blanketing, and water spraying are methods used to control spontaneous combustion. Spontaneous combustion may give rise to subsurface cavities in colliery spoil heaps, the roofs of which may be incapable of supporting a person. Burnt ashes also may cover zones that are red-hot to appreciable depths. When steam comes in contact with red-hot carbonaceous material, watergas is formed, and when the latter is mixed with air, over a wide range of concentrations, it becomes potentially explosive. If a cloud of coal dust is formed near burning spoil when reworking a heap, then this also can ignite and explode. Damping with a water spray may prove useful in the latter case.

Noxious gases are emitted from burning colliery spoil. These include carbon monoxide, carbon dioxide, sulphur dioxide and, less frequently, hydrogen sulphide. Each may be dangerous if breathed at certain concentrations that may be present at fires on spoil heaps. The rate of evolution of these gases may be accelerated by disturbing a burning heap by excavating into it or reshaping it. Carbon monoxide is the most dangerous since it cannot be detected by taste, smell or irritation and may be present in potentially lethal concentrations. By contrast, sulphur gases are readily detectable in the aforementioned ways and usually are not present in high concentrations. Even so, they still may cause distress to persons with respiratory ailments.

Old spoil heaps represent the most notable form of dereliction associated with subsurface mining. They are particularly conspicuous and can be difficult to rehabilitate into the landscape. For instance, the uppermost slopes of a spoil heap frequently are devoid of near-surface moisture and so are often barren of vegetation. In fact, in order to support vegetation a spoil heap also should have a stable surface in which roots can become established, must be non-toxic, and contain an adequate and available supply of nutrients. These derelict areas need to be restored to sufficiently high standards to create an acceptable environment. The restoration of such derelict land requires a preliminary reconnaissance of the site, followed by a site investigation. The investigation provides essential input for the design of remedial measures. Restoration of a spoil heap is an exercise in large scale earthmoving and invariably involves spreading the waste over a larger area so that gradients on the existing site can be reduced by the transfer of spoil to adjacent land. Water courses may have to be diverted, as may services, notably roads. After an old spoil heap has been regraded the actual surface still needs restoring. This is not so important if the area is to be built over as it is if it is to be used for amenity or recreational purposes. In the case where buildings are to be erected, however, the ground must be adequately compacted so that bearing capacities are acceptable and buildings are not subjected to adverse settlement. On the other hand, where the land is to be used for amenity or recreational purposes, then soil fertility must be restored so that the land can be grassed and trees planted.

6.4.2 Tailings ponds or lagoons

Tailings are fine-grained slurries that result from crushing rock that contains ore. In addition, they are produced by the washeries at collieries (Fig. 6.13). The water in tailings may contain certain chemicals associated with the metal recovery process – such as cyanide in tailings from gold mines, and heavy metals in tailings from copper-lead-zinc mines. Tailings also may contain sulphide minerals like pyrite that can give rise to acid mine drainage. Accordingly, contaminants carried in the tailings represent a source of pollution for both ground and surface water, as well as soil.

Fig. 6.13 A tailings impoundment just south of Salt Lake City, Utah.

Tailings are deposited as slurry, generally in specially constructed tailings dams. Tailings dams usually consist of raised embankments, that is, the construction of the dam is staged over the build-up of the impoundment. These dams are constructed in a similar manner to earth-fill dams used to impound reservoirs. The design of tailings dams obviously must pay due attention to their stability. Failure of a dam can lead to catastrophic consequences. The most dangerous situation occurs when the ponded water on top of the slurry erodes cracks in the dam to form pipes that may emerge on the outer slopes of the dam. The failure of the Merriespruit gold mine tailings dam in South Africa in 1994 resulted in 17 deaths (Fig. 6.14). There may have been insufficient freeboard provision and poor pool control leading to overtopping of the dam. Also, high void ratios in some parts of the dam could have meant that these zones were in a metastable condition, and were exposed by overtopping and erosion. This resulted in liquefaction of the tailings and consequent flow failure.

Fig. 6.14 The Merriespruit gold mine tailings dam.

Seepage losses from tailings lagoons that contain toxic materials can have an adverse effect on the environment. The rate at which seepage occurs from a tailings lagoon is governed by the permeability of the tailings and the ground beneath the impoundment. One of the most cost effective methods of controlling seepage loss from a tailings lagoon is to cover the whole floor of the impoundment with tailings, provided this is of low permeability, from the start of the operation. This forms a liner. Clay liners also represent an effective method of reducing seepage from a tailings lagoon. Filter drains may be placed at the base of tailings to convey water to collection ponds.

Rehabilitation objectives for tailings impoundments include their long-term mass stability, long-term stability against erosion, prevention of environmental contamination and a return of the area to productive use. Normally, when the discharge of tailings comes to an end, the level of the phreatic surface in the embankment falls as water replenishment ceases. This results in an enhancement of the stability of embankment slopes. However, where tailings impoundments are located on slopes, excess run-off into the impoundment may reduce embankment stability or overtopping may lead to failure by erosion of the downstream slope. In such cases judicious siting is called for when locating an impoundment in order to minimize inflow due to run-off. Diversion ditches can cater for some run-off but have to be maintained, as do abandoned spillways and culverts.

Accumulation of water may be prevented by capping the impoundment, the capping sloping towards the boundaries. Erosion by water or wind can be impeded by placing rip-rap on slopes and by the establishment of vegetation on the waste. The latter also helps to return the impoundment to some form of productive use. Particular precautions need to be taken where long-term potential for environmental contamination exists. For example, as the water level in the impoundment declines, the rate of oxidation of any pyrite present in the tailings increases, reducing the pH and increasing the potential for heavy

Table 6.1 Composition of acid mine water from a South African coalfield.

Determinand (mg l^{-1})	Sample 1	Sample 2	Sample 3	Sample 4	Sample 5	Sample 6
TDS	4844	2968	3202	2490	3364	3604
Suspended solids	33	10.4	12	10.0	7.6	
EC (mS m^{-1})*	471	430	443	377	404	340
pH value	1.9	2.4	2.95	2.9	2.3	2.8
Turbidity as NTU	5.5	0.6	2.0	0.9	1.7	
Nitrate NO$_3$ as N	0.1	0.1	0.1	0.1	0.1	0.1
Chlorides as Cl	310	431	406	324	353	611
Fluoride	0.6	0.5	0.33	0.6	0.6	0.84
Sulphate as SO$_4$	3250	1610	1730	1256	2124	1440
Total hardness		484	411	576	585	377
Calcium hardness as CaCO$_3$		285	310	327	282	
Magnesium hardness as MgCO$_3$		199	101	249	305	
Calcium	173.8	114.0	124	131	113	84
Magnesium	89.4	48.4	49.5	60.5	49.3	31
Sodium	247.0	326.0	311	278	267	399
Potassium	7.3	9.4	8.9	6.4	3.8	
Iron	248.3	128	140	87	89.9	193
Manganese	17.9	15	9.9	13.4	13.9	9.3
Aluminium		124		112	204	84

* EC = electrical conductivity

Fig. 6.15 Vegetation decimated by acid mine drainage issuing from a shallow abandoned mine, Witbank Coalfield, South Africa. Note the white patches where salts have been precipitated.

metal contamination. In the case of tailings from uranium mining, radioactive decay of radium gives rise to radon gas. Diffusion of radon gas does not occur in saturated tailings but after abandonment radon reduction measures may be necessary. In such cases a clay cover can be placed over the tailings impoundment to prevent leaching of contaminants and/or to reduce emission of radon gas.

Once the discharge of tailings ceases, the surface of the impoundment is allowed to dry. Drying of the decant pond may take place by evaporation and/or by drainage to an effluent plant. Desiccation and consolidation of the slimes may take a considerable time. Stabilization can begin once the surface is firm enough to support equipment and normally involves the establishment of a vegetative cover.

6.5 Acid mine drainage

Acid mine drainage is produced when natural oxidation of sulphide minerals occurs in mine rock or waste that are exposed to air and water. This is a consequence of the oxidation of sulphur in the mineral concerned to a higher oxidation state, with the formation of sulphuric acid and sulphate, and if aqueous iron is present and unstable, to the precipitation of ferric iron with iron hydroxide. Acid mine drainage is responsible for problems of water pollution in mining areas around the world. However, it does not occur if the sulphide minerals are non-reactive or if the host rock contains sufficient alkaline material to neutralize the acidity. If acid mine drainage is not controlled it can pose a serious threat to the environment since acid generation can lead to elevated levels of heavy metals and sulphate in the water affected that obviously have a detrimental effect on its quality (Table 6.1). This can have a notable impact on the aquatic environment, as well as vegetation (Fig. 6.15). The development of acid mine drainage is time dependent and at some mines may evolve over a period of years.

6.5.1 Causes of acid mine drainage

Generally, acid mine drainage from underground mines occurs as point discharges. A major source of acid mine drainage may result from the closure of a mine. When a mine is abandoned and dewatering by pumping ceases, the groundwater level rebounds. However, the workings often act as drainage systems so that the groundwater does not rise to its former level. Consequently, a residual dewatered zone remains that is subject to continuing oxidation. Groundwater may drain to the surface from old drainage adits, river bank mine mouths, faults, springs and shafts that intercept rock in which water is under artesian pressure. The large areas of fractured rock exposed in opencast coal mines or open pits can give rise to large volumes of acid mine drainage. Acid generation tends to occur in the surface layers of spoil heaps where air and water have access to sulphide minerals. Tailings deposits that have a high content of sulphide represent another potential source of acid generation. However, the low permeability of many tailings deposits, together with the fact that they commonly are flooded, means that the rate of acid generation and release is limited but it can continue to take place long after a tailings deposit has been abandoned. Mineral stockpiles also may represent sources of acid mine drainage. Major acid flushes commonly occur during periods of heavy rainfall after long periods of dry weather. Heap-leach operations at metalliferous mines include, for example, cyanide leach for gold recovery and acid leach for base metal recovery. Spent leach heaps can represent sources of acid mine drainage, especially those associated with low pH leachates.

Certain conditions including the right combination of mineralogy, water and oxygen are necessary for the development of acid mine drainage. Such conditions do not always exist. Consequently, acid mine drainage is not found at all mines with sulphide bearing minerals. The ability of a particular mine rock or waste to generate net acidity depends on the relative content of acid-generating minerals and acid-consuming or neutralizing minerals. Acid waters produced by sulphide oxidation of mine rock or waste may be neutralized by contact with acid-consuming minerals. As a result the water draining from the parent material may have a neutral pH value and negligible acidity despite ongoing sulphide oxidation. If the acid-consuming minerals are dissolved, washed out or surrounded by other minerals, then acid generation continues. Where neutralizing carbonate minerals are present, metal hydroxide sludges, such as iron hydroxides and oxyhydroxides are formed. In fact, the development of acid mine drainage is a complex combination of inorganic and sometimes organic processes and reactions. In order to generate severe acid mine drainage (pH<3) sulphide minerals must create an optimum micro-environment for rapid oxidation and must continue to oxidize long enough to exhaust the neutralization potential of the rock.

6.5.2 Control of acid mine drainage

The objective of acid mine drainage control is to satisfy environmental requirements using the most cost effective techniques. Prediction of the potential for acid generation involves the collection of available data and carrying out various tests. There are three key strategies in acid mine drainage management, namely, control of the acid generation process, control of acid migration, and collection and treatment of acid mine drainage. Control of acid mine drainage may require different approaches, depending on the severity of potential acid generation, the longevity of the source of exposure and the sensitivity of the receiving waters. Mine water treatment systems installed during operation may be adequate to cope with both operational and long-term post-closure treatment with little maintenance. On the other hand, in many mineral operations, especially those associated with abandoned workings, the long-term method of treatment may be different from that used while a mine was operational. Hence, there may have to be two stages involved with the design of a system for treatment of acid mine drainage, one for during mine operation and another for after closure.

Obviously, the best solution is to control acid generation, if possible. Source control of acid mine drainage involves measures to prevent or inhibit oxidation, acid generation or contaminant leaching. If acid generation is prevented, then there is no risk of the contaminants entering the environment. Such control methods involve the removal or isolation of sulphide material, or the exclusion of water or air. The latter is much more practical and can be achieved by the placement of a cover over acid-generating material such as waste or air-sealing adits in mines. Migration control is considered when acid generation is occurring and cannot be inhibited. Since water is the transport medium, control relies on the prevention of water entry to the source of acid mine drainage. Release control is based on measures to collect and treat either or both ground and surface acid mine drainage. In some cases, especially at working mines, this is the only practical option available. Treatment processes have concentrated on neutralization to raise the pH and precipitate metals. Lime or limestone commonly are used, although offering only a partial solution to the problem. More sophisticated processes (active treatment methods) involve osmosis (waste removal through membranes), electrodialysis (selective ion removal through membranes), ion exchange (ion removal using resin), electrolysis (metal recovery with electrodes) and solvent extraction (removal of specific ions with solvents).

Active treatment of mine water requires the installation of a treatment plant. Acid mine drainage water treated with active systems tends to produce a solid residue that has to be disposed of in tailings lagoons. This sludge contains metal hydroxides. Passive treatment involves engineering a combination of low maintenance biochemical systems (e.g. anoxic limestone drains, aerobic and anaerobic wetlands and rock filters). Such treatment does not produce large volumes of sludge, the metals being precipitated as oxides or sulphides in the substrate materials.

6.6   Coal mine effluents

Different types of waste waters and process effluents are produced as a result of coal mining and may arise due to the extraction process, by the subsequent preparation of coal, from the disposal of colliery spoil or from coal stockpiles. The strata from which the groundwater involved is derived, the mineralogical character of the coal and the colliery spoil, and the washing processes employed all affect the type of effluent produced.

Generally, the major pollutants associated with coal mining are suspended solids, dissolved salts, acidity and iron compounds. Elevated levels of suspended matter are associated with most coal mining

Table 6.2   Average quality characteristics of coal mining effluents in the Nottinghamshire coalfield, England.

Type of effluent	BOD (ATU)(mg l⁻¹)	Suspended solids (mg l⁻¹)	Chloride (mg l⁻¹)	Electrical conductivity (μS cm⁻¹)	Minimum pH	Other potential contaminants
Minewaters	2.1	57	4 900	14 000	3.5	Iron, barium, nickel, aluminium, sodium sulphate
Drainage from coal stocking sites	2.6	128	600	2 200	2.2	Iron, zinc
Spoil tip drainage	3.1	317	1 600	4 100	2.7	Iron, zinc
Coal preparation plant discharges	2.1	39	1 500	4 200	3.2	Oil from flotation chemicals
Slurry lagoon discharges	2.4	493	2 000	6 100	3.8	

effluents, with occasionally high values being recorded. Hence, the most common pollutant of a receiving stream near a coal mine is the increased concentration of suspended solids. In particular, the blanketing effect of coal slurry particles on the bed of a stream is unacceptable both in terms of appearance, and its influence on the flora and fauna in the stream. The turbidity of some streams into which colliery waste is discharged may restrict the penetration of light into the water and so have adverse effects on aquatic plant life present. In general, good fisheries are unlikely to be found where average concentrations of suspended solids exceed about 80 mg l^{-1}.

The character of drainage from coal mines varies from area to area and from coal seam to coal seam. Hence, mine drainage waters are liable to vary in both quality and quantity, sometimes unpredictably, as mine workings develop. Although not all mine waters are ferruginous, and in fact some are of the highest quality and can be used for potable supply, they are commonly high in iron and sulphates, low in pH and high in acidity (Table 6.2). Ferruginous discharges can give rise to disastrous conditions in receiving streams and can affect many kilometres of otherwise good quality water. A high level of mineralization is typical of many coal mining discharges and is reflected in the high values of electrical conductivity. The range of dissolved salts encountered in mine water is variable, with electrical conductivity values up to 335 000 μS cm^{-1} and chloride levels of 60 000 mg l^{-1} being recorded. The principal groups of salts in mine discharge waters are chlorides and sulphates, and to a lesser extent sodium and potassium salts. The more saline waters may contain significant concentrations of barium, strontium, ammonium and manganese ions.

The prevention of pollution of groundwater by coal mining effluent is of particular importance. Movement of pollutants through strata often is very slow and is difficult to detect. Hence, effective remedial action is either impractical or prohibitively expensive. Because of this there are few successful recoveries of polluted aquifers. Pollution of an aquifer by surface water can be alleviated by lining the beds of influent streams that flow across the aquifer or by providing pipelines to convey mine discharge to less sensitive water courses that do not flow across the aquifer concerned.

6.7   Heap leaching

Heap leaching involves low grade ores – such as finely disseminated gold deposits, from which the metal cannot be extracted by conventional methods – being placed on bases that have low permeability and then being sprayed with a solvent to extract the metal. For example, cyanide commonly is used to dissolve gold. The solution and dissolved metal are collected in a plastic lined pond and then treated to recover the metal. Heap leach projects can be developed on permanent bases, that is, the spent ore is left on the base after leaching, the latter acting as a liner. Conversely, if the leached ore is disposed of after the metal is extracted, then the base can be used again.

The chemistry of cyanide is complex and consequently numerous forms of cyanide are present in the leached heaps and associated tailings lagoons at gold mines. Toxicity primarily arises from free cyanide and generally metal-cyanide compounds are less toxic. Toxicity also is dependent upon the degree to which these compounds dissociate to release free cyanide. The different forms of cyanide vary widely in their rates of decay and their potential toxicity. Cyanide leaching of gold from ore usually is undertaken at a pH value of 10.3 or above so that most of the free cyanide in solution exists in the stable anion form, in this way reducing to a minimum the loss of cyanide by volatilization. When the waste is disposed of in a tailings lagoon, the pH value of the decant decreases to around 7, at which point most of the free cyanide occurs as hydrocyanic acid and is volatile. The greater the depth of water in the impoundment, the slower is the rate of loss of free cyanide by volatilization.

Cyanide forms compounds with many metals including cadmium, cobalt, copper, iron, mercury, nickel, silver and zinc. As such, these cyanide compounds occur in the waste and effluents at gold mines that use the heap leach process. Generally, the toxicity of these cyanide–metal compounds is related to their stability, in that the more stable the compound is, the less toxic it is, especially to aquatic life. For instance, stability of zinc cyanide is weak, copper cyanide and nickel cyanide are moderately strong and iron cyanide is very strong, the latter being more or less non-toxic. On the other hand, the weak and moderately stable metal–cyanide compounds can break down to form highly toxic free cyanide forms.

In addition to copper cyanide and zinc cyanide, the principal weak acid dissociable metal–cyanide species present in cyanide leach material are nickel, cadmium and silver cyanide. The breakdown of these metal–cyanide species in an impoundment involves dissociation of the cyanide ion and free metal ion. Hydrolysis of the cyanide ions then leads to the formation of hydrocyanic acid that subsequently is lost by volatilization from the water in the impoundment. The general rate of decay of cyanide in the water in an impoundment therefore depends upon the rate at which metal–cyanide species dissociate and the rate of volatilization of free cyanide.

6.8   Spontaneous combustion

All coals spontaneously ignite if the right conditions exist for the particular coal, that is, oxidation of coal exposed to air can lead to spontaneous combustion. Some coals ignite more easily than others, for example, high rank coals are less prone to spontaneous combustion than those of lower rank. The larger the surface area exposed, the greater is the opportunity for oxidation of the coal. If a coal seam has a lower than normal moisture content and then the moisture content rises, this leads to the liberation of heat. The likelihood of spontaneous combustion occurring increases as the temperature increases so that with increasing depth, and consequently increasing geothermal gradient, coal has a greater tendency to ignite. If the pyrite content in coal exceeds 2% this helps spontaneous combustion since the oxidation of pyrite is also an exothermic reaction.

When air gains access to shallow coal workings via surface fissures or due to partial collapse of the workings, conditions conducive to self-ignition may exist. Where the workings are old and abandoned, remaining coal within a worked seam may be highly fractured or weakened due to weathering and fine coal may be strewn in the roadways. Hence, a large surface area of coal is available for oxidation and, with the existence of a limited amount of air, the exothermic reaction produces a rise in temperature that eventually becomes self-generating. The flow of air through partially collapsed workings is unlikely to have a high enough velocity to convey away the heat generated. If such occurrences are not detected early and controlled adequately, large areas of coal can be destroyed by self-combustion and surface areas seriously affected. For instance, partially burnt pillars can collapse leading to subsidence and ground fissuring, which in old abandoned shallow mines can further accentuate the problem by allowing greater access of air to the workings. Air also can gain access to workings via the development of crown holes at the surface or via poorly sealed shafts. Gases such as steam, carbon dioxide, carbon monoxide and sulphur dioxide may escape from the fissures or crown holes.

If the coal seam affected by spontaneous combustion is at shallow depth, the spread of a fire may be limited by excavating trenches into the coal seam and then backfilling them. Old workings sometimes have been flooded to extinguish fires. This means that the water pumped into the area of the mine concerned must be impounded by dams or pumped into the mine more quickly than it can be discharged. The water must remain in place for a sufficient length of time to cool the coal and surrounding strata otherwise re-ignition will occur when the water level is lowered. However, neither of these techniques always proves successful. When grouting is used to extinguish a burning coal seam, drillholes initially are sunk to intercept the workings at the lowest area where burning occurs. Holes are drilled at closely spaced intervals and systematic filling takes place moving towards the higher levels of the mine. Inert gases such as nitrogen, carbon dioxide, steam and combustion by-products also have been used in attempts to extinguish mine fires, as have foams. As in the case of grouts, they are introduced into the mine via drillholes. Any fissures or crown holes at the surface that allow entry of air to a mine should be sealed prior to treatment. Thermocouples are used to determine the extent of the fire, its subsequent movement and the success of the treatment.

6.9   Gases (see also Section 3.9)

Methane and carbon dioxide are produced by the breakdown of organic matter and commonly are associated with coal bearing strata. Methane can be oxidized during migration to form carbon dioxide.

However, carbon dioxide can be generated both microbially and inorganically in a number of ways that do not involve methane. Gases may be dissolved in groundwater depending on the pressure, temperature or concentration of other gases or minerals in water. Dissolved gases may be advected by groundwater and only when the pressure is reduced and the solubility limit of the gas in water exceeded, do they come out of solution and form a separate gaseous phase. Such pressure release occurs when coal is removed during mining. It is essential that such degassing is not allowed to occur in confined spaces where an explosive mixture could develop. Gas problems are not present in every coal mine. Nonetheless, gas may accumulate in abandoned mines and methane, in particular, may escape from old workings via shafts, and via crown holes where the workings are at shallow depth.

6.10   Mineral dusts

As far as mineral dusts are concerned, they can take the form of fumes, particulates or fibres. Inhalation of mineral dusts can damage lung tissue, leading to illness and death. The diseases that result from occupational exposure to mineral dusts are referred to as pneumoconiosis. These diseases are caused by the accumulation of isometric dust particles in the lung. Particles are retained in the lung if they are small enough not to fall onto the wall of an airway before they reach the alveolar (gas exchange) part of the lung. Consequently, isometric particles are retained if they are smaller than about 5 μm in diameter and larger than about 0.5 μm. Isometric particles of the size that are retained in the lung are removed by scavenger cells known as macrophages. Unless the particles damage the cells, as happens in the case of quartz dust, the dust is removed either to the lymphatic system or up the airways leaving the lung undamaged.

A significant quantity of dust is required in the lungs to cause disease. For example, a coal miner needs to retain 100 g or more of dust, a slate quarryman around 10 to 15 g, and someone exposed to quartz dust around 5 g. The quantity of dust required depends partly on the amount of crystalline quartz it contains, on the surface area of the particles, and on the presence of other compounds such as iron oxides that may modify the effect of quartz. Nonetheless, dust related diseases can be prevented by measures that suppress dust (wet processing), by adequate ventilation, or by the use of masks that filter out dust.

Although exposure to quartz dust produces severe scarring of the lungs, that is, silicosis, it rarely causes death by itself. The most common cause of death among individuals who contract silicosis is tuberculosis. Unlike quartz dust, coal dust has very little effect on the macrophages (macrophages engulf the dust in the air spaces of the lungs, quartz is able to kill these cells without any damage to itself). Hence, macrophages can ingest large quantities of coal dust with very little deleterious effect. Other materials in coal dust such as mica, kaolinite and small amounts of quartz, may play a part in lungs that are so full of dust and engorged macrophages that the normal methods of clearance are overwhelmed.

Elongated particles and fibres have different aerodynamic properties from isometric particles, and tend to fall at a speed that is related to their cross-sectional diameter and is independent of their length. Length, however, becomes a limiting factor when the particles are longer than the diameter of the airways. Consequently, fibres are retained if they are less than 3 μm in diameter and less than 50 to 100 μm long. The lower limit of retention may be around 0.1 μm in diameter and 5 μm long. Unfortunately, elongated particles cannot be engulfed and removed by macrophages. Fibres exceeding 10 to 12 μm in length remain in the alveolar part of the lung. Smaller fibres move into the lung tissue and, although some may enter the lymphatic system, others reach the surface of the lung and the pleural space. In fact, the way in which particle shape influences clearance from the lung determines relative particle potency as regards causing disease and the location at which the disease develops.

Fibrous minerals like asbestos can cause extensive lung scarring (asbestosis), primary lung cancer and a cancer of the pleura or peritoneum called a mesothelioma. Severe disease can be associated with the retention of less than 1 g of asbestos, and mesothelioma with less than 1 mg. Amphibole fibres are straight and stiff, and can split along their longitudinal cleavage. They can have an extremely fine diameter. Of the common commercial types of asbestos, crocidolite (that has a diameter of < 0.1 μm) can penetrate the lung parenchyma through the conducting airways. Amphiboles do not undergo leaching or disintegrate in the tissue. The longer amphibole fibres become coated with an iron-containing protein complex to form asbestos bodies that probably are inert. It is probably the smaller uncoated fibres that cause the damage and may continue to do so for a lifetime. By contrast, although chrysotile [$Mg_3(OH)_4Si_2O_5$] has a diameter that is less than that of crocidolite [$Na_2Mg_3Fe_2Si_8O_{22}(OH)_2$], because of its coiled configuration, it presents a far larger aerodynamic profile and so most of these fibres tend to become caught and immobilized, and do not penetrate the smaller conducting airways. Consequently, only the shortest fibres can migrate through the lung and pleura. Once in the tissue, the magnesium in the fibres is leached out and the fibres eventually disintegrate.

6.11 Contamination due to mining

Many abandoned mine sites are heavily contaminated. For instance, crude metal extraction techniques used in the past meant that tailings contained high levels of associated metals. The large amounts of waste associated with mining represent a source of contamination of land and both ground and surface water. These wastes normally possess chemical and physical characteristics that prevent the re-establishment of plants without some form of prior remedial treatment.

Soil may be contaminated by mining waste from old workings. For example, lead contamination up to 522 mg kg^{-1} has been found in some garden soils in parts of the Tamar Valley in south west England and even higher levels occur around Ceredigion in Wales. Indeed, the prevalence of dental caries in children has been associated with high levels of lead in soil. However, a number of factors decrease the solubility of lead in waste. These include the mineral composition, the degree of encapsulation in pyrite of silicate matrices, the nature of the alteration rinds and the particle size. As such they decrease the bioavailability of lead in soils derived from or contaminated by mining wastes.

Gold mining in South Africa has meant that huge amounts of tailings have been impounded within tailings dams. Poor construction and management, in particular, of some of the older tailings dams resulted in seepage loss that adversely affected both ground and groundwater. Some tailings dams have been partially or totally reclaimed thereby leaving behind contaminated footprints. The topsoil in such areas has been highly acidified and contains heavy metals. As such, it poses a serious threat to the underlying dolostone aquifers. Soil management measures such as liming may be used to prevent the migration of contaminants from the topsoil into the subsoil and groundwater, and to aid the establishment of vegetation. Because of the cost, the removal of contaminated soil can only be undertaken in situations where small volumes are involved.

As mentioned above, acid mine drainage is responsible for contamination of ground, it giving rise to elevated levels of heavy metals in the drainage water, which pollutes natural waters and can be precipitated on or in sediments. The accumulation of zinc, copper, nickel, manganese, chromium and lead in material precipitated from streams polluted by acid mine drainage could be due to adsorption by iron and sulphate minerals, which are characteristically associated with acid mine drainage. For example, jarosite $KFe_3(OH)_6(SO_4)_2$ acts as an important trace element accumulator. The high concentrations of lead in some of the sediments concerned may be the result of sorption to goethite $FeO_2(OH)$. High concentrations of heavy metals in surface water and soils can decimate vegetation, and can seriously affect the health of animals.

6.12 Mining, dereliction and restoration (see also Chapter 10)

Some of the worst dereliction has been associated with past mineral workings and mining activities. Several industrial processes are associated with mining such as smelting and the production of coal gas, which also have an adverse impact upon the environment. Such derelict sites represent wasted resources, having a blighting effect on the surrounding area and can deter new development. Rehabilitation therefore is highly desirable. In addition, such a site may have been contaminated by the disposal of waste, and the ground may be disturbed by the presence of old building foundations and subsurface structures. Derelict sites may require varying amounts of filling, levelling and regrading.

Abandoned quarries, pits and spoil heaps are particularly conspicuous, and can be difficult to rehabilitate into the landscape. Like other forms of dereliction such features have a blighting affect on both the local environment and economy. Reclamation strategies for hard rock quarries have been surprisingly limited. In the past quarries have been screened from view by planting trees around them or, where conditions allow, used for waste disposal. Landform replication represents the best method of quarry reclamation and involves the construction of landforms and associated habitats similar to those of the surrounding environment. The extraction of sand and gravel deposits in low-lying areas along the flanks of rivers frequently means that the workings eventually extend beneath the water table. On restoration it is not necessary to fill the flooded pits completely. Partial filling and landscaping can convert such sites into recreational areas offering such facilities as sailing, fishing and other water sports. Abandoned clay pits frequently are used for disposal of domestic waste.

Restoration of opencast coal sites can begin before a site is closed, indeed this usually is the more convenient method. Worked-out areas behind the excavation front are filled with rock waste. This means that the final contours can be designed with less spoil movement than if the two operations were undertaken separately. Furthermore, more soil for spreading can be conserved when restoration and coal working are carried out simultaneously. Because of high stripping ratios (often 15:1 to 25:1), there usually is enough spoil to more or less fill the void. The restored land generally is used for agriculture or forestry but it can be used for country parks, golf courses, etc. Restoration of former open workings may involve a programme of regrading of slopes, with the addition of soil, so that trees can be planted. Spoil and smelter waste can be subjected to large-scale earthmoving to produce contoured landforms that blend more readily into the surroundings. Recreational and amenity areas may be created, and areas of interesting industrial archaeology can be turned into tourist attractions (Fig. 6.16).

An illustration of those industries associated with mining and their impact on the environment can be provided by former manufactured gas plants (FMGPs). The latter roasted coal to drive off gas and in

Fig. 6.16 A golf course constructed over smelter waste (note the black bunkers) in the centre of the photograph. The chimney in the upper part was for the smelter, note the waste heaps to the right. It and the ruins in the bottom left, which were the refining works, are sites of industrial archaeology, Anaconda, Montana. Anaconda in the late nineteenth and early twentieth century was one of the largest producers of copper in the world.

the process produced toxic wastes, notably tar. Unfortunately, most tar residuals and gas oils are highly resistant to natural degradation or attenuation in the environment and therefore potential problems associated with tar could persist for centuries. The polycyclic aromatic hydrocarbons (PAHs) are of particular concern as they are suspected of being carcinogenic. The predominant semi-volatile organic compounds (SVOCs) associated with gas manufacture are highly viscous and have a low solubility in water, and so come to rest in the vadose zone. In addition, a gas manufacturing site produced significant amounts of solid, as well as liquid waste. For instance, three tonnes of brick were removed from, and replaced, at each generator set per year. Other waste solids included ash, clinker, slag, scurf (hard carbon deposits formed on the interior surfaces of retorts and generators), spent lime, spent wood chips, spent iron spirals (for capturing sulphur) and retort and bench fragments. The solid material, some of which may be contaminated, was disposed of in dumps, which usually were located around the periphery of the plant, along an adjacent stream or in topographic hollows. Because dumps had high void ratios, toxic wastes and sludges may have been disposed within them. Adjacent low land often was used as unlined tar ponds. Once in the ground most manufactured gas wastes become immobile. Such sites are among the most demanding in terms of reclamation because they normally are severely contaminated. Contaminated material may have to be removed, which is expensive, or be encapsulated by a specially designed cover system, composed of clay reinforced with geogrid, with its impermeability enhanced by the inclusion of geomembranes. When such sites are redeveloped they often need to be compacted. This can be brought about by vibrocompaction or dynamic compaction (see Chapter 11) prior to the placement of the cover system if the latter is used.

7

Waste, Contamination and the Environment

7.1   Introduction

With increasing industrialization, technical development and economic growth, the quantity of waste produced by society has increased immensely. Many types of waste material are produced of which domestic waste, commercial waste, industrial waste, radioactive waste, and mining waste (see Chapter 6) are probably the most notable. Over and above this, waste can be regarded as non-hazardous or hazardous, the latter posing health risks and environmental problems if not managed properly. Waste may take the form of solids, sludges, liquids, gases or any combination of these. Furthermore, deposited waste can undergo changes that result in dangerous substances being produced.

Since waste products differ considerably from one another, the storage facilities they require also differ. Highly toxic, non-degradable wastes should be disposed of underground if they cannot be burnt in order to provide long-term isolation from the environment. Solid, unreactive, immobile inorganic wastes can be disposed of at above-ground disposal sites.

The best method of disposal is determined by the type and amount of waste on the one hand, and the geological and hydrogeological conditions of the waste disposal site on the other. Initially, a desk study is undertaken to help locate a disposal site. The primary task of the site exploration that follows is to determine the geological and hydrogeological conditions. Chemical analysis of the groundwater, together with mineralogical analysis of the rocks or soils at a potential site, may afford information about the future development of a site if used for disposal. At the same time the leaching capacity of the groundwater is determined, which allows prediction of reactions between wastes and soil or rock. If groundwater must be protected or highly mobile toxic or very slowly degradable substances are present in wastes, then impermeable liners should be used to inhibit infiltration of leachate into the surrounding ground.

7.2   Domestic refuse and sanitary landfills

Domestic refuse is a heterogeneous collection of almost anything (i.e. waste food, garden rubbish, paper, plastic, glass, rubber, cloth, ashes, building waste, metals, etc.), much of which is capable of reacting with water to give a liquid rich in organic matter, mineral salts and bacteria, referred to as leachate. Leachate is formed from rainfall which infiltrates a landfill and dissolves the soluble fraction of the waste, as well as from the soluble products formed as a result of the chemical and biochemical processes occurring within decaying wastes. The organic carbon content of waste is especially important since this influences the growth potential of pathogenic organisms. Any assessment of the state of a landfill and its environment must take into consideration the substances present in the landfill, their mobility now and in the future, the potential pathways along which pollutants can travel and the targets potentially at risk from the substances involved. One of the difficulties in predicting the effect of leachate on groundwater is the continual change in the characteristics of leachate as a landfill ages.

7.2.1   Leachate generation

Leachate is formed by the action of liquids, primarily water, within a landfill (Fig. 7.1). The generation of leachate occurs once the absorbent characteristics of the refuse are exceeded. The waste in a landfill

Fig. 7.1   Leachate seeping from a landfill near Durban, South Africa.

site generally has a variety of origins. Many of the organic components are biodegradable and, initially, decomposition of the waste is aerobic. Bacteria flourish in moist conditions and waste contains varying amounts of liquid, which may be increased by infiltration of precipitation. Once decomposition starts the oxygen in the waste rapidly becomes exhausted, and so the waste becomes anaerobic.

There are basically two processes by which anaerobic decomposition of organic waste takes place. Initially, complex organic materials are broken down into simpler organic substances, which are typified by various acids and alcohols. The nitrogen present in the original organic material tends to be converted into ammonium ions, which are readily soluble and may give rise to significant quantities of ammonia in the leachate. The reducing environment converts oxidized ions such as those in ferric salts to the ferrous state. Ferrous salts are more soluble and therefore iron is leached from a landfill. The sulphate in a landfill may be reduced biochemically to sulphides. Although this may lead to the production of small quantities of hydrogen sulphide, the sulphide tends to remain in the landfill as highly insoluble metal sulphides. In a young landfill the dissolved salt content may exceed 10 000 mg l^{-1}, with relatively high concentrations of sodium, calcium, chloride, sulphate or iron, whereas as a landfill ages the concentration of inorganic materials usually decreases (Table 7.1). Suspended particles may be present in leachate due to washout of fine material from a landfill. The second stage of anaerobic decomposition involves the formation of methane. In other words, methanogenic bacteria use the end products from the first stage of anaerobic decomposition to produce methane and carbon dioxide.

Physical and chemical processes are involved in the attenuation of leachate in soils. These include precipitation, ion exchange, adsorption or filtration. Insoluble heavy metal sulphides and soluble iron sulphide are formed where organic pollution is greatest, and ferric and inorganic hydroxides are precipitated between the oxidizing and reducing zones. The ion exchange and adsorption properties of a soil or rock mass primarily are attributable to the presence of clay minerals. The humic material in soil also has a high ion exchange capacity. Adsorption is susceptible to changes in pH value as the pH affects the surface charge on a colloid particle or molecule. Hence, at low pH values the removal rates due to adsorption can be reduced significantly. In many porous soils or rocks filtration of suspended matter occurs within short distances. On the other hand, a rock mass containing open discontinuities such as some limestones may transmit leachate for several kilometres. Pathogenic micro-organisms, which may

Table 7.1 Typical composition of leachates from recent and old domestic wastes.

Determinand	Leachate from recent wastes	Leachate from old wastes
pH	6.2	7.5
Chemical oxygen (mg l^{-1})	23800	1160
Biochemical oxygen (mg l^{-1})	11900	260
Total organic carbon (mg l^{-1})	8000	465
Fatty acids (mg m^{-1})	5688	5
Ammoniacal-N (mg l^{-1})	790	370
Orthophosphate (mg l^{-1})	0.73	1.4
Chloride (mg l^{-1})	1315	2080
Sodium (Na) (mg l^{-1})	960	1300
Magnesium (Mg) (mg l^{-1})	252	185
Potassium (K) (mg l^{-1})	780	590
Calcium (Ca) (mg l^{-1})	1820	250
Manganese (Mn) (mg l^{-1})	27	2.1
Iron (Fe) (mg l^{-1})	540	23
Nickel (Ni) (mg l^{-1})	0.6	0.1
Copper (Cu (mg l^{-1})	0.12	0.3
Zinc (Zn) (mg l^{-1})	21.5	0.4
Lead (Pb) (mg l^{-1})	8.4	0.14

be found in a landfill, do not usually travel far in the soil because of the changed environmental conditions. In fact, pathogenic bacteria normally are not present within a few tens of metres of a landfill.

7.2.2 Methanogenesis

The biochemical decomposition of domestic and other putrescible refuse in a landfill produces gas consisting primarily of methane (CH_4), with smaller amounts of carbon dioxide (CO_2) and volatile organic acids. In the initial weeks or months after placement, a landfill is aerobic and gas production is mainly CO_2, but it also contains oxygen (O_2) and nitrogen (N_2). As a landfill becomes anaerobic, the evolution of O_2 declines to almost zero and N_2 to less than 1%. The principal gases produced during the anaerobic stage are CO_2 and CH_4, with CH_4 production increasing slowly as methanogenic bacteria establish themselves. The factors that influence the rate at which gas is produced include the character of the waste, the moisture content, temperature and pH of a landfill. If toxic chemicals are present in a landfill, then bacterial activity, and especially methanogenesis, may be inhibited. A moisture content of 40% or higher is desirable for optimum gas production. Generally, the rate of gas production increases with increasing temperature. The pH value of a landfill should be around 7.0 for optimum production of gas, methanogenesis tending to cease below a pH of 6.2. Although the amount of gas generated by domestic refuse varies appreciably, between 2.2 and 250 litres per kilogram dry weight may be produced.

Methane production can constitute a dangerous hazard because methane is combustible, and in certain concentrations explosive (i.e. 5–15% by volume in air), as well as asphyxiating. Appropriate safety precautions must be taken during site operation. In many instances landfill gas is able to disperse safely to the atmosphere from the surface of a landfill. However, when a landfill is completely covered with a soil capping of low permeability, the potential for gas to migrate along unknown pathways increases and there are cases of hazard arising from methane migration. In such instances, proper closure of a landfill site can require gas management to control methane gas by venting to the atmosphere or for collection. Alternatively, a geocomposite of adequate transmissivity, along with a perforated pipe collection system, can be used for gathering the gas. Impermeable barriers (clay, bentonite, geomembranes or cement) can be used as gas cut-offs. An impermeable barrier should extend to the base of the fill or the water table, whichever is the higher. Unfortunately, there are cases on record of explosions occurring in buildings due to the ignition of accumulated methane derived from landfills near to or on which they were built. The source of such gas should be identified so that remedial action can be taken. Ideally, planners of residential developments should avoid landfill sites. Monitoring of landfill gas is an important aspect of safety. Instruments usually monitor methane, as this is the most important component of landfill gas.

7.2.3 Site selection

Selection of a landfill site for a particular waste or a mixture of wastes involves a consideration of economic and social factors, as well as geological and hydrogeological conditions. Consideration also needs to be given to the availability of construction materials and subsequent site rehabilitation. The ideal landfill site should be hydrogeologically acceptable, posing no potential threat to water quality; be free from running or static water; and have a sufficient store of soil material suitable for covering each individual layer of waste. It also should be situated at least 200 m away from any residential development.

When assessing the suitability of a site, two of the principal considerations are the ease with which any potential pollutant can be transmitted through the substrata and the distance it is likely to spread from the site. Consequently, the primary and secondary permeabilities of the formations underlying a possible landfill area are of major importance. Most argillaceous sedimentary, massive igneous and metamorphic rock formations have low permeabilities and therefore are likely to afford the most protection to water supply. By contrast, the least protection is provided by rock masses intersected by open discontinuities or in which solution features are developed. Sandy soils may act as filters leading to dilution and decontamination of leachate. Hence, sites for disposal of domestic refuse can be chosen where decontamination has the maximum chance of reaching completion and where groundwater sources are located far enough away to enable dilution to be effective.

The position of the water table is important as it determines whether wet or dry tipping will take place. Generally, unless waste is inert, wet tipping should be avoided. The position of the water table also determines the location at which flow is discharged. The hydraulic gradient determines the direction and velocity of the flow of leachate when it reaches the water table and also influences the amount of dilution that leachate undergoes. Aquifers that contain potable supplies of water must be protected. If near a proposed landfill site, a thorough hydrogeological investigation is necessary to ensure that site operations will not pollute aquifers. If pollution is a possibility, the site must be designed to provide some form of artificial protection, otherwise the proposal should be abandoned.

7.2.4 Landfill design

The design of a landfill site is influenced by the physical and biochemical properties of the wastes. The need for control of leachate production at a particular landfill is dependent on the extent of any possible pollution problems at that site. Site selection therefore has an important influence on the need for leachate control. Nonetheless, the use of leachate control and/or treatment methods may permit unsuitable sites to be used for the disposal of solid wastes. Leachate control should be planned before landfill development, especially if control techniques are to be installed beneath the waste. The most common means of controlling leachate is to minimize the amount of water infiltrating the site by encapsulating the waste in impermeable material. Hence, well designed landfills usually possess a cellular structure, as well as a lining and a cover, that is, the waste is contained within a series of cells formed of clay soil (Fig. 7.2). The cells are covered at the end of each working day with a layer of clay soil and compacted. Compaction is important since it reduces settlement and hydraulic conductivity, while

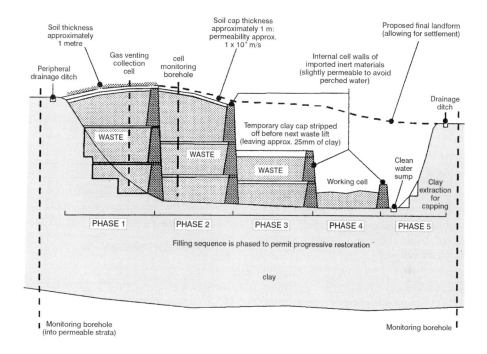

Fig. 7.2   Cellular construction of a landfill with kind permission of DEFRA, Crown Copyright.

increasing shear strength and bearing capacity. Furthermore, the smaller the quantity of air trapped within landfill waste, the lower is the potential for spontaneous combustion. Locally available soil can be mixed with the waste. As far as landfill stability is concerned, the potential for slope failure in a landfill is related to compaction control during disposal and the heavier the roller that is used for compaction the better. Even so, conventional compaction techniques do not always achieve effective results, especially with highly non-uniform waste. In such instances, dynamic compaction has been used with good results.

Since all natural materials possess some degree of permeability, total containment only can be achieved if an artificial lining is provided over the bottom of a site. However, there is no guarantee that clay, soil-cement, asphalt or plastic linings will remain impermeable permanently. Thus, the migration of materials from a landfill site into the substrata will occur eventually, although the length of time before this happens may be subject to uncertainty. In some instances the delay will be sufficiently long for the pollution potential of the leachate to be greatly diminished. One of the methods of tackling the problem of pollution associated with landfills is by dilution and dispersal of the leachate. Otherwise leachate can be collected by internal drains within the landfill and conveyed away for treatment.

Landfill liners are constructed from a wide variety of materials (Fig. 7.3). Adequate site preparation is necessary if a lining system is to perform satisfactorily. Nonetheless, no liner system, even if perfectly designed and constructed, will prevent all seepage losses. For instance, no liner, no matter how rigid or highly reinforced, can withstand large differential settlement without eventually leaking or possibly failing. Clay, bentonite, geomembrane, soil-cement, or bitumen-cement can be placed beneath a landfill to inhibit movement of leachate into the soil. Attempts should be made to ensure that the liner does not crack, or geomembranes tear. A double liner system is intended to prevent leakage. In this system the upper geomembrane acts as the primary liner for leachate collection at the base of the landfill. A perforated

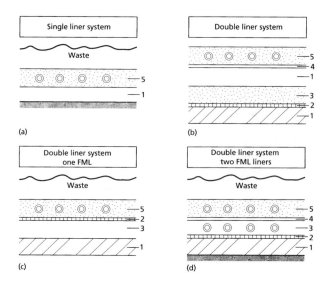

Fig. 7.3 Landfill liner systems. 1 = compacted low permeability clay; 2 = flexible membrane liner (FML); 3 = leachate collection/detection system; 4 = FML; 5 = primary leachate collection system. ◎ = collection pipes.

pipe collector system is located below the primary liner and is bedded within a crushed stone or sand drainage blanket. A secondary geomembrane liner occurs beneath the drainage blanket, and only functions if leakage from the primary liner occurs. Leachate drainage systems should be designed to collect the anticipated volume of leachate likely to be produced by the landfill at any time during its life. Drainage systems from a landfill commonly lead into a sump from which leachate can be conveyed for treatment.

The principal function of a cover is to minimize infiltration of precipitation into a landfill, and hence minimize the formation of leachate. The nature of the soil used for covering waste materials is very important. At many sites, however, the quality of borrow material is less than the ideal and so some blending with imported soil may be required. Cover systems may consist of a number of components as shown in Fig. 7.4. The key geotechnical factors in cover design are its stability and its resistance to cracking. Cracking of a clay cover may be brought about by desiccation or by the build-up of gas pressure beneath, if a venting system is not functional or not installed. It also may be difficult to maintain the integrity of the cover if large differential settlements occur in a landfill. A hydraulic conductivity of 10^{-9} m s^{-1} usually is specified for clay covers but perhaps never attained because of cracking.

The slope of the surface of the landfill influences infiltration. Water tends to collect on a flat surface and subsequently infiltrates, whereas it tends to run off steeper slopes. However, the possibility exists that surface run-off will erode the cover when the surface slope exceeds about 8%. Surface water should be collected in ditches and routed from the site.

The major function of all cut-off systems is to isolate wastes from the surrounding environment, so offering protection to soil and groundwater from contamination. Hence, the hydraulic and gas conductivities of cut-off systems are of paramount importance. Vertical cut-off walls may surround a site. Where an impermeable horizon exists at shallow depth beneath the site, the cut-off wall should be keyed into it. A very deep cut-off is required where no impermeable stratum exists. If leachate flows from an existing landfill, then the construction of a seepage cut-off system provides a solution. Steel sheet piles, geosynthetic sheet piles and secant piles are used as cut-off walls, as are grout curtains. The

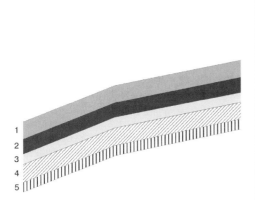

Layer	Description of layer	Typical materials	Typical thickness (m)
1	Surface layer	Topsoil; geosynthetic erosion control layer; cobbles; paving material	0.15
2	Protection layer	Soil; recycled or reused waste material; cobbles	0.3-1
3	Drainage layer	Sand or gravel; geonet or geocomposite	0.3
4	Barrier layer	Compacted clay; geomembrane; geosynthetic clay liner; waste material; asphalt	0.3-1
5	Gas collection layer and/or foundation layer	Sand or gravel; soil; geonet or geotextile; recycled or reused waste material	0.3

Fig. 7.4 Basic components of a landfill cap or cover.

latter are more likely to be constructed by jet grouting than by injection. Barriers also can be constructed as trenches excavated under bentonite slurry and filled with soil-bentonite or cement-bentonite. Plastic concrete cut-off walls are similar to cement-bentonite cut-offs except that they contain aggregate. A geomembrane sheeting-enclosed cut-off wall consists of a geomembrane that is fabricated to form a U-shaped envelope that fits the dimensions of a slurry trench. Ballast is placed within the geomembrane in order to sink it into the trench, after which the envelope is filled with wet sand. Monitoring wells can be placed in the sand.

7.2.5 Effect of leachate on water

Leachates contain many contaminants that may have a deleterious effect on surface water. If, for example, leachate enters a river, then oxygen is removed from the river by bacteria as they break down the organic compounds in the leachate. In cases of severe organic pollution, the river may be completely denuded of oxygen with disastrous effects on aquatic life. The principal inorganic pollutants in leachate that may cause problems are ammonia, iron, heavy metals and, to a lesser extent, chloride, sulphate, phosphate and calcium. Ammonia can be present in landfill leachates up to several hundred milligrams per litre, whereas unpolluted rivers have a very low content of ammonia. Discharge of leachate high in ammonia into a river exerts an oxygen demand on the receiving water. In addition, ammonia is toxic to fish, and as ammonia is a fertilizer it may alter the ecology of the river. Leachate that contains ferrous iron is particularly objectionable in a river since ochreous deposits are formed by chemical or biochemical oxidation of the ferrous compounds to ferric compounds. The turbidity caused by the oxidation of ferrous iron can lower the amount of light and so reduce the number of flora and fauna. Heavy metals can be toxic to fish at relatively low concentrations. They also can affect the organisms on which fish feed.

Physically, leachate affects river quality in terms of suspended solids, colour, turbidity and temperature. Suspended solids, colour and turbidity reduce light intensity in a river. This can affect the food chain and the lack of photosynthetic activity by plants reduces oxygen replacement in a river. The suspended solids may settle on the bed of a river in significant quantities. This can destroy plant and animal life.

If leachate enters the phreatic zone, it mixes and moves with the groundwater. The organic carbon content in the leachate leads to an increase in the biochemical oxygen demand (BOD) in the groundwater that may increase the potential for reproduction of pathogenic organisms. If anaerobic conditions develop, then metals such as iron and manganese may dissolve in the water causing further problems. If the groundwater has a high buffering capacity, then the effects of mixing acidic or alkaline leachate with the groundwater are reduced. The worst situation occurs in discontinuous rock masses where groundwater movement is dominated by fissure flow, or where groundwater movement is slow in a shallow water table so that little dilution occurs. The velocity of groundwater flow is important in that a high velocity gives rise to more dispersion in the direction of flow and relatively less laterally so that the leachate plume forms a narrow cone in the direction of groundwater flow. Low groundwater velocity leads to a wider plume. Ideally, a groundwater monitoring programme should be established prior to tipping and continue for anything up to 20 years after completion of the site.

7.3   Hazardous wastes

A hazardous waste can be regarded as any waste or combination of wastes of inorganic or organic origin that, because of its quantity, concentration, physical, chemical, toxicological or persistency properties, may give rise to acute or chronic impacts on human health and/or the environment, when improperly treated, stored, transported or disposed of. Such waste can be generated from a wide range of commercial, industrial, agricultural and domestic activities and can take the form of liquid, sludge or solid. The characteristics of the waste not only influence its degree of hazard but also are of great importance in the choice of a safe and environmentally acceptable method of disposal. Hazardous wastes may involve one or more risks such as explosion, fire, infection, chemical instability or corrosion, or acute toxicity. An assessment of the risk posed to health and/or the environment by hazardous waste must take into consideration its biodegradability, persistency, bioaccumulation, concentration, volume production, dispersion and potential for leakage into the environment. Hazardous wastes therefore require special treatment and cannot be released into the environment, added to sewage or be stored in a situation that is either open to the air or from which aqueous leachate can emanate.

Assessment of the suitability of a site for hazardous waste disposal is a complex matter that involves the use of models to predict the chemical behaviour of the waste in the ground, and the potential for mobilization and migration in groundwater. The quantity of waste involved, the manner and conditions of use, and the susceptibility of people or other living things to a certain waste can be used to determine its degree of hazard. Hazard ratings can be categorized as extreme, high, moderate or low. On the one hand, waste that contains significant concentrations of extremely hazardous material, including certain carcinogens, teratogens, and infectious substances, is of primary concern. On the other, the low category of hazardous waste contains potentially hazardous constituents but in concentrations that represent only a limited threat to health or the environment. If the hazard rating is less than the low category, then the waste can be regarded as non-hazardous and disposed of as general waste.

The form and rates of release of waste to the environment, together with the time and place at which releases occur can be determined with reasonable degrees of confidence. To translate these results into a risk assessment requires a parallel prediction of the consequences of a release of waste. Risk associated with the disposal of wastes involves the probability of an event or process that leads to release occurring within a given time period, multiplied by the consequences of that release. Alternatively, the risk can be regarded as the probability that an individual or group will be exposed to a pollutant, multiplied by the probability that this exposure will give rise to a serious health effect. The risk can be calculated given adequate epidemiological data on the health effects of toxic substances.

The minimum requirements for the treatment and disposal of hazardous waste involve ensuring that certain classes of waste are not disposed of without pretreatment. The objective of treating a waste is to reduce or destroy the toxicity of the harmful components in order to minimize the impact on the environment. In addition, waste treatment can be used to recover materials during waste minimization programmes. The method of treatment chosen is influenced by the physical and chemical characteristics of the waste. Physical treatment methods are used to remove, separate and concentrate hazardous and toxic materials. Chemical treatment is used in the application of physical treatment methods and to lower the toxicity of a hazardous waste by changing its chemical nature. This may produce essentially non-hazardous materials. In biological treatments, microbial activity is used to reduce or destroy the toxicity of a waste. The principal objective of such processes as immobilization, solidification or encapsulation is to convert hazardous waste into an inert mass with very low leachability. Macro-encapsulation involves the containment of waste in drums or other approved containers within a reinforced concrete cell that is stored within a landfill. Incineration can be regarded as both a means of treatment or of disposal.

Safe disposal of hazardous waste is the ultimate objective of waste management, disposal being in a landfill, by burial, by incineration or marine disposal. When landfill is chosen as the disposal option, the capacity of the site to accept certain substances without exceeding a specified level of risk has to be considered. The capacity of a site to accept waste is influenced by the geological and hydrogeological conditions, the degree of hazard presented by the waste, the leachability of the waste and the design of the landfill. Certain hazardous wastes may be prohibited from disposal in a landfill, such as explosive wastes or flammable gases. Protection of groundwater from the disposal of toxic waste in landfills can be brought about by containment. A number of containment systems have been developed that isolate wastes and include compacted clay barriers, slurry trench cut-off walls, geomembrane walls, sheet piling, grout curtains and extraction wells. Monitoring of hazardous waste repositories forms an inherent part of the safety requirements governing their operational and post-operational periods.

Disposal of liquid hazardous waste has been undertaken by injection into deep wells located in rock masses below fresh water aquifers, thereby ensuring that contamination or pollution of groundwater supplies does not occur. In such instances, the waste generally is injected into a permeable bed of rock several hundreds or even thousands of metres below the surface, which is confined by relatively impervious formations. However, even where geological conditions are favourable for deep-well disposal, the space for waste disposal frequently is restricted and the potential injection zones usually are occupied by groundwater. Accordingly, any potential formation into which waste can be injected must possess sufficient porosity, permeability, volume and confinement to guarantee safe injection. The piezometric pressure in the injection zone influences the rate at which the reservoir can accept liquid waste. A further point to consider is that induced seismic activity has been associated with the disposal of fluids in deep wells. Two important factors relating to the cost of construction of a well are its depth and the ease with which it can be drilled.

Monitoring is especially important in deep-well disposal that involves toxic or hazardous materials. A system of observation wells sunk into the subsurface reservoir concerned, in the vicinity of the disposal well, allows the movement of liquid waste to be monitored. In addition, shallow wells sunk into fresh groundwater aquifers permit monitoring of water quality so that any upward migration of the waste can be noted readily. Effective monitoring requires that the geological and hydrogeological conditions are accurately evaluated and mapped before the disposal programme is started.

Low-level hazardous waste can be co-disposed, that is, mixed with very much larger quantities of 'non-hazardous' domestic waste and buried in suitable disused quarries or clay pits. The objectives of co-disposal are to absorb, dilute and neutralize liquids, and to provide a source of biodegradable materials to encourage microbial activity.

7.4 Radioactive waste

Low-level radioactive waste contains small amounts of radioactivity and so does not present a significant environmental hazard if properly dealt with. It can be disposed of safely by burying in carefully controlled and monitored sites where the hydrogeological and geological conditions severely limit the migration of radioactive material. The trend in low-level radioactive waste disposal in the United States is to provide concrete barriers in the disposal facility, which prevent the short-term release of contamination. Such disposal facilities include below-ground vaults, above-ground vaults, modular concrete canisters placed in trenches, earth-mounded concrete bunkers, augered holes and mined cavities.

High-level radioactive waste unfortunately cannot be made non-radioactive and so disposal has to take account of the continuing emission of radiation. Furthermore, as radioactive decay occurs, the resulting daughter product is chemically different from the parent product. The daughter product also may be radioactive but its decay mechanism may differ from that of the parent. This is of particular importance in the storage and disposal of radioactive material. The half-life of a radioactive substance determines the time that a hazardous waste must be stored for its activity to be reduced by half. However, if a radioactive material decays to form another unstable isotope and if the half-life of the daughter product is long, as compared with that of the parent, then although the activity of the parent will decline with time that of the daughter will increase. Consequently, the storage time of these radioactive wastes also must consider the half-lives of the products that result from decay.

7.4.1 Disposal of radioactive waste

In general, two types of high-level radioactive waste are produced, namely, spent fuel rods from nuclear reactors and reprocessed waste. At present, both kinds of waste are adequately isolated from the environment in container systems. However, because much radioactive waste remains hazardous for hundreds or thousands of years it should be disposed of far from the surface environment and where it will require no monitoring. Ice sheets have been suggested as a repository for isolating high-level radioactive waste. The presumed advantages are disposal in a cold remote area, in a material that would entomb the wastes for many thousands of years. The high cost, adverse climate and uncertainties of ice dynamics are factors that do not favour such a means of disposal. Disposal on the deep sea bed is another suggestion and involves emplacement in sedimentary deposits at the bottom of the sea. Such deposits have sorptive capacity for many radionuclides that might leach from breached waste containers and if any radionuclides escaped they would be diluted by dispersal. Currently, however, disposal of radioactive waste beneath the sea floor is prohibited by international convention. Disposal of radioactive waste on a remote island involves the placement of wastes within deep stable geological formations. It also relies on the unique hydrological system associated with island geology. The rock melt concept involves the direct placement of liquids or slurries of high-level wastes or dissolved spent fuel in underground cavities. After evaporation of the slurry water, the heat from radioactive decay would melt the surrounding rock. In about 1000 years the waste rock mixture would resolidify, trapping the radioactive material in a relatively insoluble matrix deep underground. The rock melt concept, however, is suitable only for certain types of wastes. Moreover, since solidification takes about 1000 years, the waste is most mobile during the period of greatest fission product hazard.

Placement of encapsulated nuclear waste in drill holes as deep as 1000 m in stable rock formations cannot dispose of high volumes of waste. Similarly, injection of liquid waste into porous or fractured strata, at depths from 1000 to 5000 m, can only accommodate limited quantities of waste. Such waste must be suitably isolated by relatively impermeable overlying strata and relies on the dispersal and diffusion of the liquid waste through the host rock. The limits of diffusion need to be well defined. Alternatively, thick beds of shale, at depths between 300 and 500 m, can be fractured by high pressure injection and then waste, mixed with cement or clay grout, can be injected into the fractured shale and allowed to solidify in place. The fractures need to be produced parallel to the bedding planes. This requirement limits the depth of injection. The concept is applicable only to reprocessed wastes or to spent fuel that has been processed in liquid or slurry form.

The most favoured method, since it probably will give rise to the least problems subsequently, is disposal in chambers excavated deep within the Earth's crust in geologically acceptable conditions (Fig. 7.5). Deep disposal of high-level radioactive waste involves the multiple barrier concept that is based upon the principle that uncertainties in performance can be minimized by conservation in design. These barriers can include waste encapsulation, waste containers, engineered barriers such as backfills, and geological host rocks that are of low permeability. A deep disposal repository should consist of a large underground system located at least 200 m, and preferably 300 m or more, beneath the ground surface, in which there is a complex of horizontally connected tunnels for transportation, ventilation and emplacement of high-level radioactive waste. It also requires a series of inclined tunnels and vertical shafts to connect the repository with the surface. Ideally, the waste should be so well entombed that none will reappear at the surface or if it does so, in amounts minute enough to be acceptable.

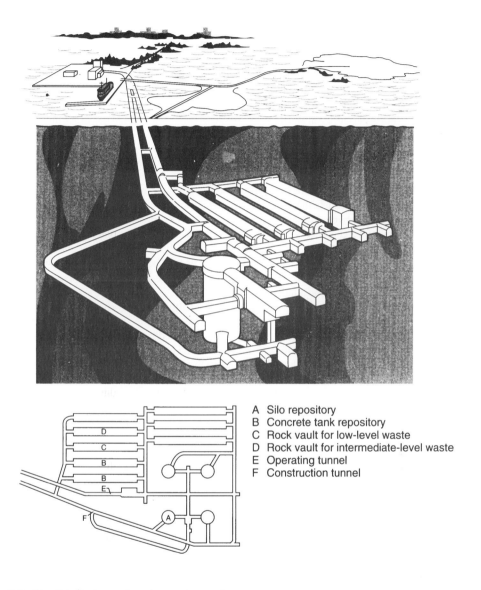

A Silo repository
B Concrete tank repository
C Rock vault for low-level waste
D Rock vault for intermediate-level waste
E Operating tunnel
F Construction tunnel

Fig. 7.5 Possible final repository for reactor waste.

The necessary safety of a permanent repository for radioactive wastes has to be demonstrated by a site analysis, which takes account of the site geology, the type of waste and their interrelationship. The site analysis must assess the thermo-mechanical load capacity of the host rock so that disposal strategies can be determined. It must determine the safe dimensions of an underground chamber and evaluate the barrier systems to be used. According to the multi-barrier concept, the geological setting for a waste repository must be able to make an appreciable contribution to the isolation of the waste over a long period of time. Hence, the geological and tectonic stability (e.g. mass movement or earthquakes), the load bearing capacity (e.g. settlement or cavern stability), geochemical and hydrogeological development (e.g. groundwater movement and potential for dissolution of rock) are important aspects of safety development. Disposal would be best undertaken in a geological environment with little or no groundwater circulation, as groundwater is the most probable means of moving waste from a repository to the biosphere. The geological system should be complemented by multiple engineered barriers such as a waste container, buffer materials and backfill. For example, radioactive material sealed in metal canisters could be stored in deep underground caverns excavated in relatively impermeable rock types in geologically stable areas (i.e., areas that do not experience volcanic activity) in which there is a minimum risk of seismic disturbance and that are not likely to undergo significant erosion. Deep structural basins are considered as possible locations.

7.4.2 Geological conditions and disposal

Stress redistribution due to subsurface excavation and possible thermally induced stresses should not endanger the state of equilibrium in the rock mass concerned and should not give rise to any unacceptable convergence or support damage during the operative period. The long-term integrity of the rock mass must be assured. Therefore, it is necessary to determine the distribution of stress and deformation in the host rock of the repository. This may involve consideration of the temperature-dependent rheological properties of the rock mass in order to compare them with its load bearing capacity. Ideally, substantial strength is necessary for engineering construction of subsurface repository facilities, especially in maintaining the integrity of underground openings.

Completely impermeable rock masses are unlikely to exist although many rock types may be regarded as practically impermeable, such as large igneous rock massifs, some thick sedimentary sequences, metamorphic rocks and rock salt. The permeability of a rock mass depends mainly on the discontinuities present, their width, amount of infill and their intersections. The repository needs to be watertight to prevent the transport of radionuclides by groundwater to the surface. Test pumping of the groundwater system, and sealing by injection techniques may be necessary.

Rock types such as thick deposits of salt or shale, or granites or basalts at depths of 300 to 500 m are regarded as the most feasible in which to excavate caverns for disposal of high-level radioactive waste. Once a repository is fully loaded, it can be backfilled, with the shafts being sealed to prevent the intrusion of water. Once sealed, the system can be regarded as isolated from the human environment. Thick deposits of salt have certain advantages. Salt has a high thermal conductivity and so rapidly dissipates any heat associated with high-level nuclear waste. It is 'plastic' at proposed repository depths so that any fractures that may develop due to construction operations will 'self-heal'. Salt also possesses gamma-ray protection similar to concrete, undergoes only minor changes when subjected to radioactivity, and tends not to provide paths of escape for fluids or gases. The attractive feature of deep salt deposits is the lack of water and the inability of water from an external source to move through the salt. These advantages may be compromised if the salt contains numerous interbedded clay or mudstone horizons, open cavities containing brine, or faults cutting the salt beds so providing conduits for external water. The solubility of salt requires that sources of unsaturated water are totally isolated from underground openings in beds of salt by watertight linings, isolation seals, cut-offs and/or by collection systems. If suitable precautions are not taken in more soluble horizons in salt, any dissolution that occurs can lead to the irregular development of a cavern being excavated. Any water that does accumulate in salt will be a concentrated brine that will be corrosive to metal canisters. The potential for heavy groundwater inflows during shaft sinking requires the use of grouting or ground freezing. Rock salt is a visco-plastic material that exhibits short-term and long-term creep. Hence, caverns in rock salt are subject to convergence as a result of plastic deformation of the salt. The rate of convergence increases with increasing temperature and stress in the surrounding rock mass. Since temperature and stress increase with depth, convergence also increases with depth. If creep deformation is not restrained or not compensated for by other means, excessive rock pressures can develop on a lining system that may approach full overburden pressure.

Not all shales are suitable for the excavation of underground caverns in that soft compacted shales present difficulties in terms of wall and roof stability. Caverns also may be subject to floor heave. Caverns could be excavated in competent cemented shales. Not only do these possess low permeability, they also could adsorb any ions likely to move through them. A possible disadvantage is that if temperatures in a cavern exceeded 100°C, then clay minerals could lose water and therefore shrink. This could lead to the development of fractures. In addition, the adsorption capacity of the shale would be reduced.

Granite is less easy to excavate than rock salt or shale but is less likely to offer problems of cavern support. It provides a more than adequate shield against radiation and will disperse any heat produced by radioactive waste. The quantity of groundwater in granite masses is small and its composition is generally non-corrosive. However, fissure and shear zones do occur within granites along which copious quantities of groundwater can flow. Discontinuities tend to close with depth and faults may be sealed. For location and design of a repository for spent nuclear fuel, the objective is to find 'solid blocks' that are large enough to host the tunnels and caverns of the repository. The Precambrian shields represent stable granite-gneiss regions.

The large thicknesses of lava flows in basalt plateaux mean that such successions also could be considered for disposal sites. Like granite, basalt also can act as a shield against radiation and can disperse heat. Frequently, the contact between flows is tight and little pyroclastic material is present. Joints may not be well developed at depth and strengthwise basalt should support a cavern. However, the durability of basalts on exposure may be suspect, which could give rise to spalling from the perimeter of a cavern. Furthermore, such basalt formations can be interrupted by feeder dykes and sills that may be associated with groundwater.

7.5 Contamination

In many of the industrialized countries of the world, one of the legacies of the past two centuries is the contamination of land and groundwater as a result of industry and society having disposed of their waste with little regard for future consequences. Contaminated land, because of its nature or former uses, is land that contains substances that, when present in sufficient quantities or concentrations, are likely to cause harm, directly or indirectly, to people and/or to the environment, or to give rise to hazards likely to affect a proposed form of development. The degree of hazard depends upon the mobility of the contaminant(s) within the ground and different types of soils have different degrees of reactivity to compounds that are introduced. The risk is influenced by the future use of a site. Hence, when contaminated sites are cleared for redevelopment they pose problems. Contamination can take many forms and can be variable in nature across a site.

Ideally, when contaminated land is redeveloped, the clean-up involved should attempt to return the land to a standard that would allow any future use of the site in question. However, the concept of total clean-up is a standard of excellence that in practice usually is cost prohibitive. Hence, standards of relevance become a necessary prerequisite in order to avoid negative land values that mean that remedial action would not take place. A more practical strategy is the 'suitable for use' approach to the control and treatment of contaminated land. This concept supports sustainable development by reducing damage from past activities and permits contaminated land to be kept in, or returned to, beneficial use wherever possible. Such an approach only requires remedial action where the contamination poses unacceptable risks to health or the environment and where there are appropriate cost-effective means available to do so. Guideline values for concentrations of contaminants in soil may be used to indicate that a possibility of significant harm exists. However, any guidelines should attempt to relate hazards to land uses, thereby differing from assessment systems based only on concentration limits. In other words, they should recognize lower thresholds for certain uses such as residential developments with gardens than for hard cover areas. When using guideline values, it must be demonstrated that the values or assumptions underlying them are appropriate for use, and characteristic of the land and the ecosystems in question. Even so, in any given circumstances the hazards that arise from contaminated land will be peculiar to the site and will differ in significance.

In fact, the presence of potentially harmful substances at a site may not require remedial action, if it can be demonstrated that they are inaccessible to living things or materials that may be detrimentally affected. However, consideration must always be given to the migration of soluble substances. The migration of soil-borne contaminants primarily is associated with groundwater movement, and the effectiveness of groundwater to transport contaminants is dependent largely upon their solubility. The quality of water can provide an indication of the mobility of contamination and the rate of dispersal. For example, in an alkaline environment, the solubility of heavy metals becomes mainly neutral due to the formation of insoluble hydroxides. Provided groundwater conditions remain substantially unchanged during the development of a site, then the principal agent likely to bring about migration will be percolating surface water. On many sites the risk of migration off-site is of a very low order because the compounds have low solubility, and frequently most of their potential for leaching has been exhausted. Liquid and gas contaminants, of course, may be mobile. Obviously, care must be taken on site during working operations to avoid the release of contained contaminants (e.g. liquors in buried tanks) into the soil. Where methane has been produced in significant quantities, it can be oxidized by bacteria as it migrates through the ground with the production of carbon dioxide. However, this process does not necessarily continue.

7.5.1 Investigation of contaminated sites

If any investigation of a site that is suspected of being contaminated is going to achieve its purpose, its objectives must be defined and the level of data required determined – the investigation being designed to meet the specific needs of the project concerned. This is of greater importance when related to potentially contaminated land than an ordinary site. Investigations of potentially contaminated sites should be approached in a staged manner. This allows for communication between interested parties, and helps minimize costs and delays by facilitating planning and progress of the investigation. After the completion of each stage, an assessment should be made of the degree of uncertainty and of acceptable risk in relation to the proposed new development. Such an assessment should be used to determine the necessity for and type of further investigation.

The first stage in any investigation of a site suspected of being contaminated is a desk study that provides data for the design of the subsequent ground investigation. The desk study should be supplemented by a site inspection. This often is referred to as a land quality appraisal. The desk study should identify past and present uses of the site, and the surrounding area, and the potential for and likely forms of contamination. The objectives of the desk study are to identify any hazards and the primary targets likely to be at risk; to provide data for health and safety precautions for the site investigation; and to identify any other factors that may act as constraints on development. Hence, the desk study should attempt to provide information on the layout of the site, including structures below ground; its physical features; the geology and hydrogeology of the site; the previous history of the site; the nature and quantities of materials to be handled and the processing involved; health and safety records; and methods of waste disposal. It should allow a preliminary risk assessment to be made and the need for further investigation to be established.

The preliminary investigation should formulate objectives so that the work is cost effective. Such an investigation provides the data for planning the main field investigation, which includes the personnel and equipment needs, the sampling and analytical requirements, and the health and safety requirements. In other words, it is a fact-finding stage that should confirm the chief hazard and identify any additional ones so that the main investigation can be carried out effectively. The preliminary investigation also should refer to any short-term or emergency measures that are required on the site before the commencement of full-scale operations.

Just as a normal site investigation, one that is involved with the exploration for contamination needs to determine the nature of the ground. In addition, it needs to assess the ability of the ground to transmit any contaminants either laterally or upward by capillary action. Permeability testing therefore is required. The investigation must establish the location of any perched water tables and aquifers, and any linkages between them, as well as determining the chemistry of the groundwater on site. Investigation of contaminated sites frequently requires the use of a team of specialists and without expert interpretation many of the benefits of site investigation may be lost. Most sites require careful interpretation of the ground investigation data. An assessment should be made as to whether protective clothing should be worn by site operatives.

The exploratory methods used in the main site investigation can include manual excavation, trenching and the use of trial pits, light cable percussion boring, power auger drilling, rotary drilling, and water and gas surveys. Excavation of pits and trenches are widely used techniques for investigating contaminated land. Visually different materials should be placed in different stockpiles as a trench or pit is dug, or borehole advanced, to facilitate sampling. One of the factors that should be avoided is cross-contamination. This is the transfer of materials by the exploratory technique from one depth into a sample taken at a different depth. Consequently, a high specification of cleaning operation should be carried out on equipment between both exploratory and sampling locations.

Sampling procedures are of particular importance and the value of the data obtained from them is related to how representative the samples are. Some materials can change as a result of being disturbed when they are obtained or during handling. Hence, sampling procedure should take account of the areas of a site that require sampling; the pattern, depth, types and numbers of samples to be collected; their handling, transport and storage; as well as sample preparation and analytical methods. A sampling plan should specify the objectives of the investigation, as well as sample locations and frequency. Historical data should be used to ensure that any potential 'hot spots' are sampled satisfactorily. A sampling grid should be used in such areas.

Volatile contaminants or gas-producing material can be determined by sampling the soil atmosphere by using a hollow gas probe, inserted to the required depth. The probe is connected to a small vacuum pump and a flow of soil gas induced. A sample is recovered by using a syringe. Care must be taken to determine the presence of gas at different horizons by the use of sealed response zones. Analysis usually is conducted on site by portable gas chromatography or photo-ionization. Standpipes can be used to monitor gases during the exploration work.

To ensure that the site investigation is conducted in the manner intended and the correct data recorded, the work should be carefully specified in advance. As the investigation proceeds, it may become apparent that the distribution of material about a site is not as predicted by the desk study or preliminary investigation. Hence, the site investigation strategy would need to be adjusted. The data obtained during the investigation must be accurately recorded in a manner whereby it can be understood subsequently. The testing programme should identify the types, distribution and concentration (severity) of contaminants, and any significant variations or local anomalies. Comparisons with surrounding uncontaminated areas can be made.

Once completed, the site characterization process, when considered in conjunction with the development proposals, will enable the constraints on development to be identified. These constraints, however, cannot be based solely on the data obtained from the site investigation but must take account of financial and legal considerations. If hazard potential and associated risk are regarded as too high, then the development proposals will need to be reviewed. For instance, it recently was proposed to drive a sewer tunnel in part of Glasgow, Scotland, through superficial deposits at about 10 m below ground surface (Fig. 7.6). However, after it was discovered that the ground was contaminated, radical changes were proposed, it being decided to locate the tunnel in a sandstone at an average invert level of 11 m below rockhead, or about 22 m below ground surface. When the physical constraints and hazards have been assessed, then a remediation programme can be designed that allows the site to be economically and safely developed. It is at this stage when clean-up standards are specified in conjunction with the assessment of the contaminative regime of the surrounding area.

7.5.2 Remediation of contaminated sites

A wide range of technologies is available for the remediation of contaminated sites and the applicability of a particular method depends partly on the site conditions, the type and extent of contamination, and the extent of the remediation required. The effectiveness of the pollution control measures needs to be monitored throughout the remediation programme. In this way any parts of the site that fail to meet the

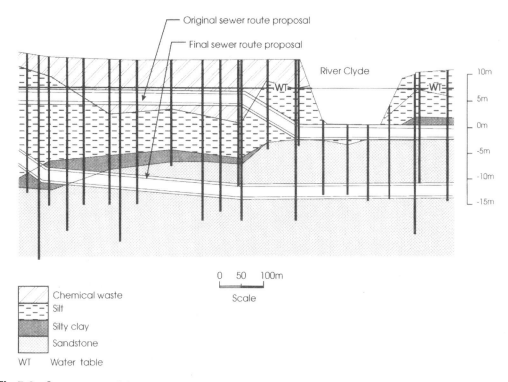

Fig.7.6 Cross-section of the planned route of a sewer tunnel in Glasgow, Scotland. It was relocated at greater depth because the site investigation revealed contaminated ground at the initial planned position. Boreholes shown as vertical black lines.

clean-up criteria can be dealt with immediately, so improving the construction schedule. Furthermore, a further sampling and testing programme is required in order to verify that the remediation operation has complied with the clean-up acceptance criteria for the site.

Most of the remediation technologies differ in their applicability to treat particular contaminants in the soil. Landfill disposal and containment are the exceptions in that they are capable of dealing with most soil contamination problems. Highly contaminated areas of a site may be covered with a drain and seal system, for example, consisting of geosynthetics and clay cover, to avoid immediate contact with contaminated soil, and to prevent the percolation of rain water through contaminated ground that would cause migration of contaminants into the saturated zone beneath. In addition, observation wells can be installed for sampling groundwater and can be used as recovery wells to extract and clean groundwater if it becomes contaminated subsequently. The rehabilitation of former gas manufacturing sites often involves excavation of contaminated areas and replacement by coarse material. However, the removal of contaminated soil from a site for disposal in a special landfill facility transfers the problem from one location to another. Hence, the in-place treatment of sites, where feasible, is the better course of action. Containment is used to isolate contaminated sites from the environment by the installation of a barrier system such as a cut-off wall.

Fixation or solidification processes reduce the availability of contaminants to the environment by chemical reactions with additives (fixation) or by changing the phase (e.g. liquid to solid) of the contaminant by mixing with another medium. Such processes usually are applied to concentrated sources of hazardous materials. Various cementing materials such as Portland cement and quicklime can be used to immobilize heavy metals.

Some contaminants can be removed from the soil by heating. For instance, soil can be heated to between 400° and 600°C to drive off or decompose organic contaminants such as hydrocarbons, solvents, volatile organic compounds and pesticides. Mobile units can be used on site, the soil being removed, treated and then returned as backfill.

Incineration, whereby soils are heated to between 1500° and 2000°C, is used for dealing with hazardous soils containing halogenated organic compounds such as polychlorinated biphenyls (PCBs) and pesticides that are difficult to remove by other techniques. Incineration involves removal of the soil and then it usually is crushed and screened to provide fine material for firing. The ash that remains may require additional treatment since heavy metal contamination may not have been removed by incineration. It then is disposed of in a landfill.

In situ vitrification transforms contaminated soil into a glassy mass. It involves electrodes being inserted around the contaminated area and sufficient electric current being supplied to melt the soil (the required temperatures can vary from 1600° to 2000°C). The volatile contaminants are either driven off or destroyed, while the non-volatile contaminants are encapsulated in the glassy mass when it solidifies.

Solvents can be used to remove contaminants from the soil. For example, soil flushing makes use of water, water-surfactant mixtures, acids, bases, chelating agents, oxidizing agents and reducing agents to extract semi-volatile organics, heavy metals and cyanide salts from the vadose zone of the soil. The technique is used in soils that are sufficiently permeable (not less than 10^{-7} m s^{-1}) to allow the solvent to permeate and the more homogeneous the soil is the better. The solvents are injected into the soil and the contaminated extractant removed by pumping.

Soil washing involves using particle size fractionation; aqueous-based systems employing some type of mechanical and/or chemical process; or counter current decantation with solvents for organic contaminants; and acids/bases or chelating agents for inorganic contaminants, to remove contaminants from excavated soils. Steam injection and stripping can be used to treat soils in the vadose zone contaminated with volatile compounds. One of the disadvantages is that some steam turns to water on cooling, which means that some contaminated water remains in the soil.

Vacuum extraction involves the removal of contaminants by the use of vacuum extraction wells. It can be applied to volatile organic compounds residing in the unsaturated soil or to volatile light non-aqueous phase liquids (LNAPLs) resting on the water table.

Bioremediation involves the use of microbial organisms to bring about the degradation or transformation of contaminants so that they become harmless. The micro-organisms involved in the process either occur naturally or are artificially introduced into the ground. In the case of the former, the microbial action is stimulated by optimizing the conditions necessary for growth. The principal use of bioremediation is in the degradation and destruction of organic contaminants, although it also has been used to convert some heavy metal compounds into less toxic states. Bioremediation can be carried out on ground *in situ* or ground can be removed for treatment.

7.5.3 Remediation of contaminated groundwater

Contaminated groundwater either can be treated *in situ* or it can be abstracted and treated. The solubility in water and volatility of contaminants influence the selection of the remedial technique used. Some organic liquids are only slightly soluble in water and are immiscible. These are known as non-aqueous phase liquids (NAPLs), when dense they are referred to as DNAPLs or 'sinkers' and when light as LNAPLs or 'floaters'. Examples of the former include many chlorinated hydrocarbons such as trichloroethylene and trichloethane; while petrol, diesel oil and paraffin provide examples of the latter. The permeability of the ground influences the rate at which contaminated groundwater moves, and therefore the ease and rate at which it can be extracted.

The pump-and-treat method is the most widely used means of remediation of contaminated groundwater. The water is abstracted from the aquifer concerned by wells, trenches or pits and treated at the surface. It is then injected back into the aquifer. The pump-and-treat method proves most successful when the contaminants are highly soluble and are not readily adsorbed by clay minerals in the ground.

Methods of treatment that are used to remove contaminants that are dissolved in water, once it has been abstracted, include standard water treatment techniques, air stripping of volatiles, carbon adsorption, microfiltration and bioremediation. In the case of LNAPLs, they can be separated from the groundwater either by using a skimming pump in a well or at the surface using oil–water separators. It usually does not prove possible to remove all light oil in this way so that other techniques may be required to treat the residual hydrocarbons. Oily substances and synthetic organic compounds normally are much more difficult to remove from an aquifer. In fact, the successful removal of DNAPLs is impossible at the present. As such, they can be dealt with by containment.

Active containment refers to the isolation or hydrodynamic control of contaminated groundwater. The process makes use of pumping and recharge systems to develop 'zones of stagnation' or to alter the flow pattern of the groundwater. Cut-off walls are used in passive containment to isolate the contaminated groundwater.

Air sparging is a type of *in situ* air stripping in which air is forced under pressure through an aquifer in order to remove volatile organic contaminants. It also enhances desorption and bioremediation of contaminants in saturated soils. The air is removed from the ground by soil venting systems. The injection points, especially where contamination occurs at shallow depth, are located beneath the area affected. The air that is vented may have to be collected for further treatment as it could be hazardous.

In situ bioremediation makes use of microbial activity to degrade organic contaminants in the groundwater so that they become non-toxic. Oxygen and nutrients are introduced into an aquifer to stimulate activity in aerobic bioremediation whereas methane and nutrients may be introduced in anaerobic bioremediation.

7.6   Contamination in estuaries

Estuaries often act as receptacles for contaminants as they frequently are the sites of harbours and industrial activity, with large centres of population. Hence, many estuaries in the industrialized world are accumulation sinks of contaminants and so have associated ecological problems. Sediments, in particular the fine and organic fractions, are regarded as carriers and possible sources of heavy metal and organic contaminants. Large-scale sediment transport patterns in a particular estuary may concentrate contaminants in specific areas not related to their point of introduction. Furthermore, sediment transport in estuaries may be inland or seawards due to variations in tidal and freshwater flow, and sediment may spend long periods stored in intertidal zones. In medium to high tidal range estuaries the sediment and associated contaminant patterns are further complicated by the existence of turbidity maxima. Turbidity maxima is the accumulation of fine muddy material in the upper reaches of estuaries around the limit of tidal influence.

When trace metals are released into the water column they can be transferred rapidly to the sediment phase by adsorption onto suspended particulate matter (SPM) followed by sedimentation. However, some heavy metals have a particular affinity to particulates whereas other heavy metals preferentially stay in solution. For example, lead has a strong affinity to particulates whereas zinc has no particular affinity and prefers to remain in solution. Intertidal flats may be considered as important trace metal sinks since they accumulate large amounts of SPM. In polluted estuaries the deposition of SPM on intertidal flats therefore may result in contaminated intertidal sediment. An understanding of the distribution and movement of contaminated SPM in estuaries and over intertidal flats is important in any determination of the transport and fate of particle-bound contaminants, and also can be relevant to dredging and spoil dumping.

The content of organic matter in estuarine sediments has the ability to concentrate trace metals in sediments. For instance, organic matter appears to have a strong association with zinc, lead, arsenic and copper, and a weaker association with chromium, tin and nickel. In addition, the water in estuaries surrounded by industrial areas is oxygen deficient and under anoxic conditions bacterial sulphate reduction allows dissolved metals in the water column to be deposited as sulphides.

8

Land Evaluation and Site Assessment

8.1 Introduction

Land evaluation and site assessment are undertaken to help determine the most suitable use of land in terms of planning and development or for construction purposes. In the process, the impact of a particular project on the environment may have to be determined, this is especially the case as far as large projects are concerned. Obviously, there has to be a geological input into these processes. The impact of land development usually is most notable in urban areas where the human pressures on land are greatest.

Investigations in relation to land-use planning and development obviously can take place at various scales, from regional, to local, to site investigations. Regional and local investigations generally are undertaken at state, county or municipal level, and may be involved with the location of routeways, the location and use of mineral resources, with problems due to previous land use or with land capability studies and zoning for future land use. Local site investigations tend to be undertaken for specific reasons such as the location of a suitable site for a landfill, of a site for the construction of a reservoir and dam, or the development of a gravel pit. In such cases investigations are necessary to obtain the relevant information (including geological) for the planning processes – which in many countries may include a public enquiry. Regional investigations frequently entail the production of engineering geomorphological and environmental geological maps, including hazard maps, with associated reports. Site investigations tend to involve the production of engineering geological (or geotechnical) maps and reports.

Any investigation begins with the formulation of aims, what does it want to achieve and which type of information is of relevance to the particular project in question? Once the pertinent questions have been posed, the nature of the investigation can be defined and the process of data collection can begin. The amount of detail required depends largely on the purpose of the investigation. For instance, less detail is required for a feasibility study for a project than is required by engineers for the design and construction of that project. Various methodologies are employed in data collection. These may include the use of remote sensing imagery, aerial photography, existing literature and maps, fieldwork and mapping, subsurface exploration by boring and drilling, sample collection, geophysical surveying and *in situ* testing. In some instances, geochemical data, notably when water or ground is polluted or contaminated, may need to be gathered, or monitoring programmes carried out. Once the relevant data has been obtained it must be interpreted and evaluated, and then, along with the conclusions, embodied in a report, which contains maps and/or plans. Geographical information systems (GIS) may be used to help process the data.

8.2 Remote sensing

Remote sensing imagery and aerial photographs have proved valuable aids in land evaluation, especially in those underdeveloped regions of the world where good topographic maps do not exist. They commonly represent one of the first stages of data collection in the process of land assessment. However, the amount of useful information obtainable from imagery and aerial photographs depends upon their characteristics, as well as the nature of the terrain they portray. Remote imagery and aerial photographs prove of most value during the planning and reconnaissance stages of a project. The information they provide can be transposed to a base map and this is checked during fieldwork. This information not only allows the fieldwork programme to be planned much more effectively, but it also helps to shorten the period spent in the field.

Remote sensing involves the identification and analysis of phenomena on the Earth's surface by using devices borne by aircraft or spacecraft. Most techniques used in remote sensing depend upon recording energy from part of the electromagnetic spectrum, ranging from gamma rays, through the visible spectrum to radar. The scanning equipment used measures both emitted and reflected radiation, the employment of suitable detectors and filters permitting the measurement of certain spectral bands. Signals from several bands of the spectrum can be recorded simultaneously by multi-spectral scanners.

8.2.1 Infrared linescanning

Infrared linescanning is dependent upon the fact that all objects emit electromagnetic radiation generated by the thermal activity of their component atoms. Emission is greatest in the infrared region of the electromagnetic spectrum for most materials at ambient temperature. It involves scanning a succession of parallel lines across the track of an aircraft with a scanning spot. The radiation is picked up by a detector that converts it to electrical signals that are transformed into visible light, thereby enabling a record to be made on film or magnetic tape. The data can be processed in colour, as well as black and white. Unfortunately, prints are increasingly distorted with increasing distance from the line of flight, which limits the total useful angle of scan to about 60° on either side. In order to reduce the distortion along the edges of the imagery, flight lines have a 50 to 60% overlap. Although temperature differences of 0.1°C can be recorded by infrared linescan, these do not represent differences in the absolute temperature of the ground but in emission of radiation. Careful calibration therefore is needed in order to obtain absolute values. Emitted radiation is determined by the temperature of the object and its emissivity, which can vary with surface roughness, soil type, moisture content and vegetative cover.

The use of infrared linescan depends on clear calm weather. Also, the time of the flight is important as thermal emissions vary significantly throughout the day. From the geological point of view, pre-dawn flying proves most suitable for thermal infrared linescan. This is because radiant temperatures are fairly constant and reflected energy is not important, whereas during a sunny day radiant and reflected energy are roughly equal so that the latter may obscure the former.

A grey scale can be used to interpret the imagery, it being produced by computer methods from linescan data that have been digitized. This enables maps of isoradiation contours to be produced. Colour enhancement also has been used to produce isotherm contour maps, with colours depicting each contour interval. Identification of grey tones is the most important aspect as far as the interpretation of thermal imagery is concerned, since these provide an indication of the radiant temperatures of a surface. Warm areas give rise to light tones, and cool areas to dark tones. Relatively cold areas are depicted as purple and relatively hot areas as red on a colour print. Thermal inertia is important in this respect since rocks with high thermal inertia, such as dolostone or quartzite, are relatively cool during the day and warm at night. Rocks and soils with low thermal inertia, for example, shale, gravel or sand, are warm during the day and cool at night. In other words, the variation in temperature of materials with high thermal inertia during the daily cycle is much less than those with low thermal inertia. Because clay soils possess relatively high thermal inertia they appear warm in pre-dawn imagery whereas sandy soils, because of their relatively low thermal inertia, appear cool.

The moisture content of soil influences the image produced, that is, soils that possess high moisture content may mask differences in soil types. Fault zones often are picked out because of their higher moisture content. Similarly, the presence of old landslides frequently can be detected due to their moisture content differing from that of their surroundings. Free-standing bodies of water usually are readily visible on thermal imagery, however, the high thermal inertia of highly saturated organic deposits may approach that of water masses, the two therefore may prove difficult to distinguish at times. Texture also can help interpretation. For instance, outcrops of rock may have a rough texture due to the presence of bedding or jointing, whereas soils usually give rise to a relatively smooth texture. However, where soil cover is less than 0.5 m, the rock structure usually is observable on the imagery since deeper moister soil occupying discontinuities gives a darker signature.

8.2.2 Side-looking airborne radar

In side-looking airborne radar (SLAR), short pulses of energy, in a selected part of the radar waveband, are transmitted sideways to the ground from antennae on both sides of an aircraft. The pulses of energy strike the ground along successive range lines and are reflected back at time intervals related to the height of the aircraft above the ground. The reflected pulses are transformed into black and white images. Returning pulses cannot be accepted from any point within 45° from the vertical so that there is a blank space under the aircraft along its line of flight. Also, the image becomes increasingly distorted towards the track of the aircraft. The belt covered by normal SLAR imagery varies from 2 to 50 km and, although the scanning is oblique, the system converts it to an image that is more or less planimetric.

Although variations in vegetation produce slightly different radar responses, a SLAR image depicts the ground devoid of vegetation. Displacements of relief are to the side towards the imaging aircraft, and radar shadows fall away from the flight line and are normal to it. The shadows form black areas that yield no information. However, because the wavelengths used in SLAR are not affected by cloud cover, imagery can be obtained at any time. This is particularly important in equatorial regions, which are rarely free of cloud. Imagery recorded by radar systems can provide appreciable detail of landforms as they are revealed due to the low angle of incident illumination. In addition, lateral overlap of radar cover can give a stereoscopic image, which offers a more reliable assessment of the terrain. Mosaics are suitable for the identification of regional geological features and for preliminary identification of terrain units.

8.2.3 Satellite imagery

In many parts of the world a LANDSAT image may provide the only form of base map available. The large areas of the ground surface that satellite images cover give a regional physiographic setting and permit the distinction of various landforms according to their characteristic photo-patterns. Accordingly, such imagery can provide a geomorphological framework from which a study of the component landforms is possible. The character of the landforms may afford some indication of the type of material of which they are composed and geomorphological data aid the selection of favourable sites for investigation on larger scale aerial surveys. Small scale imagery may enable regional geological relationships and structures to be identified that are not noticeable on larger scale imagery or mosaics.

Later generation LANDSAT satellites carry an improved imaging system called thematic mapper (TM), as well as a multispectral scanner (MSS). The TM is a cross-track scanner with an oscillating scan mirror and arrays of 16 detectors for each of the visible and reflected infrared bands. Thermatic mapper images have a spatial resolution of 30 m and excellent spectral resolution. Generally, TM bands are processed as normal and infrared colour images. Data gathered by LANDSAT TM are available as computer-compatible tapes or as CD-ROMS, which can be read and processed by computers. The weakest point in the system is the lack of adequate stereovision capability, however, a stereomate of a TM image can be produced with the help of a good digital elevation model. The French SPOT satellite is equipped with two sensor systems that cover adjacent paths, each with a swath width of 60 m. Potentially higher temporal resolution is provided by the sideways viewing option since the satellite can observe a location not directly under the orbital path. SPOT senses the terrain in a single wide panchromatic band and in three narrower spectral bands corresponding to the green, red and near infrared parts of the spectrum. The spatial resolution in the panchromatic mode is 10 m and the three spectral bands have a spatial resolution of 20 m. Images can be produced for stereoscopic purposes. Better resolution can be produced by the newest generation of high-resolution scanners on satellites (pixel sizes are down to 1 m). Radar satellite images are available from the European ERS-1 and the Japanese JERS.

The images are reproduced for four spectral bands plus two false colour composites. The infrared band is probably the best for geological purposes. Because separate images within different wavelengths are recorded at the same time, the likelihood of recognizing different phenomena is enhanced significantly. Since the energy emitted and reflected from objects commonly varies according to wavelength, its characteristic spectral pattern or signature in an image is determined by the amount of energy transmitted to the sensor within the wavelength range in which that sensor operates. As a consequence, a unique tonal signature frequently may be identified for a feature if the energy that is being emitted and/or reflected from it is broken into specially selected wavelength bands. Differentiation can be made between rock types if reflected energy from the shorter and longer ends of the visible spectrum is recorded separately. The ability to distinguish between different materials increases when imagery is recorded by different sensors outside the visible spectrum, the spectral characteristics then being influenced by the atomic composition and molecular structure of the materials concerned.

Satellite images may be interpreted in a similar manner to aerial photographs, although the images do not come in stereopairs. Nevertheless, a pseudostereoscopic effect may be obtained by viewing two different spectral bands (band-lap stereo) of the same image or by examining images of the same view taken at different times (time-lap stereo). There is also a certain amount of side-lap, which improves with latitude. This provides a true stereographic image across a restricted strip of a print, however, significant effects are only produced by large relief features. Interpretation of satellite data also may be accomplished by automated methods using digital data directly or by using interactive computer facilities with visual display devices.

The value of space imagery is important where existing map coverage is inadequate. For example, it can be of use for the preparation of maps of terrain classification, for engineering soil maps, for maps used for route selection, for inventories of construction materials, and for inventories of drainage networks and catchment areas. A major construction project is governed by the terrain – optimum location requiring minimum disturbance of the environment. In order to assess the ground conditions it is necessary to make a detailed study of all the photo-pattern elements that comprise the landforms on the satellite imagery. Important evidence relating to soil types, or surface or subsurface conditions may be provided by erosion patterns, drainage characteristics or vegetative cover. Engineering soil maps frequently are prepared on a regional basis for both planning and location purposes in order to minimize construction costs, the soils being delineated for the landforms within the regional physiographic setting.

8.3 Aerial photographs and photogeology

Generally, aerial photographs are taken from an aeroplane that is flying at an altitude of between 800 and 9000 m, the height being governed by the amount of detail that is required. Photographs may be taken at different angles ranging from vertical to low oblique (excluding horizon) to high oblique (including horizon). Vertical photographs, however, are the most relevant for photogeological purposes. Oblique photographs occasionally have been used for survey purposes but, because their scale of distortion from foreground to background is appreciable, they are not really suitable. Nevertheless, because they offer a graphic visual image of the ground they constitute good illustrative material.

There are four main types of film used in normal aerial photography, namely, black and white, infrared monochrome, true colour and false colour. Black and white film is used for topographic survey work and for normal interpretation purposes. The other types of film are used for special purposes. For example, infrared monochrome film makes use of the fact that near-infrared radiation is strongly absorbed by water. Accordingly, it is of particular value when mapping shorelines, the depth of shallow underwater features and the presence of water on land, as for instance, in channels, at shallow depths underground or beneath vegetation. Furthermore, it is more able to penetrate haze than conventional photography. True colour photography displays variation of hue, value and chroma, rather than tone only and generally offers much more refined imagery. As a consequence, colour photographs are better than black and white for photogeological interpretation, in that there are more subtle changes in colour in the former than in grey tones in the latter. Hence, they record more geological information. However, colour photographs are more expensive and it is difficult to reproduce slight variations in shade consistently in processing. Another disadvantage is the attenuation of colour in the atmosphere, with the blue end of the spectrum suffering a greater loss than the red end. Even so, at the altitudes at which photographs normally are taken the colour differentiation is reduced significantly. Obviously, true colour is primarily of value if it is closely related to the geology of the area shown on the photograph. False colour refers to infrared colour photography, and provides a more sensitive means of identifying exposures of bare grey rocks than any other type of film. Lineaments, variations in water content in soils and rocks, and changes in vegetation that may not be readily apparent on black and white photographs often are depicted clearly by false colour. A summary of the types of geological information that can be obtained from aerial photographs is given in Table 8.1.

Normally, vertical aerial photographs have 60% overlap on consecutive prints on the same run, and adjacent runs have a 20% overlap or side-lap. As a result of tilt (the angular divergence of the aircraft from a horizontal flight path) no photograph is ever exactly vertical but the deviation is almost invariably less than 1°. Scale distortion away from the centre of the photograph represents another source of error.

Table 8.1 Types of photogeological investigation.

Structural geology	Mapping and analysis of folding. Mapping of regional fault systems and recording any evidence of recent fault movements. Determination of the number and geometry of joint systems.
Rock types	Recognition of the main lithological types (crystalline and sedimentary rocks, unconsolidated deposits).
Soil surveys	Determining main soil type boundaries, relative permeabilities and cohesiveness, periglacial studies.
Topography	Determination of relief and landforms. Assessment of stability of slopes, detection of old landslides.
Stability	Slope instability (especially useful in detecting old failures which are difficult to appreciate on the ground) and rock fall areas, quick clays, loess, peat, mobile sand, soft ground, features associated with old mine workings.
Drainage	Outlining of catchment areas, steam divides, surface run-off characteristics, areas of subsurface drainage such as karstic areas, especially of cavernous limestone as illustrated by surface solution features; areas liable to flooding. Tracing swampy ground, perennial or intermittent streams, and dry valleys. Levées and meander migration. Flood control studies. Forecasting effect of proposed obstructions. Run-off characteristics. Shoals, shallow water, stream gradients and widths.
Erosion	Areas of wind, sheet and gully erosion, excessive deforestation, stripping for opencast work, coastal erosion.
Groundwater	Outcrops and structure of aquifers. Water-bearing sands and gravels. Seepages and springs, possible productive fracture zones. Sources of pollution. Possible recharge sites.
Reservoirs and dam sites	Geology of reservoir site, including surface permeability classification. Likely seepage problems. Limit of flooding and rough relative values of land to be submerged. Bedrock gulleys, faults and local fracture pattern. Abutment characteristics. Possible diversion routes. Ground needing clearing. Suitable areas for irrigation.
Materials	Location of sand and gravel, clay, rip-rap, borrow and quarry sites with access routes.
Routes	Avoidance of major obstacles and expensive land. Best graded alternatives and ground conditions. Sites for bridges. Pipe and power line reconnaissance. Best routes through urban areas.
Old mine workings	Detection of shafts and shallow abandoned workings, subsidence features.

8.3.1 Topographical interpretation

Examination of consecutive pairs of aerial photographs with a stereoscope allows observation of a three-dimensional image of the ground surface. The three-dimensional image means that heights can be determined and contours can be drawn, thereby producing topographic maps. However, the relief presented in this image is exaggerated so that slopes appear steeper than they actually are. Nonetheless, this helps detection of minor changes in slope and elevation. Unfortunately, exaggeration proves a disadvantage in mountainous areas, as it becomes difficult to distinguish between steep and very steep slopes.

Aerial photographs may be combined in order to cover larger regions. The simplest type of combination is the uncontrolled print lay-down that consists of photographs, laid along side each other, which have not been accurately fitted into a surveyed grid. Photomosaics represent a more elaborate type of print lay-down, requiring more care in their production, and controlled photomosaics are based on a number of geodetically surveyed points. They can be regarded as having the same accuracy as topographic maps.

A study of aerial photographs allows the area concerned to be divided into topographical and geological units. It also enables the geologist to plan fieldwork and to select locations for sampling, resulting in a shorter, more profitable period in the field. When a detailed interpretation of aerial photographs is required, the photographs can be enlarged up to approximately twice the scale of the final map to be produced.

8.3.2 Photogeological interpretation

The image perceived when stereopairs of aerial photographs are observed, represents a combination of variations in both relief and tone. However, relief and tone on aerial photographs are not absolute quantities for particular rock types. For instance, relief represents the relative resistance of rocks to erosion, as well as the amount of erosion that has occurred. Tone is important since small variations may be indicative of different types of rock. Unfortunately, tone is affected by light conditions, which vary with weather, time of day, season and processing. Nonetheless, basic intrusions normally produce darker tones than acid intrusions. Quartzite, quartz schist, limestone, chalk and sandstone tend to give light tones; while slates, micaceous schists, mudstones and shales give medium tones; and basalts, dolerites and amphibolites give dark tones. Regional geological structures frequently are easier to recognize on aerial photographs, which provide a broad synoptic view, than they are in the field.

Care must be exercised in the interpretation of the dip of strata from stereopairs of aerial photographs. For example, dips of 50° or 60° may appear almost vertical, and dips between 15° and 20° may look more like 45° because of vertical exaggeration. However, with practice dips can be estimated reliably in the ranges, less than 10°, 10–25°, 25–45°, and over 45°. Moreover, displacement of relief makes all vertical structures appear to dip towards the central or principal point of a photograph. Because relief displacement is much less in the central areas of photographs than at their edges, it is obviously wiser to use the central areas when estimating dips. It also must be borne in mind that the topographic slope need bear no relation to the dip of the strata composing the slope. However, scarp slopes do reflect the dip of rocks, and as dipping rocks cross interfluves and river valleys they produce crescent and V-shaped traces respectively. The pointed end of the V always indicates the direction of dip and the sharper the angle of the V, the shallower the dip. It may be possible to estimate the dip from bedding traces if there are no dip slopes. Vertical beds are independent of relief.

The axial trace of a fold can be plotted, and the direction and amount of its plunge can be assessed when the direction and amount of dip of the strata concerned can be estimated from aerial photographs. Steeply plunging folds have well-rounded noses and the bedding can be traced in a continuous curve. On the other hand, gently plunging folds occur as two bedding lineaments meeting at an acute angle (the nose) to form a single lineament. In addition, the presence of repeated folding sometimes may be recognized by plotting bedding plane traces on aerial photographs.

Lineaments are any alignment of features on an aerial photograph. The various types recognized include topographic, drainage, vegetative and colour alignments. Bedding is portrayed by lineaments that usually are few in number and occur in parallel groups. If a certain bed is more resistant than those flanking it, then it forms a clear topographic lineament. Even if bedding lineaments are interrupted by streams they usually are persistent and can be traced across the disruptive feature. Foliation may be indicated by lineaments. It often can be distinguished from bedding since parallel lineaments that represent foliation tend to be both numerous and impersistent.

Straight lineaments that appear as slight negative features on aerial photographs usually represent faults or master joints. In order to identify the presence of a fault there should be some evidence of movement. Usually, the termination or displacement of other structures provides such evidence. Faults may be less obvious in areas of thick soil or vegetation cover. Also, faults running parallel to the strike of strata may be difficult to recognize. Of course, joints show no evidence of displacement. Jointing patterns may assist the recognition of certain rock types, as for example, in limestone or granite terrains.

Dykes and veins also give rise to straight lineaments, which are at times indistinguishable from those produced by faults or major joints. If, however, dykes or veins are wide enough, they may offer a relief or tonal contrast with the country rock and then are distinctive. Basic dykes often are responsible for dark lineaments, and acid dykes and quartz veins for light coloured lineaments. Even so, because relative tone depends very much on the nature of the country rock, positive identification cannot be made from aerial photographs alone.

If the area portrayed by the aerial photographs is subject to active erosion, then it frequently is possible to differentiate between different rock masses, although it usually is not possible to identify the rock types. Normally, only general rather than specific rock types are recognizable from aerial photographs, for example, superficial deposits, sedimentary rocks, metamorphic rocks, intrusive rocks and extrusive rocks. Superficial deposits can be grouped into transported and residual categories. Transported superficial deposits can be recognized by their blanketing effect on the geology beneath; by their association with their mode of transport and with diagnostic landforms such as meander belts, river terraces, drumlins, eskers, sand dunes, etc. and their relatively sharp boundaries. Residual deposits generally do not blanket the underlying geology completely and in places there are gradational boundaries with rock outcrops. Obviously, no mode of transport can be recognized. It usually is possible to distinguish between metamorphosed and unmetamorphosed sediments, as metamorphism tends to make rocks more similar as far as resistance to erosion is concerned. Metamorphism also should be suspected when rocks are tightly folded and associated with multiple intrusions. By contrast, rocks that are horizontally bedded or gently folded, and are unaffected by igneous intrusions are unlikely to be metamorphic. As noted above, acid igneous rocks give rise to light tones on aerial photographs and they may display evidence of jointing. The recognition of volcanic cones indicates the presence of extrusive rocks.

8.4   Site investigation

The general objective of a site investigation is to determine the suitability of a site for a proposed development. It primarily involves gathering data on ground conditions at and below the surface, and is a prerequisite for the successful and economic design of engineering structures and earthworks. A site investigation should attempt to foresee and provide against difficulties that may arise during construction operations due to ground and/or other local conditions, and should continue after construction begins. It is essential that the assessment of ground conditions that constitute the basic design assumption is checked as construction proceeds and designs should be modified accordingly if conditions differ from those predicted. An investigation of a site for an important project requires both the exploration and sampling of strata likely to be affected. Data relating to the groundwater conditions, the discontinuity pattern in rock masses, the extent of weathering of rock masses, geohazards and past land use also are required for many projects.

8.4.1   Desk study and preliminary reconnaissance

The desk study normally is the first stage in a site investigation, the objective being to examine available records and data relevant to the area or site concerned to obtain a general idea of the existing geological conditions prior to a field investigation. In addition, a desk study can be undertaken in order to determine the factors that affect a proposed development for feasibility assessment and project planning purposes. The terms of reference for a desk study need to be defined clearly in advance of the commencement of the study. The effort expended in a desk study depends upon the type of project, the geotechnical complexity of the area or site, and the availability of relevant information.

A desk study for the planning stage of a project can encompass a range of appraisals from the preliminary rapid response to the comprehensive. Nonetheless, there are a number of common factors that need to be taken into account. These are summarized in Table 8.2, from which it can be concluded that an appraisal report typically includes a factual and interpretative description of the surface and geological conditions, information on previous site usage, a preliminary assessment of the suitability of the site for the planned development, an identification of any potential hazards, and provisional recommendations with regard to ground engineering aspects. However, a desk study should not be regarded as an alternative to a ground investigation for a construction project.

Detailed searches for information can be extremely time consuming and may not be justified for small schemes at sites where the ground conditions are relatively simple or well known. In such cases a study of the relevant topographical, and geological maps and memoirs, and possibly aerial photographs may suffice. On large projects literature and map surveys may save time, and thereby reduce the cost of the associated site exploration. The data obtained during such searches should help the planning of the site exploration and should prevent duplication of effort. A desk study also can reduce the risk of encountering unexpected ground conditions that could adversely affect the financial viability of a project.

The preliminary reconnaissance involves a walk over the site noting, where possible, the distribution of the soil and rock types present, the relief of the ground, the surface drainage and associated features, actual or likely landslip areas, ground cover and obstructions, earlier uses of the site such as tipping, or evidence of underground workings, etc. The inspection should not be restricted to the site but should examine adjacent areas to see how they affect or will be affected by development of the site in question. The importance of the preliminary investigation is that it should assess the suitability of the site for the proposed works and, if suitable, it helps form the basis for planning the site exploration. The preliminary reconnaissance also allows a check to be made on any conclusions reached by the desk study.

8.4.2 Site exploration

The aim of a site exploration is to try to determine, and thereby understand, the nature of the ground conditions on and surrounding a site. The extent to which this is carried out depends upon the complexity of the ground conditions, and the size and importance of the project. A report embodying the findings of an investigation can be used for design purposes and should contain geological plans and models of the site with accompanying sections, thereby conveying a three-dimensional picture of the subsurface strata.

The scale of any field mapping depends on the particular requirement, the complexity of the geology, and the staff and time available. Rock and soil types should be mapped according to their lithology and, if possible, presumed physical behaviour. Particular attention should be given to the nature of any superficial deposits and, where present, made-over ground. Geomorphological conditions, hydro-geological conditions, landslips, subsidences, borehole and field test information all can be recorded on maps.

Subsurface exploration can be carried out in a number of ways, that is, by digging pits and/or trenches, and by boring or drilling holes in soils or rocks respectively. Holes can be bored by hand or mechanically. In the case of boreholes or drill holes, there are no given rules regarding their location or the depth to which they should be sunk. This depends upon the geological conditions and the type of project concerned. The information provided by the preliminary reconnaissance and from any trial trenches aids the initial planning and layout of boreholes. Holes should be located so as to detect the geological sequence and structure. Obviously, the more complex this is, the greater the number of holes needed. In some instances, it may be as well to start with a widely spaced network of holes. As information is obtained, further holes

Table 8.2 Summary contents of engineering geological desk study appraisals.*

Item	Content and main points of relevance
Introduction	Statement of terms of reference and objectives, with indication of any limitations. Brief description of nature of project and specific ground-orientated proposals. Statement of sources of information on which appraisal is based.
Ground conditions	Description of relevant factual information.Identification of any major features which might influence scheme layout, planning or feasibility.
Site description and topography	Descriptions of existing surface conditions from study of topographic maps and actual photographs, and also from site walkover inspection (if possible).
Engineering history	Review of information on previous surface conditions and usage (if different from present) based on study of old maps, photographs, archival records and related to any present features observed during site walkover. Identification of features such as landfill zones, mine workings, pits and quarries, sources of contamination, old water courses, etc.
Engineering geology	Description of subsurface conditions, including any information on groundwater, from study of geological maps and memoirs, previous site investigation reports and any features or outcrops observed during site walkover. Identification of possible geological hazards, e.g. buried channels in alluvium, solution holes in chalk and limestone, swelling/shrinkable clays.
Provisional assessment of site suitability	Summary of main engineering elements of proposed scheme, as understood. Comments on suitability of site for proposed development, based on existing knowledge.
Provisional land classification	Where there is significant variation in ground conditions or assessed level of risk, subdivision of the site into zones of high and low risk, and any intermediate zones.Comparison of various risk zones with regard to the likely order of cost and scope of subsequent site exploration requirements, engineering implications, etc.
Provisional engineering comments	Statement of provisional engineering comments on such aspects as: foundation conditions and which method(s) appears most appropriate for structural foundations and ground slabs; road pavement subgrade conditions; drainage; excavatability of soils and rocks; suitability of local borrow materials for use in construction; slope stability considerations; nature and extent of any remedial works; temporary problems during construction
Recommendations for further work	Proposals for phased ground investigation, with objectives, requirements, and estimated budget costs.

*Herbert, S.M., Roche, D.P. and Card, G.B. (1987) The value of engineering geological desk study appraisals in scheme planning. In *Planning in Engineering Geology, Engineering Geology Special Publication No 4,* Culshaw, M.G., Bell, F.G., Cripps, J.C. and O'Hara, M. (Eds.), The Geological Society, London, 151–154. With permission of the Geological Society.

can be put down, if and where necessary. Exploration should be carried out to a depth that includes all strata likely to be affected significantly by structural loading.

The results from a borehole or drill hole should be documented on a log (Fig. 8.1). Apart from the basic information such as number, location, elevation, date, client, contractor, and engineer responsible,

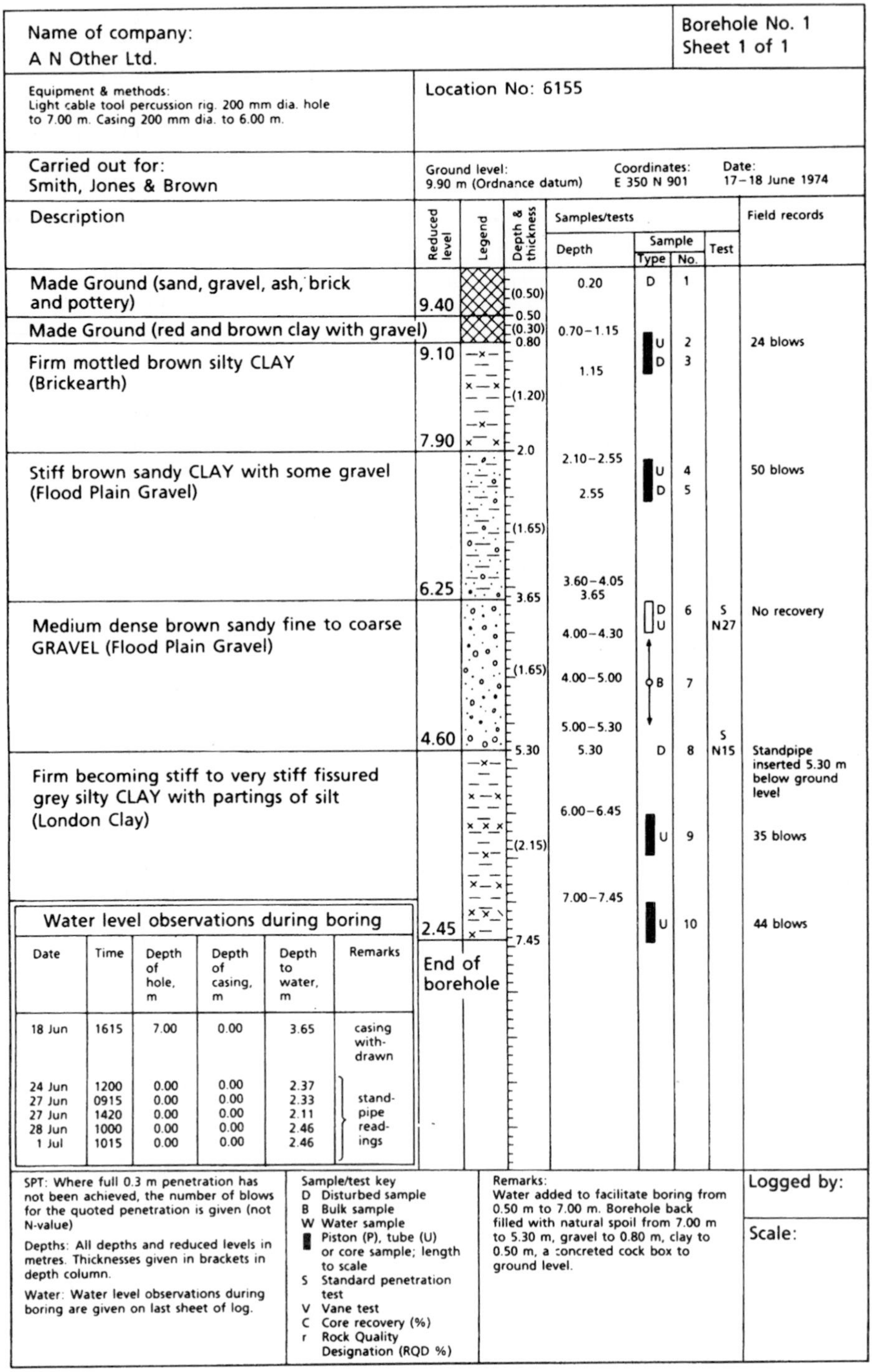

Name of company: A N Other Ltd.		Borehole No. 1 Sheet 1 of 1
Equipment & methods: Light cable tool percussion rig. 200 mm dia. hole to 7.00 m. Casing 200 mm dia. to 6.00 m.	Location No: 6155	
Carried out for: Smith, Jones & Brown	Ground level: 9.90 m (Ordnance datum) Coordinates: E 350 N 901 Date: 17–18 June 1974	

Description	Reduced level	Legend	Depth & thickness	Samples/tests				Field records
				Depth	Sample Type	No.	Test	
Made Ground (sand, gravel, ash, brick and pottery)	9.40		(0.50)	0.20	D	1		
Made Ground (red and brown clay with gravel)	9.10		0.50 (0.30) 0.80	0.70–1.15	U	2		24 blows
Firm mottled brown silty CLAY (Brickearth)	7.90		(1.20) 2.0	1.15	D	3		
Stiff brown sandy CLAY with some gravel (Flood Plain Gravel)	6.25		(1.65) 3.65	2.10–2.55 2.55 3.60–4.05 3.65	U D	4 5		50 blows
Medium dense brown sandy fine to coarse GRAVEL (Flood Plain Gravel)	4.60		(1.65) 5.30	4.00–4.30 4.00–5.00 5.00–5.30 5.30	D U B D	6 7 8	S N27 S N15	No recovery Standpipe inserted 5.30 m below ground level
Firm becoming stiff to very stiff fissured grey silty CLAY with partings of silt (London Clay)	2.45		(2.15) 7.45	6.00–6.45 7.00–7.45	U U	9 10		35 blows 44 blows

Water level observations during boring

Date	Time	Depth of hole, m	Depth of casing, m	Depth to water, m	Remarks
18 Jun	1615	7.00	0.00	3.65	casing with-drawn
24 Jun	1200	0.00	0.00	2.37	
27 Jun	0915	0.00	0.00	2.33	stand-pipe read-ings
27 Jun	1420	0.00	0.00	2.11	
28 Jun	1000	0.00	0.00	2.46	
1 Jul	1015	0.00	0.00	2.46	

End of borehole

| SPT: Where full 0.3 m penetration has not been achieved, the number of blows for the quoted penetration is given (not N-value)

Depths: All depths and reduced levels in metres. Thicknesses given in brackets in depth column.

Water: Water level observations during boring are given on last sheet of log. | Sample/test key
D Disturbed sample
B Bulk sample
W Water sample
▮ Piston (P), tube (U) or core sample; length to scale
S Standard penetration test
V Vane test
C Core recovery (%)
r Rock Quality Designation (RQD %) | Remarks:
Water added to facilitate boring from 0.50 m to 7.00 m. Borehole back filled with natural spoil from 7.00 m to 5.30 m, gravel to 0.80 m, clay to 0.50 m, a concreted cock box to ground level. | Logged by:

Scale: |

Fig. 8.1 A borehole log (with permission of the Geological Society).

the fundamental requirement of a borehole/drill hole log is to show how the sequence of strata changes with depth. Individual soil or rock types are presented in symbolic form on a log. The material recovered must be adequately described, and in the case of rocks frequently includes an assessment of the degree of weathering, fracture index, and relative strength (e.g. by point load testing core). The type of boring or drilling equipment should be recorded, the rate of progress made being a significant factor. The water level in the hole and any water loss, when it is used as a flush during rotary drilling, should be noted, as these reflect the mass permeability of the ground. If any *in situ* testing is done during boring or drilling operations, then the type(s) of test and the depth at which it/they were carried out must be recorded. The depths from which samples are taken also must be recorded.

Direct observation of strata, discontinuities and cavities in drill holes can be undertaken by cameras or closed-circuit television equipment, and drill holes can be viewed either radially or axially. Remote focusing for all heads and rotation of the radial head through 360° are controlled from the surface. Television heads have their own light source. However, if the drill hole is deflected from the vertical, variations in the distribution of light may result in some lack of picture definition.

The simplest method whereby data relating to subsurface conditions in soils can be obtained is by hand augering. This is used principally in fine-grained soils. Any soil samples that are obtained by augering are disturbed and invariably some amount of mixing of soil types occurs. Critical changes in the ground conditions are unlikely to be located accurately. Power augers are available as solid stem or hollow stem, both having an external continuous helical flight. The latter is used in those soils in which the borehole does not remain open. Solid stem augers are used in stiff clay soils that do not need casing. Disturbed samples taken from auger holes often are unreliable. In favourable ground conditions, such as firm and stiff homogeneous clay soils, auger rigs are capable of high output rates. The development of large earth augers have made it is possible to sink 1 m diameter boreholes in soils more economically than previously. The ground conditions can be inspected directly from such holes. Depending on the ground conditions, the boreholes may be unlined, lined with steel mesh or cased with steel pipe. In the latter case, windows are provided at certain levels for inspection and sampling.

Trenches and pits allow the ground conditions in soils and highly weathered rocks to be examined directly, although they are limited as far as their depth is concerned. Trenches, to a depth of some 5 m, can provide a flexible, rapid and economic method of obtaining information. Groundwater conditions and stability of the sides obviously influence whether or not they can be excavated, and safety must be observed at all times. This may necessitate shoring the sides. Pits are used when the initial subsurface survey has revealed any areas of special difficulty. The soil conditions in pits and trenches can be mapped and/or photographed, and undisturbed and/or disturbed samples collected.

The light cable and tool boring rig is used for investigating soils (Fig. 8.2). The hole is sunk by repeatedly dropping one of the tools into the ground. The basic tools are the shell and the claycutter, which are essentially open-ended steel tubes to which cutting shoes are attached, and are used in coarse-grained soils and fine-grained soils respectively. A power winch is used to lift the tool, suspended on a cable wire, and by releasing the clutch of the winch the tool drops and cuts into the soil. Once a hole is established it is lined with casing, the drop tool operating within the casing. This type of rig usually is capable of penetrating about 60 m of soil. Disturbed samples are obtained from the shell and the claycutter.

In the wash boring method the hole is advanced by the combined use of chopping bits and jetting the soil or weak rock, the cuttings thereby produced being washed from the hole by the water used for jetting. The method cannot be used for sampling and therefore its primary purpose is to sink the hole between sampling positions. When a sample is required, the bit is replaced by a sampler. Some indication of the type of ground penetrated may be obtained from the cuttings carried to the surface by the wash water, however, it may be difficult to identify strata with certainty. Wash boring may be used in cased and uncased holes, casing being used in coarse-grained soils to avoid collapse of the hole.

As far as soils are concerned, samples may be divided into disturbed and undisturbed types. An undisturbed sample can be regarded as one that is removed from its natural condition without disturbing its structure, density or moisture content. Undisturbed samples may be obtained by hand from surface

Fig. 8.2 A cable and tool rig.

exposures, pits and trenches. Careful hand trimming is used to produce a regular block. Block samples are waxed, together with reinforcing layers of thin cloth, before being placed in containers.

The fundamental requirement of any undisturbed sampling tool is that on being forced into the ground it should cause as little displacement of the soil as possible. The amount of displacement is influenced by a number of factors, such as the cutting edge of the sampler. A thin cutting edge and sampling tube minimizes displacement but is easily damaged, and it cannot be used in gravelly and hard soils. The standard sampling tube used with a light cable and tool rig for obtaining samples from fine-grained soils is referred to as the U100. It has a diameter of 100 mm, a length of approximately 450 mm and walls 1.2 mm thick (Fig. 8.3). The upper end of the tube is fastened to a check valve, which allows air or water to escape during driving, the cutting shoe being attached to the lower end. On withdrawal from

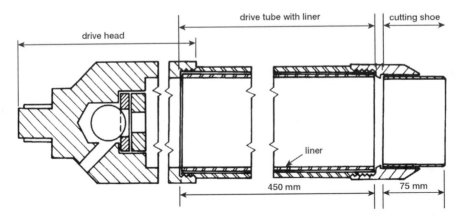

drive tube with liner

cutting shoe

drive head

liner

450 mm

75 mm

Fig. 8.3 A U100 sampling tube.

the borehole the check valve and cutting shoe are removed, and the sample is sealed in the tube with paraffin wax and end caps screwed on.

A thin-walled piston sampler is used for obtaining clay soils with a shear strength of less than 50 kPa, since soft clays tend to expand into the sample tube. Expansion is reduced by a piston in the sampler, the thin-walled tube being jacked down over a stationary internal piston, which when sampling is complete, is locked in place and the whole assembly then is pulled. Where continuous samples are required, particularly from rapidly varying or sensitive soils, a Delft sampler may be used. This can obtain a continuous sample from ground level to depths of about 20 m. The core is retained in a self-vulcanising sleeve as the sampler is continuously advanced into the soil.

Rotary-percussion rigs are designed for rapid drilling in rock. The rock is subjected to rapid high speed impacts whilst the bit rotates. The technique is most effective in brittle materials since it relies on chipping the rock. The rate at which drilling proceeds depends upon the type of rock, particularly its strength, hardness and fracture index; the type of drill and drill bit; the flushing medium (usually water or compressed air) and the pressures used; as well as the experience of the drilling crew. Drill flushings should be sampled at regular intervals, at changes in the physical appearance of the flushings and at significant changes in penetration rates. This method of drilling sometimes is used as a means of advancing a hole at low cost and high speed between intervals where core drilling is required.

For many exploration purposes a solid and as near as possible continuous rock core is required for examination and testing. In this case a rotary drilling machine is used with a coring bit attached to a core barrel. Both the bit and core barrel are attached by rods to the drill, by which they are rotated (Fig. 8.4). The flush is pumped through the drill rods and discharged at the bit. The flushing agent, usually water, serves to cool the bit and to remove the cuttings from the drillhole. The core is cut with the bit and housed in the core barrel. The bit is set with diamonds or tungsten carbide inserts (Fig. 8.5). Tungsten bits are not suitable for drilling in very hard rocks. Core bits vary in size and accordingly core sticks range between 17.5 and 165 mm diameter. Generally, the larger the bit, the better is the core recovery.

Fig. 8.4 A light drilling rig.

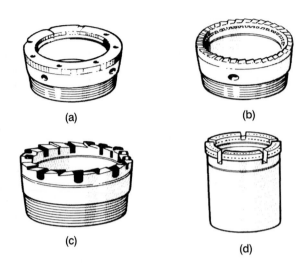

Fig. 8.5 Some common types of coring bits (a) surface set diamond bit (bottom discharge) (b) stepped sawtooth bit (c) tungsten carbide bit (d) impregnated diamond bit.

A variety of core barrels is available for rock sampling. The simplest type of core barrel is the single tube barrel but because it is suitable only for use in hard massive rocks, it is rarely used. In the single tube barrel, the barrel rotates the bit and the flush moves over the core. In double tube barrels the flush passes between the inner and outer tubes. Double tubes may be of the rigid or swivel type. The disadvantage of the rigid barrel is that both the inner and outer tubes rotate together and in soft rock this can break the core as it enters the inner tube. It is therefore only suitable for hard rock formations. In the double tube swivel-type core barrel the inner tube remains stationary (Fig. 8.6). It is suitable for use in medium and hard rocks, and gives improved core recovery in soft friable rocks. The face-ejection barrel is a variety of the double tube swivel-type in which the flushing fluid does not affect the end of the core. This type of barrel is a minimum requirement for coring badly shattered, weathered and soft rock formations. Triple tube barrels have been developed for obtaining cores from very soft rocks and from highly jointed and cleaved rock. Most rock cores should be removed by hydraulic extruders whilst the tube is held horizontal.

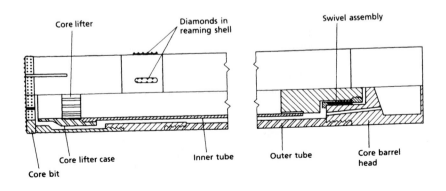

Fig. 8.6 Double tube swivel type core barrel.

The inner tube of a double core barrel can be lined with a plastic sleeve before drilling commences in order to reduce disturbance during extrusion.

If casing is used for diamond drilling operations, then it is drilled into the ground using a tungsten carbide or diamond tipped casing shoe with air, water or mud flush. The casing may be inserted down a hole drilled to a larger diameter to act as conductor casing when reducing and drilling ahead in a smaller diameter. Alternatively, it may be drilled or reamed in a larger diameter than the initial hole to allow continued drilling at the same diameter.

The weakest strata are generally of the greatest interest but these are the very materials that are most likely to be lost during drilling or to deteriorate after extraction. Therefore, the percentage core recovery in a lithological unit and the maximum intact core length should be recorded. Zones of core loss also should be identified. Shales and mudstones are particularly prone to deterioration once sampled and some may disintegrate completely if allowed to dry. Ideally, such material should be tested as soon as possible after recovery. Deterioration of suspect material may be reduced by wrapping cores with aluminium foil or plastic sheets. Core sticks can be photographed before they are removed from the site.

8.5 Geophysical exploration

A geophysical exploration can provide subsurface information over a large area. The information obtained may help eliminate less favourable alternative sites, may aid the location of test holes in critical areas and may prevent unnecessary repetitive boring or drilling in fairly uniform ground. A geophysical survey also may help detect variations in subsurface conditions between test holes, as these only provide information about the strata where they are sunk, telling nothing about the ground between. On the other hand, boreholes or drill holes are required to aid interpretation and correlation of geophysical results. Therefore, an appropriate combination of direct exploratory and geophysical methods often can yield a high standard of data.

The actual choice of method used for a particular survey may not be difficult to make. When dealing with layered rocks, provided their geological structure is not too complex, seismic methods have a distinct advantage in that they give more detailed, precise and unambiguous information than any other method. Electrical methods may be preferred for small-scale work where the geological structures are simple. Magnetic and gravimetric methods generally are used to delineate lateral changes or vertical structures. On occasions, more than one method may be used. Seismic methods also can be used to provide an assessment of the elastic properties of rock masses and their rock mass quality. Resistivity methods have been used in groundwater exploration.

8.5.1 Seismic methods

The sudden release of energy from the detonation of an explosive charge in the ground or the mechanical pounding of the surface generates shock waves that radiate out in a hemi-spherical wave front from the point of release. The waves generated are compressional (P), dilational shear (S) and surface waves. The velocities of the shock waves generally increase with depth below the surface since the elastic moduli increase with depth. The compressional waves travel faster, and are more easily generated and recorded than shear waves. They therefore are used almost exclusively in seismic exploration. The shock wave velocity depends on many variables, including rock fabric, mineralogy and pore water. In general, velocities in crystalline rocks are high to very high (Table 8.3). Velocities in sedimentary rocks increase concomitantly with consolidation and decrease in pore fluids, and with increase in the degree of cementation and diagenesis. Unconsolidated sedimentary deposits have maximum velocities varying as a function of the volume of voids, either air filled or water filled, mineralogy and grain size.

When seismic waves pass from one layer to another some energy is reflected back towards the surface whilst the remainder is refracted. Thus, two methods of seismic surveying can be distinguished, namely, seismic reflection and seismic refraction. Measurement of the time taken from the generation of the shock waves until they are recorded by detector arrays forms the basis of the two methods. The seismic reflection method is the most extensively used of all geophysical techniques, its principal employment

Table 8.3 Velocities of compressional waves of some common rocks.

V_p (km s^{-1})		V_p (km s^{-1})	
Igneous rocks		*Sedimentary rocks*	
Basalt	5.2–6.4	Gypsum	2.0–3.5
Dolerite	5.8–6.6	Limestone	2.8–7.0
Gabbro	6.5–6.7	Sandstone	1.4–4.4
Granite	5.5–6.1	Shale	2.1–4.4
Metamorphic rocks		*Unconsolidated deposits*	
Gneiss	3.7–7.0	Alluvium	0.3–0.6
Marble	3.7–6.9	Sands and gravels	0.3–1.8
Quartzite	5.6–6.1	Clay (wet)	1.5–2.0
Schist	3.5–5.7	Clay (sandy)	2.0–2.4

being in the oil industry. In this technique the depth of investigation is large compared with the distance from the shock source to detector array. This is to exclude refraction waves. In seismic refraction one ray approaches the interface between two rock types at a critical angle, which means that if the ray is passing from a low (V_0) to a high velocity (V_1) layer it will be refracted along the upper boundary of the latter layer. After refraction, the pulse travels along the interface with velocity V_1. The material at the boundary is subjected to oscillating stress from below. This generates new disturbances along the boundary that travel upwards through the low velocity rock and eventually reach the surface. At short distances from the point where the shock waves are generated the geophones record direct waves whilst at a critical distance both the direct and refracted waves arrive at the same time. Beyond this point, because the rays refracted along the high velocity layer travel faster than those through the low velocity layer above, they reach the geophones first. In refraction work the object is to develop a time–distance graph, which involves plotting arrival times against geophone spacing (Fig. 8.7). The depth (Z) to the high velocity layer, then can be obtained from the graph by using the expression:

$$Z = \frac{x}{2} \frac{\sqrt{(V_1 - V_0)}}{\sqrt{(V_1 + V_0)}} \tag{8.1}$$

where V_0 is the speed in the low velocity layer, V_1 is the speed in the high velocity layer and x is the critical distance. The method also works for multi-layered rock sequences if each layer is sufficiently thick and transmits seismic waves at higher speeds than the one above it. However, in the refraction method a low velocity layer underlying a high velocity layer usually cannot be detected as in such an inversion the pulse is refracted into the low velocity layer. Also a layer of intermediate velocity between an underlying refractor and overlying layers can be masked as a first arrival on the travel–time curve. The latter is known as a blind zone. The position of faults also can be estimated from the time–distance graphs. Thus, the distance between geophones, together with the total length and arrangement of the array, has to be carefully chosen to suit each particular situation.

8.5.2 Resistivity methods

In the resistivity method an electric current is introduced into the ground by means of two current electrodes and the potential difference between two potential electrodes is measured. Since most of the principal rock forming minerals are practically insulators, the resistivity of rocks and soils is determined

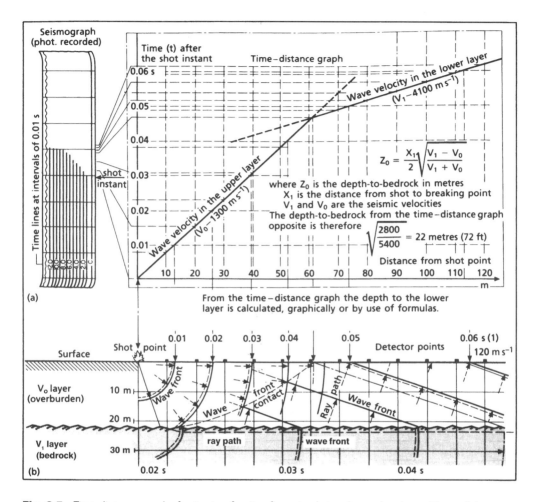

Fig. 8.7 Time-distance graph of seismic refraction for a simple two-layered system with parallel interfaces.

by the amount of conducting mineral constituents and the content of mineralized water in their pores. The latter condition is by far the dominant factor and, in fact, most rocks and soils conduct an electric current only because they contain water. The widely differing resistivity values of the various types of impregnating water can cause variations in the resistivity of rocks ranging from a few tenths of an ohm-metre to hundreds of ohm-metres (Ω-m), as can be seen from Table 8.4.

The resistivity method is based on the fact that any subsurface variation in conductivity alters the pattern of current flow in the ground and therefore changes the distribution of electric potential at the surface. Since the electrical resistivity of unconsolidated deposits and bedrock differ from each other, the resistivity method may be used in their detection, and to give their approximate thicknesses, relative positions and depths (Table 8.5). The first step in any resistivity survey involves conducting a resistivity depth sounding at the site of a borehole in order to establish a correlation between resistivity and lithological layers. If a correlation cannot be established, then an alternative method is required.

The electrodes are normally arranged along a straight line, the potential electrodes being placed

211

Table 8.4 Resistivity of some types of natural water.

Type of water	Resistivity in Ω-m
Meteoric water, derived from precipitation	30–1000
Surface waters, in districts of igneous rocks	30–500
Surface waters, in districts of sedimentary rocks	10–100
Groundwater, in areas of igneous rocks	30–150
Groundwater, in areas of sedimentary rocks	>1
Sea water	about 0.2

Table 8.5 Resistivity values of some common rock types.

Rock type	Resistivity (Ω-m)
Topsoil	5–50
Peat and clay	8–50
Clay, sand and gravel mixtures	40–250
Saturated sand and gravel	40–100
Moist to dry sand and gravel	100 –3 000
Mudstones, marls and shales	8–100
Sandstones and limestones	100 –1 000
Crystalline rocks	200 –10 000

inside the current electrodes and all four are symmetrically disposed with respect to the centre of the configuration. The configurations of the symmetric type that are used most frequently are those introduced by Wenner and by Schlumberger. Other configurations include the dipole-dipole and the pole-dipole arrays. In the Wenner configuration the distances between all four electrodes are equal (Fig. 8.8a). The spacings can be progressively increased, keeping the centre of the array fixed or the whole array, with fixed spacings, can be shifted along a given line. In the Schlumberger arrangement the potential electrodes maintain a constant separation about the centre of the station whilst if changes with depth are being investigated the current electrodes are moved outwards after each reading (Fig. 8.8b). The expressions used to compute the apparent resistivity (ρ_a) for the Wenner and Schlumberger configurations are as follows:

Wenner:

$$\rho_a = 2\pi aR \tag{8.2}$$

Schlumberger:

$$\rho_a = \pi \, \frac{(L^2 - l^2)}{2l} \, R \tag{8.3}$$

where a, L and l are explained in Figure 8.8 and R is the resistance reading.

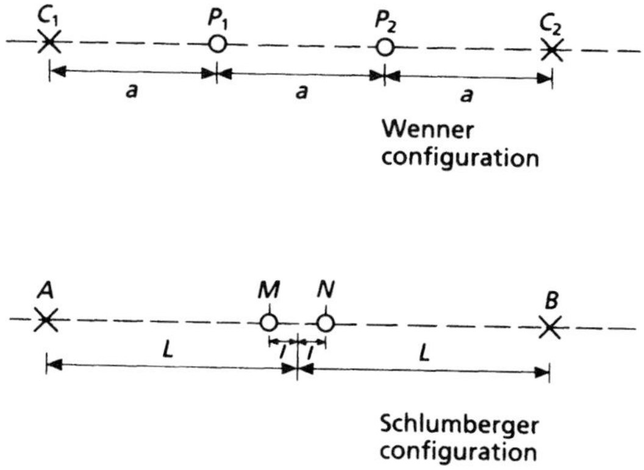

Fig. 8.8   The Wenner and Schlumberger configurations.

Horizontal profiling is used to determine variations in apparent resistivity in a horizontal direction at a pre-selected depth. For this purpose an electrode configuration, with fixed inter-electrode distances, is moved along a straight traverse, resistivity determinations being made at stations at regular intervals. The length of the electrode configuration must be carefully chosen because it is the dominating factor regarding depth penetration. The data of a constant separation survey, consisting of a series of traverses arranged in a grid pattern, may be used to construct a contour map of lines of equal resistivity. These maps often are extremely useful in locating areas of anomalous resistivity such as gravel pockets in clay soils and the trend of buried channels. Even so, interpretation of resistivity maps in terms of the delineation of lateral variations is mainly qualitative.

Electrical sounding furnishes information concerning the vertical succession of different conducting zones and their individual thicknesses and resistivities. In electrical sounding the midpoint of the electrode configuration is fixed at the observation station while the length of the configuration is increased gradually. As a result, the current penetrates deeper and deeper, the apparent resistivity being measured each time the current electrodes are moved outwards. The readings therefore become increasingly affected by the resistivity conditions at advancing depths. The Schlumberger configuration is preferable to the Wenner configuration for depth sounding. The data obtained usually is plotted as a graph of apparent resistivity against electrode separation, in the case of the Wenner array, or half the current electrode separation for the Schlumberger array. The electrode separation at which inflection points occur on the graph provide an idea of the depth of interfaces. The apparent resistivities of the different parts of the curve provide some idea of the relative resistivities of the layers concerned. Generally, it is not possible to determine the depths to more than three or four layers.

8.5.3   Electromagnetic methods

In most electromagnetic methods, electromagnetic energy is introduced into the ground by inductive coupling and is produced by passing an alternating current through a coil. The receiver also detects its signal by induction, for example, the terrain conductivity meter measures the conductivity of the ground by such an inductive method. The conductivity meter is carried along traverse lines across a site and can provide a direct continuous readout. Hence, surveys can be carried out rapidly. Conductivity values are taken at positions set out on a grid pattern. One of the disadvantages of the very low frequency (VLF) method is that wave penetration is limited. Hence, measurements of resistivity based on electromagnetic energy primarily are applied to soil cover and fill. Attenuation of electromagnetic energy is a problem in the electromagnetic pulse technique but it has been used with success to investigate very dry sites and low porosity rocks.

The ground-probing radar method is based upon the transmission of pulsed electromagnetic waves. In this method the travel times of the waves reflected from subsurface interfaces are recorded as they arrive at the surface and the depth (Z) to an interface is derived from:

$$Z = vt/2 \qquad (8.4)$$

where v is the velocity of the radar pulse and t is its travel time. The conductivity of the ground imposes the greatest limitation on the use of radar probing, that is, the depth to which radar energy can penetrate depends upon the effective conductivity of the strata being probed. This is governed chiefly by the water content and its salinity, of the strata. Furthermore, the value of effective conductivity is also a function of temperature and density, as well as the frequency of the electromagnetic waves being propagated. The least penetration occurs in saturated clayey materials or where the moisture content is saline. For example, attenuation of electromagnetic energy in wet clay and silt mean that depth of penetration frequently is less than 1 m. The technique appears to be reasonably successful in sandy soils and rocks in which the moisture content is non-saline. Rocks like limestone and granite can be penetrated for distances of tens of metres and in dry conditions the penetration may reach 100 m. Dry rock salt is radar-translucent, permitting penetration distances of hundreds of metres.

8.5.4 Magnetic methods

All rocks are magnetized to a lesser or greater extent by the Earth's magnetic field, so in magnetic prospecting accurate measurements are made of the anomalies produced in the local geomagnetic field by this magnetization. The intensity of magnetization and hence the amount by which the Earth's magnetic field is changed locally, depends on the magnetic susceptibility of the material concerned. In addition to the magnetism induced by the Earth's field, rocks possess a permanent magnetism that depends upon their history. The strength of the magnetic field is measured in nanoTeslas (nT). The average strength of the Earth's magnetic field is about 50 000 nT but the variations associated with magnetized rock formations are very much smaller than this.

Aeromagnetic surveying has almost completely supplanted ground surveys for regional reconnaissance purposes. Accurate identification of the plan position of the aircraft for the whole duration of the magnetometer record is essential. The object is to produce an aeromagnetic map, the base map with transcribed magnetic values being contoured at 5 to 10 nT intervals.

The aim of most ground surveys is to produce isomagnetic contour maps of anomalies to enable the form of the causative magnetized body to be estimated (Fig. 8.9). Profiles are surveyed across the trend of linear anomalies with stations, if necessary, at intervals of as little as 1 m. A base station is set up beyond the anomaly where the geomagnetic field is uniform. The reading at the base station is taken as zero and all subsequent readings are expressed as plus-or-minus differences. Large metallic objects like pylons are a serious handicap to magnetic exploration and must be kept at a sufficient distance, as it is difficult to correct for them.

A magnetometer may be used for mapping geological structures, for example, in some thick sedimentary sequences it is sometimes possible to delineate the major structural features because the succession includes magnetic horizons. These may be ferruginous sandstones or shales, tuffs or basic lava flows. In such circumstances anticlines produce positive and synclines negative anomalies. Faults and dykes are indicated on isomagnetic maps by linear belts of somewhat sharp gradient or by sudden swings in the trend of the contours.

8.5.5 Gravity methods

The Earth's gravity field varies according to the density of the subsurface rocks but at any particular locality its magnitude also is influenced by latitude, elevation, neighbouring topographical features and the tidal deformation of the Earth's crust. The effects of these latter factors have to be eliminated in any gravity survey where the object is to measure the variations in acceleration due to gravity precisely. This information then can be used to construct a contoured gravity map. In survey work, modern practice is to measure anomalies in gravity units (g.u. = 10^{-6} m s^{-2}). Modern gravity meters used in exploration

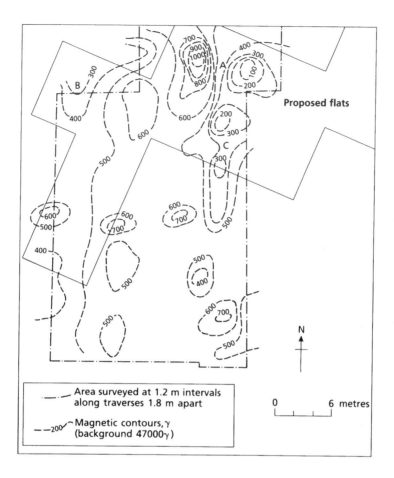

Fig. 8.9 Isomagnetic map of a site proposed for development in which mine shafts occurred at A, B and C.

measure not the absolute value of the acceleration due to gravity but the small differences in this value between one place and the next.

A gravity survey is conducted from a local base station at which the value of the acceleration due to gravity is known with reference to a fundamental base where the acceleration due to gravity has been accurately measured. The way in which a gravity survey is carried out largely depends on the objective in view. Gravity methods can be used in regional reconnaissance surveys to reveal anomalies that subsequently may be investigated by other methods. Such large-scale surveys covering hundreds of square kilometres, carried out in order to reveal major geological structures, are done by vehicle or helicopter with a density of only a few stations per square kilometre. For more detailed work such as the location of faults, the spacing between stations may be as small as 20 m. The location and elevation of stations must be established with very high precision because gravity differences large enough to be of geological significance are produced by changes in elevation of several millimetres and of only 30 m in north-south distance.

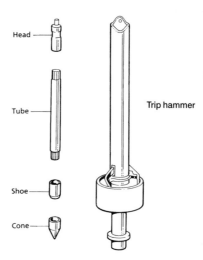

Fig. 8.10 Standard penetration test equipment.

8.6 In situ testing

Under certain circumstances, especially when samples are difficult to obtain, data may be obtained directly by carrying out tests in the field. This is particularly the case with coarse-grained soils and sensitive clay soils. Also, the data obtained by direct testing in place is likely to give more reliable results than those obtained from laboratory tests.

8.6.1 Penetrometer tests

There are two types of penetrometer tests, that is, dynamic tests and static tests. Both methods measure the resistance offered by the soil at any particular depth to penetration of a conical point. Penetration of the cone creates a complex shear failure and thus provides an indirect measure of the *in situ* shear strength of the soil.

The most widely used dynamic method is the standard penetration test (SPT). This empirical test, which makes use of a light cable and tool rig, consists of driving a split-spoon sampler, with an outside diameter of 50 mm, into the soil at the base of a borehole (Fig. 8.10). Drivage is accomplished by a trip hammer, weighing 63 kg, falling freely through a distance of 750 mm onto the drive head, which is fitted at the top of the rod assembly. First, the split-spoon is driven 150 mm into the soil at the bottom of the borehole. It then is driven a further 300 mm and the number of blows required to drive this distance is recorded. The blow count is referred to as the *N* value from which the relative density of a sandy soil can be assessed (Table 8.6). Refusal is regarded as 100 blows. The results obtained from the standard penetration test provide an evaluation of the degree of compaction of sands and the *N* values may be related to the values of the angle of internal friction (ϕ) and the allowable bearing capacity. The SPT also can be employed in clay soils, weak rocks and in the weathered zones of harder rocks.

The Dutch cone penetrometer is the most widely used static method (Fig. 8.11). It is particularly useful in soft clay soils and loose sands where boring operations tend to disturb the soil. In this technique a tube and inner rod with a conical point at the base are hydraulically advanced into the ground. The cone has a cross-sectional area of 1000 mm^2 with an angle of 60°. At approximately every 300 mm depth the cone is advanced ahead of the tube a distance of 50 mm and the maximum resistance is noted. The tube is advanced to join the cone after each measurement and the process repeated. The resistances are plotted against their corresponding depths so as to give a profile of the variation. Recordings also can be

Table 8.6 Relative density and consistency of soil.

(a) Relative density of sand and SPT values, and relationship to angle of friction

SPT(N)	Relative density (D_r)	Description of compactness	Angle of internal friction (φ)
4	0.2	Very loose	Under 30°
4–10	0.2–0.4	Loose	30°–35°
10–30	0.4–0.6	Medium dense	35°–40°
30–50	0.6–0.8	Dense	40°– 40°
over 50	0.8–1.0	Very dense	over 45°

(b) N-values, consistency and unconfined compressive strength of cohesive soils

N	Consistency	Unconfined compressive strength (kPa)
Under 2	Very soft	Under 20
2 – 4	Soft	20–40
5 – 8	Firm	40–75
9 – 15	Stiff	75–150
16 – 30	Very stiff	150–300
over 30	Hard	over 300

$$D_r = \frac{e_{max} - e}{e_{max} - e_{min}} \quad e = \text{void ratio}$$

taken continuously. One type of Dutch cone penetrometer has a sleeve behind the cone that can measure side friction. The ratio of sleeve resistance to that of cone resistance is higher in fine-grained than in coarse-grained soils thus affording some estimate of the type of soil involved.

In the piezocone a cone penetrometer is combined with a piezometer, the latter being located between the cone and the friction sleeve. The pore water pressure is measured by the piezometer at the same time as the cone resistance and sleeve friction. Because of the limited thickness of the piezometer (the filter is around 5 mm), much thinner layers can be determined with greater accuracy than with a conventional cone penetrometer. If the piezocone is kept at a given depth so that the pore water pressure can dissipate with time, then this allows assessment of the in situ permeability and consolidation characteristics of the soil to be made.

8.6.2   Shear vane test and in situ shear test

Because soft clays may suffer disturbance when sampled and therefore give unreliable results when tested for strength in the laboratory, a vane test often is used to measure the in situ undrained shear strength. Vane tests can be used in clay soils that have a consistency varying from very soft to firm. In its simplest form the shear vane apparatus consists of four blades arranged in cruciform fashion that are attached to the end of a rod (Fig. 8.12). All rotating parts, other than the vane, are enclosed in guide tubes in order to eliminate the effects of friction of the soil on the vane rods during the test. The vane normally is housed in a protective shoe. The vane and rods are pushed into the soil from the surface or

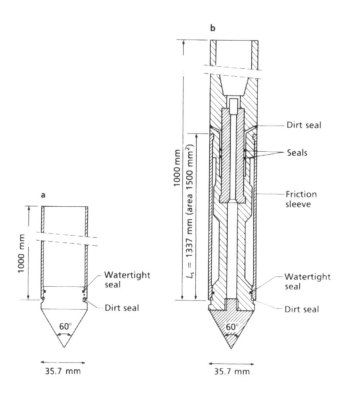

Fig. 8.11 Cone penetrometer tips (a) without friction sleeve (b) with friction sleeve.

the base of a borehole to a point 0.5 m above the required depth. Then the vane is pushed out of the protective shoe and advanced to the test position. It then is rotated at a rate of 6° to 12° per minute. Torque is applied to the vane rods by means of a torque-measuring instrument mounted at ground level. The maximum torque required for rotation is recorded. As the vane is rotated the soil fails along a cylindrical surface defined by the edges of the vane as well as along the horizontal surfaces at the top and bottom of the blades. The shearing resistance is obtained from the following expression:

$$\tau = \frac{M}{\pi(D^2 H/2 + D^3/6)} \tag{8.5}$$

where τ is the shearing resistance, D and H are the diameter and height of the vane respectively, and M is the torque.

In situ shear tests usually are performed on blocks, 700 × 700 mm, cut in rock. The block is sheared by a horizontal jack while a confining load is imposed by a vertical jack. It is advantageous to make the tests inside galleries, where reactions for the jacks are provided by the sides and roof. The tests are performed at various normal loads and give an estimate of the angle of shearing resistance and cohesion of the rock. Such shear testing may be carried out along a discontinuity in rock material in order to assess its shear strength.

8.6.3 Pressuremeter tests

The Menard pressuremeter is used to determine the *in situ* strength and deformation of soils, and weak

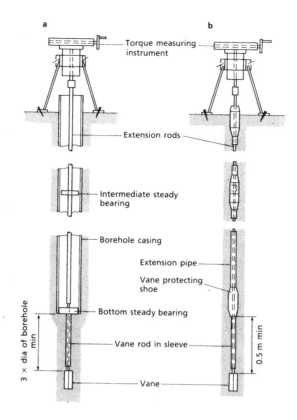

Fig. 8.12 Shear vane tests (a) borehole vane test (b) penetration vane test.

and weathered rocks. The test takes account of the influence of discontinuities and is particularly useful in those soils from which undisturbed samples cannot be obtained readily. Where possible testing is carried out in an unlined hole but if necessary a special slotted casing is used to provide support. This pressuremeter consists essentially of a probe that is placed in a borehole at the appropriate depth and then expanded. The probe consists of a cylindrical metal body over which are fitted three cylinders (Fig. 8.13). A rubber membrane covers the cylinders and is clamped between them to give three independent cells. The cells are inflated with water and a common gas pressure is applied by a volumeter located at the surface; in this way a radial stress is applied to the soil. The deformations produced by the central cell are indicated on the volumeter and allow the deformation modulus of the ground to be determined. If the test is continued until the ground fails, then this provides an estimate of its ultimate bearing capacity.

The major advantage of a self-boring pressuremeter is that a borehole is unnecessary, as a result the interaction of the probe and the soil is improved. Self-boring is brought about either by jetting or by using a chopping tool.

A dilatometer is a type of pressuremeter and is used in a drill hole to obtain data relating to the deformability of a rock mass. These instruments range up to about 300 mm in diameter and over 1 m in length, and can exert pressures of up to 20 MPa on the drill hole walls. Diamentral strains can be measured either directly along two perpendicular diameters or by measuring the amount of liquid pumped into the instrument.

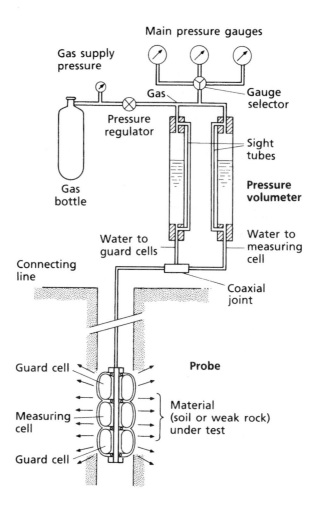

Fig. 8.13 Pressuremeter test equipment.

8.6.4 Plate load tests

A plate load test provides information by which the bearing capacity and settlement characteristics of the ground can be assessed. It is carried out in a trial pit, usually at excavation base level. Plates vary in size from 0.15 to 0.61 m in diameter, the size of plate used being determined by the spacing of any discontinuities present. The plate should be properly bedded and the test carried out on undisturbed material so that reliable results may be obtained. The load is applied by a jack, in increments, either of one-fifth of the proposed bearing pressure or in steps of 25 to 50 kPa (these are smaller in soft soils, that is, where the settlement under the first increment of 25 kPa is greater than $0.002D$, D being the diameter of the plate). Successive increments should be made after settlement has ceased. The test generally is continued up to two or three times the proposed loading, or in clay soils until settlement equal to 10 to 20% of the plate dimension is reached or the rate of increase of settlement becomes excessive. Consequently, the ultimate bearing capacity at which settlement continues without increasing the load rarely is reached. Large plate bearing tests frequently are used to determine the value of Young's modulus of the foundation rock at large construction sites.

8.6.5 Assessment of permeability

An initial assessment of the magnitude and variability of the *in situ* permeability can be obtained from tests carried out in boreholes as the hole is advanced. By artificially raising the level of water in a borehole (falling head test) above that in the surrounding ground, the flow rate from the borehole can be measured. However, in very permeable soils it may not be possible to raise the level of water in a borehole. Conversely, the water level in a borehole can be artificially depressed (rising head test) so allowing the rate of water flow into the borehole to be assessed. Wherever possible, a rising and a falling head test should be carried out at each required level and the results averaged.

The constant head method of *in situ* permeability testing is used when the rise or fall in the level of the water is too rapid for accurate recording (i.e. occurs in less than 5 min). This test normally is conducted as an inflow test in which the flow of water into the ground is kept under a sensibly constant head (e.g. by adjusting the rate of flow into a borehole so as to maintain the water level at a point on the inside of the casing near the top). The method is only applicable in permeable ground such as gravels, sands and broken rock, when there is a negligible or zero time for equalization. The rate of flow is measured once a steady flow into and out of a borehole has been attained over a period of some 10 min.

The permeability of an individual bed or section of rock can be determined by a water injection or packer test carried out in a drill hole. This is done by sealing off a length of uncased hole with packers and injecting water under pressure into the test section (Fig. 8.14). Usually, because it is more convenient, packer tests are carried out after the entire length of a hole has been drilled. Two packers are used to seal off selected test lengths and the tests are performed from the base of the hole upwards. The hole must be flushed to remove sediment prior to a test being performed. With double packer testing the variation in permeability throughout a test hole is determined. The rate of flow of water over the test length is measured under a range of constant pressures and recorded. The permeability is calculated from a flow-pressure curve. Water generally is pumped into the test section at steady pressures for periods of 15 min, readings of water absorption also being taken every 15 min. The test usually consists of five cycles at successive pressures of 6, 12, 18, 12 and 6 kPa for every metre depth of packer below the surface.

Field pumping tests allow the determination of the coefficients of permeability and storage, as well as the transmissivity, of a larger mass of ground than the aforementioned tests. A pumping test involves abstracting water from a well at known discharge rate(s) and observing the resulting water levels as drawdown occurs. At the same time the behaviour of the water table in the aquifer can be recorded in observation wells radially arranged about the abstraction well. There are two types of pumping test, namely, the constant pumping rate aquifer test and the step performance test. In the former test the rate of discharge is constant whereas in a step performance test there are a number of stages, each of equal length of time, but at different rates of discharge. The step performance test usually is carried out before the constant pumping rate aquifer test. Yield–drawdown graphs are plotted from the information obtained (Fig. 8.15).

8.6.6 Assessment of flow

A flowmeter log provides a record of the direction and velocity of groundwater movement in a drill hole. Flowmeter logging requires the use of a velocity-sensitive instrument, a system for lowering the instrument into the hole, a depth measuring device to determine the position of the flowmeter and a recorder located at the surface. The direction of flow of water is determined by slowly lowering and raising the flowmeter through a section of hole, 6 to 9 m in length, and recording velocity measurements during both traverses. If the velocity measured is greater during the downward than the upward traverse, then the direction of flow is upward and vice versa. A flowmeter log made while a drill hole is being pumped at a moderate rate or by allowing water to flow if there is sufficient artesian head, permits identification of the zones contributing to the discharge. It also provides information on the thickness of these zones

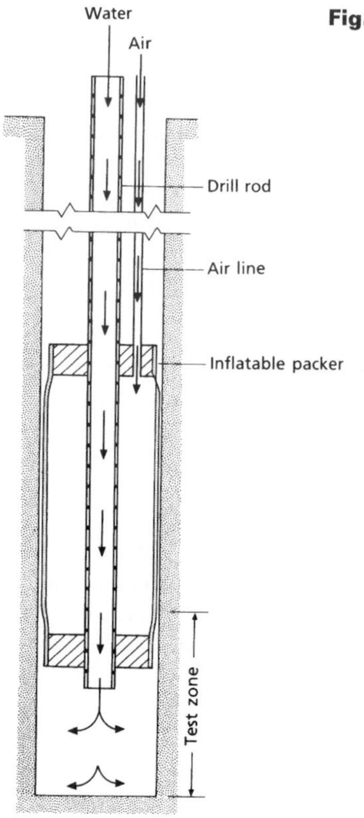

Fig. 8.14 Packer test equipment.

Water

Air

— Drill rod

— Air line

— Inflatable packer

Test zone

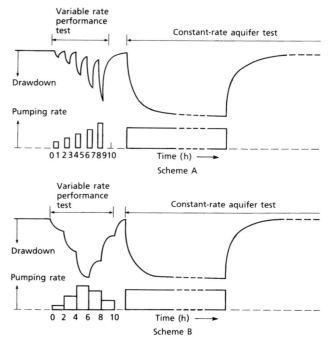

Variable rate
performance
test

Constant-rate aquifer test

Drawdown

Pumping rate

0 1 2 3 4 5 6 7 8 9 10   Time (h) ⟶

Scheme A

Variable rate
performance
test

Constant-rate aquifer test

Drawdown

Pumping rate

0 2 4 6 8 10   Time (h) ⟶

Scheme B

Fig. 8.15 Pumping test
procedure.

and the relative yield at that discharge rate. Because the yield varies approximately directly with the drawdown of water level in a well, flowmeter logs made by pumping, should be pumped at least at three different rates. The drawdown of water level should be recorded for each rate.

A number of different types of tracer have been used to investigate the movement of groundwater and the interconnection between ground and surface water. The ideal tracer should be easy to detect quantitatively in minute concentrations; it should not change the hydraulic characteristics of, or be adsorbed by the media through which it is flowing; it should react with the groundwater concerned; and it should have a low toxicity. The type of tracers in use include water soluble dyes that can be detected by colorimetry, sodium chloride or sulphate salts that can be detected chemically, and strong electrolytes that can be detected by electrical conductivity. Radioactive tracers also are used and one of their advantages is that they can be detected in minute quantities in water. Such tracers should have a useful half-life and should present a minimum of hazard. When a tracer is injected via a drill hole into groundwater it is subject to diffusion, dispersion, dilution and adsorption. Even if these processes are not significant, flow through an aquifer may be stratified or concentrated along discontinuities. Therefore, a tracer may remain undetected unless the observation drill holes intersect these discontinuities. The vertical velocity of water movement in a drill hole can be assessed by using tracers. A tracer is injected at a given depth, and the direction and rate of movement is monitored by a probe. Similarly, determination of permeability can be done by measuring the time it takes for a tracer to move between two test holes. Like pumping tests, this tracer technique is based on the assumption that the aquifer is homogeneous and that observations taken radially at the same distance from the well are comparable. The method requires that injection and observation wells are close together to avoid excessive travel time and that the direction of flow is known so that observation holes are correctly sited.

8.6.7   Pore water pressures and their assessment

Subsurface water is normally under pressure that increases with increasing depth below the water table to very high values. Such water pressures have a significant influence on the engineering behaviour of most rock and soil masses, and their variations are responsible for changes in the stresses in these masses, which affect their strength and deformation characteristics.

Piezometers are installed in the ground in order to monitor and obtain measurements of pore water pressures (Fig. 8.16). Observations should be made regularly so that changes due to external factors such as excessive precipitation, tides, the seasons, etc., are noted, it being most important to record the maximum pressures that have occurred. Standpipe piezometers allow the determination of the position of the water table and the permeability. For example, the water level can be measured with an electric dipmeter. Piezometer tips that have leads going to a constant head permeability unit enable the rate of flow through the tip to be measured. Hydraulic piezometers can be installed at various depths in a borehole where it is required to determine pore water pressures. They are connected to a manometer board that records the changes in pore water pressure. Usually, simpler types of piezometer are used in more permeable soils. When a piezometer is installed in a borehole it should be surrounded with a filter of clean sand. The sand should be sealed both above and below the piezometer to enable the pore water pressures at that particular level to be measured with a minimum of influence from the surrounding strata, since the latter may contain water at different pressures. The response of piezometers in rock masses can be very much influenced by the incidence and geometry of the discontinuities so that the values of water pressure obtained may be misleading if due regard is not given to these structures.

223

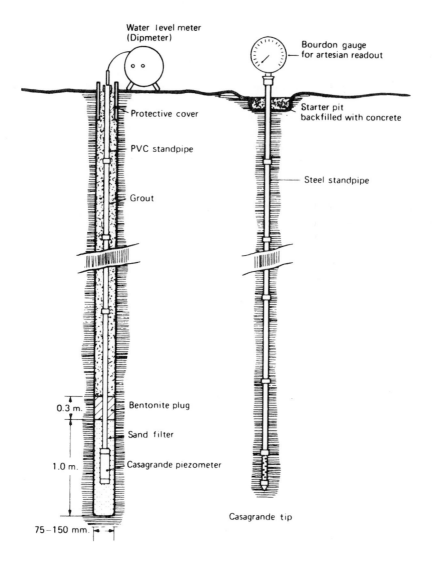

Fig. 8.16 Standard piezometers.Borehole standard piezometer (left) and drive-in standpipe piezometer (right).

9

Engineering Aspects of Soils and Rocks

9.1 Basic properties of soil

Soil consists of an assemblage of particles between which there are voids, and as such can contain three phases, that is, solids, water and air. The interrelationships of the masses and volumes of these three phases are important since they help define the character of a soil. One of the most fundamental properties of a soil is the void ratio, which is the ratio of the volume of the voids to that of the volume of the solids. The porosity is a similar property and is defined as the ratio of the volume of the voids to the total volume of the soil, expressed as a percentage (Table 9.1). Both void ratio and porosity indicate the relative proportion of void volume in a soil sample. Water content plays a fundamental part in determining the engineering behaviour of any soil and is expressed as a percentage of the mass of the solid material in the soil sample. The degree of saturation expresses the relative volume percentage of water in the voids.

The unit weight of a soil is its mass per unit volume, whilst its specific gravity, which refers to the soil particles, is the ratio of its mass to that of an equal volume of water. The density of a soil is the ratio of its mass to that of its volume. A number of types of density are distinguished. The dry density is the mass of the solid particles divided by the total volume (Table 9.1), whereas the bulk density is simply the mass of the soil (including its natural moisture content) divided by its volume. The saturated density is the density of the soil when saturated, whilst the submerged density is the ratio of effective mass to volume of soil when submerged. A useful way to characterize the density of a coarse-grained soil is by its relative density, D_r, which is defined as:

$$D_r = \frac{e_{max} - e}{e_{max} - e_{min}} \tag{9.1}$$

where e is the naturally occurring void ratio, e_{max} is the maximum void ratio and e_{min} is the minimum void ratio.

The particle size distribution expresses the size of particles in a soil in terms of percentages, by weight, of boulders, cobbles, gravel, sand, silt and clay (Table 9.2). Two methods usually are used to determine the particle size distribution in a soil, that is, sieving and sedimentation. The results of particle size analysis are given in the form of a series of fractions, by weight, of different size grades. These fractions are expressed as a percentage of the whole sample and are generally summed to obtain a cumulative percentage. Cumulative curves are then plotted on a semi-logarithmic paper to give a graphical representation of the particle size distribution. The slope of the curve provides an indication of the degree of sorting (see Fig. 1.17).

The Atterberg or consistency limits of fine-grained soils are founded on the concept that such soils can exist in any of four states depending on their moisture content. In other words, a fine-grained soil is solid when dry but as water is added, it first turns to a semi-solid, then to a plastic, and finally to a liquid state. The moisture content at the boundaries between these states is referred to as the shrinkage limit

Table 9.1 Description of dry density, porosity, void ratio and degree of saturation of soils (after International Association of Engineering Geology).

Class	Dry density (Mg m^{-3})	Description	Porosity (%)	Description
1	less than 1.4	Very low	over 50	Very high
2	1.4–1.7	Low	50–45	High
3	1.7–1.9	Moderate	45–35	Medium
4	1.9–2.2	High	35–40	Low
5	over 2.2	Very high	less than 30	Very low

Class	Void ratio	Description	Degree of saturation (%)	Description
1	over 1.8	Very high	less than 25	Naturally dry
2	1.0–1.8	High	25–50	Wet
3	0.8–0.55	Medium	50–80	Very wet
4	0.55–0.43	Low	80–95	Highly saturated
5	less than 0.43	Very low	over 95	Saturated

Table 9.2 Particle size distribution of soils.

Types of material		Sizes (mm)
Boulders		Over 200
Cobbles		60–200
Gravel	Coarse	20–60
	Medium	6–20
	Fine	2–6
Sand	Coarse	0.6–2
	Medium	0.2–0.6
	Fine	0.06–0.2
Silt	Coarse	0.02–0.06
	Medium	0.006–0.02
	Fine	0.002–0.006
Clay		less than 0.002

Table 9.3 Plasticity according to liquid limit.

Description	Plasticity	Range of liquid limit
Lean or silty	Low plasticity	less than 35
Intermediate	Intermediate plasticity	35–50
Fat	High plasticity	50–70
Very fat	Very high plasticity	70–90
Extra fat	Extra high plasticity	over 90

(SL), the plastic limit (PL) and the liquid limit (LL) respectively. The shrinkage limit is defined as the percentage moisture content of a soil at the point where it suffers no further decrease in volume on drying. The plastic limit is the percentage moisture content at which a soil can be rolled, without breaking, into a thread 3 mm in diameter, any further rolling causing it to crumble. Lastly, the liquid limit is defined as the minimum moisture content at which a fine-grained soil will flow under its own weight and is determined by a cone penetrometer or the Casagrande apparatus. Clays may be classified according to their liquid limit as shown in Table 9.3. The numerical difference between the liquid and plastic limits is termed the plasticity index (PI). This indicates the range of moisture content over which the material exists in a plastic condition. The plasticity of a soil is influenced by its clay fraction, in particular the amount and type of clay minerals present influence the quantity of attracted water held in a fine-grained soil. This, in turn, can be related to the activity of fine-grained soil, which is defined as the plasticity index divided by the percentage, by mass, finer than 0.002 mm. The following grades of activity are recognized:

1. Inactive with activity less than 0.5.
2. Inactive with activity range 0.5 to 0.75.
3. Normal with activity range 0.75 to 1.25.
4. Active with activity range 1.25 to 2.
5. Active with activity greater than 2.

The liquidity index of a soil is defined as its moisture content in excess of the plastic limit, expressed as a percentage of the plasticity index. It describes the moisture content of a soil with respect to its index limits and indicates in which part of its plastic range a soil lies, that is, its nearness to the liquid limit. A third index is the consistency index, that is, the ratio of the difference between the liquid limit and natural moisture content to the plasticity index. It can be used to classify the different types of consistency of fine-grained soils, as shown in Table 9.4.

9.2   Soil classification and description

Any engineering classification of soil involves grouping the different soil types into categories that possess similar properties and in so doing provides the engineer with a systematic method of soil description. As one of the basic properties of soils is grain size, then this parameter is used in classification, coarse-grained soils being distinguished from the fine-grained soils. Gravels and sands are the two principal types of coarse-grained soils in the American Unified Classification and are divided into subgroups on a basis of grading (Table 9.5). Well graded soils are those in which the particle size distribution extends over a wide range without excess or deficiency in any particular sizes. In uniformly graded soils the particle size distribution extends over a very limited range and in poorly or gap graded soils there is a deficiency in a particular particle size range.

A plasticity chart, in which the plasticity index is plotted against liquid limit, is used to classify fine-

Table 9.4 Consistency of fine grained soils.

Description	Consistency index	Approximate undrained shear strength(kPa)	Field identification
Hard		over 300	Indented with difficulty by thumbnail, brittle
Very stiff	above 1	150–300	Readily indented by thumbnail, still very tough
Stiff	0.75–1	75–150	Readily indented by thumb but penetrated only with difficulty. Cannot be moulded in the fingers.
Firm	0.5–0.75	40–75	Can be penetrated several centimetres by thumb with moderate effort, and moulded in the fingers by strong pressure
Soft	less than 0.5	20–40	Easily penetrated several centimetres by thumb, easily moulded
Very soft		less than 20	Easily penetrated several centimetres by fist, exudes between fingers when squeezed in fist

Table 9.5 Symbols used in the Unified Soil Classification.

Main soil type		Prefix
Coarse-grained soils	Gravel	G
	Sand	S
Fine-grained soils	Silt	M
	Clay	C
	Organic silts and clays	O
Fibrous soils	Peat	Pt

Subdivision		Suffix
For coarse-grained soils	Well graded, with little or no fines	W
	Well graded with suitable clay binder	C
	Uniformly graded with little or no fines	U
	Poorly graded with little or no fines	P
	Poorly graded with appreciable fines or well graded with excess fines	F
For fine-grained soils	Low compressibility (plasticity)	L
	Medium compressibility (plasticity)	I
	High compressibility (plasticity)	H

grained soils (see Table 9.6). The A line on the plasticity chart is taken as the boundary between organic and inorganic soils, the latter lying above the line. Each of the main soil types and subgroups are given a letter, a pair of which are combined in the group symbol, the former being the prefix, the latter the suffix (Table 9.5). Most soils consist of more than one grade size type. If boulders and cobbles are present in

Table 9.6 Simplified version of Unified Soil Classification.

Excludes particles larger than 60 mm			Typical names	Group symbols	Information required for describing soils
Coarse grained soils. More than half of material is *larger* than 0.06 mm	Gravels More than half of coarse fraction is gravel size	Clean gravels (little or no fines)	Well graded gravels, gravel-sand mixtures, little or no fine	GW	Give typical name; indicate approximate percentages of sand and gravel; maximum size; angularity, surface condition, and hardness of the coarse grains; local or geological name and other pertinent descriptive information; and symbols in parentheses
			Poorly graded gravels, gravel-sand mixtures, little or no fine	GP	
		Gravels with fines (appreciable amount of fines)	Silty gravels, poorly graded gravel–sand–clay mixtures	GM	For undisturbed soils add information on stratification, degree of compactness, cementation, moisture conditions and drainage characteristics
			Clayey gravels poorly graded gravel–sand–clay mixtures	GC	
	Sands More than half of coarse fraction is sand size	Clean sands (little or no fines)	Well graded sands, gravelly sands, little or no fines	SW	Example: Silty sand, gravelly; about 20% hard, angular gravel particles 12.5 mm maximum size; rounded and subangular sand grains coarse to fine, about 15% nonplastic fines; well compacted and moist in place; alluvial sand; (*SM*)
			Poorly graded sands, gravelly sands, little or no fines	SP	
		Sands with fines (appreciable amount of fines)	Silty sands, poorly graded sand–silt mixtures	SM	
			Clayey sands poorly graded sand–clay mixtures	SC	

Table 9.6 cont'd.

		Typical names	Group symbols	Information required for describing soils
Fine grained soils. More than half of material is *smaller* than 0.06 mm	Silts and clays liquid limit less than 50	Inorganic silts and very fine sands, rock flour, silty or clayey fine sands with slight plasticity	*ML*	Give typical name; indicate degree and character of plasticity, amount and maximum size of coarse grains; colour in wet condition, odour, if any, local or geological name and other pertinent descriptive information, and symbol in parentheses
		Inorganic clays of low to medium plasticity, gravelly clays, sandy clays, silty clays, lean clays	*CL*	For undisturbed soils add information on structure, stratification, consistency in undisturbed and remoulded states, moisture and drainage conditions
		Organic silts and organic silt–clays of low plasticity	*OL*	
	Silts and clays liquid limit greater than 50	Inorganic silts micaceous or diatomaceous fine sandy or silty soils, clastic silts	*MH*	Example: Clayey silt: brown; slightly plastic; small percentage of fine sand, numerous vertical root holes; firm and dry in place; loess (*ML*)
		Inorganic clays of high plasticity, fat clays	*CH*	
		Organic clays of medium to high plasticity	*OH*	
Highly organic soils		Peat and other highly organic soils	*Pt*	Identified by colour, odour, spongy feel and frequently fibrous texture

Plasticity chart for laboratory classification of fine-grained soils.

composite soil types, then their presence should be recorded separately in the soil description. A full description of a soil should provide data on its grading, plasticity, particle characteristics and colour, as well as on its fabric, the nature of its bedding and discontinuities, and strength.

When small amounts of organic matter occur dispersed throughout a soil, they can have a notable effect on plasticity and therefore the engineering properties. Increasing quantities of organic matter increase these effects. Soils in which the organic contents may be up to 30%, by weight, and contain moisture contents ranging up to 250% behave primarily as mineral soils, albeit with different parameters. Peat is an accumulation of plant remains that have undergone some degree of decomposition. Inorganic soil material may occur as secondary constituents in peat, and should be described, for example, as slightly clayey or sandy.

9.3 Shear strength of soil

The shear strength of a soil is regarded as the maximum resistance that it can offer to shear stress. When this maximum is reached, the strength of the soil is fully mobilized, any additional stress causing the soil to fail. However, the shear strength value determined experimentally is not a unique constant that is characteristic of the material but varies with the method of testing. Shear displacement also continues to take place after the shear strength is exceeded. The stress on any plane surface can be resolved into the normal stress, σ_n, which acts perpendicular to the surface, and the shearing stress (τ) which acts along the surface, the magnitude of the resistance being given by the Coulomb equation:

$$\tau = c + \sigma_n \tan \phi \qquad (9.2)$$

where ϕ is the angle of shearing resistance and c is the cohesion. The stress that controls changes in the volume and the strength of a soil is known as the effective stress (σ'). When a load is applied to a saturated soil it is either carried by the pore water, which gives rise to an increase in the pore water pressure (u) or by the soil skeleton, or both. The effect that a load has on a soil therefore is affected by the drainage conditions but in most practical cases the effective stress is equal to the intergranular stress and can be determined from the equation:

$$\sigma' = \sigma - u \qquad (9.3)$$

where σ is the total stress. Hence, shear strength depends upon effective stress and not total stress. Accordingly, the Coulomb equation must be modified in terms of effective stress and becomes:

$$\tau = c' + \sigma' \tan \phi' \qquad (9.4)$$

Because of the complex nature of the shearing resistance of soils several methods of testing are employed in its assessment. The principal tests are the triaxial test and the shear box test. The triaxial test is the more reliable of the two and so is used more frequently. In both tests a confining stress is imposed on the sample prior to it being sheared and several tests have to be carried out on the same soil in order to determine its cohesion and angle of shearing resistance.

The internal frictional resistance of a soil, for example, as is developed in sand, is generated by friction when the grains in the zone of shearing are caused to slide, roll and rotate against each other. Local crushing may occur at the points of contacts that undergo the highest stress. The total resistance is the sum of the behaviour of all the particles and is influenced by the confining stress, the coefficient of friction and angles of contact between the minerals, as well as their surface roughness. However, the angle of shearing resistance or internal friction does not depend solely on the internal friction between grains. A proportion of the shearing stress on the plane of failure is utilized in overcoming interlocking, that is, shearing stress also is dependent upon the initial void ratio or density of granular soil. In addition, it is influenced by the size and shape of the grains. The larger the grains, the wider the zone that is affected. The more angular the grains, then the greater is the frictional resistance to their relative movement. Electrical forces of attraction and repulsion also may be involved in shearing resistance.

In clay soils the cohesion that is developed by the molecular attractive forces between the minute soil particles is mainly responsible for the resistance offered to shearing. Because molecular attractive forces depend to a large extent on the mineralogical composition of the particles and on the type of concentration of electrolytes present in the pore water, the magnitude of the true cohesion also depends on these factors. When clay soil is strained it develops an increasing resistance (strength), but under a given effective pressure the resistance offered is limited, the maximum value corresponding to the peak strength. If testing is continued beyond the peak strength, then as displacement increases, the resistance decreases, again to a limiting value that is termed the residual strength. The development of residual strength is therefore a continuous process. In moving from peak to residual strength, cohesion falls to almost, or actually, zero but the angle of shearing resistance is reduced by only a few degrees. Under a given effective pressure, the residual strength of a clay soil is the same whether it is normally or overconsolidated (see Section 9.4). In other words, the residual shear strength of a clay soil is independent of its post-depositional history, unlike the peak undrained shear strength that is controlled by the history of consolidation, as well as diagenesis. Furthermore, the value of residual shear strength decreases as the amount of clay fraction increases in a deposit and also is influenced by the type of clay minerals present.

In the case of an undisturbed clay soil, its shear strength often is found to be greater than that obtained when it is remoulded and tested under the same conditions and at the same water content. The ratio of the undisturbed to the remoulded strength at the same moisture content is defined as the sensitivity of a clay soil. The following grades of sensitivity are recognized:

1. Insensitive clays, under 1.
2. Low sensitive clay soils, 1 to 2.
3. Medium sensitive clay soils, 2 to 4.
4. Sensitive clay soils, 4 to 8.
5. Extra-sensitive clay soils, 8 to 16.
6. Quick clay, over 16.

Clay soils with high sensitivity values have little strength after being disturbed. Indeed, if they are disturbed this may cause a material that is initially fairly strong to behave as a viscous fluid. Sensitive clays generally possess high moisture contents, frequently with liquidity indices well in excess of unity. A sharp increase in moisture content may cause a great increase in sensitivity, sometimes with disastrous results. Heavily overconsolidated clays are insensitive. The effect of remoulding on clay soils of various sensitivities is illustrated in Fig. 9.1. Some clay soils with moderate to high sensitivity show a regain in strength when, after remoulding, they are allowed to rest under unaltered external conditions. Such soils are thixotropic.

9.4 Consolidation of fine-grained soils

The theory of consolidation enables the amount and rate of settlement that are likely to occur when structures are constructed on fine-grained soils to be determined. If such a soil is saturated, then the load is carried initially by the pore water that causes a pressure, the hydrostatic excess pressure, to develop. The excess pore water pressure is dissipated at a rate that depends upon the permeability of the soil mass and the load is transferred eventually to the soil structure. The change in volume during consolidation is equal to the volume of the pore water expelled and corresponds to the change in void ratio of the soil. This is termed primary consolidation. Settlement occurs as the void ratio decreases. In clay soils, because of their low permeability, the rate of consolidation is slow. Further consolidation may occur due to a rearrangement of the soil particles. This secondary consolidation is usually much less significant. However, it should not be assumed that secondary consolidation always follows primary consolidation as the two processes can occur at the same time. The compressibility of a clay soil is related to its geological history, that is, to whether it is normally consolidated or overconsolidated. A normally consolidated clay is one that has at no time during its geological history been subjected to pressures greater than its existing overburden pressure, whereas an overconsolidated clay is one that has, in that part of its former overburden has been removed by erosion.

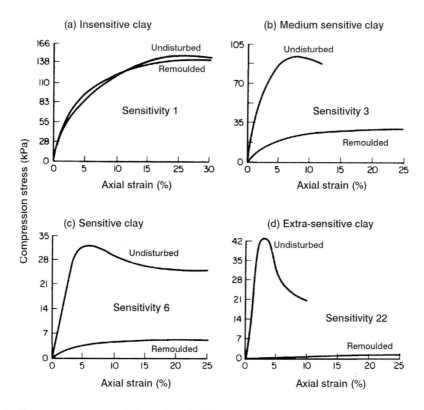

Fig. 9.1 Stress–strain curves of clay soils with different sensitivities.

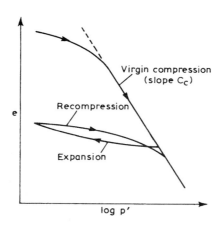

Fig. 9.2 The results from an oedometer test plotted as an e log p curve.

The relationship between unit load and void ratio for a soil can be represented by plotting the void ratio (e) against the logarithm of the unit load (p) (Fig. 9.2). In the laboratory the relationship between e and p is determined by using an oedometer. The shape of the $e \log p$ curve is related to the stress history of a particular clay soil. In other words, the $e \log p$ curve for a normally consolidated clay is linear and is referred to as the virgin compression curve. On the other hand, if the clay soil is overconsolidated the $e \log p$ curve is not straight and the preconsolidation pressure can be derived from the curve. The preconsolidation pressure refers to the maximum overburden pressure to which a deposit has been subjected. Overconsolidated clay soil is appreciably less compressible than normally consolidated clay.

The compressibility of a clay can be expressed in terms of the compression index (C_c) or the coefficient of volume compressibility (m_v). The compression index is the slope of the linear section of the $e \log p$ curve and is dimensionless. It tends to be applied to normally consolidated clays. The value of C_c for cohesive soils ranges from 0.075 for sandy clays to more than 1.0 for highly colloidal bentonic clays. The approximate range of the degree of compressibility for some clay soils is given in Table 9.7a. It can be seen that the compressibility index increases with increasing clay content and so with increasing liquid limit. Indeed, for normally consolidated clays it is related to their liquid limit, the relationship between the two being expressed as:

$$C_c = 0.009(LL - 10) \tag{9.5}$$

This relationship does not apply to highly organic clays, to where the liquid limit exceeds 100% or to where the liquid limit is exceeded by the natural moisture content. The coefficient of volume compressibility is defined as the volume change per unit volume per unit increase in load. The value of m_v for a given soil depends upon the stress range over which it is determined, for example, it usually is recommended that it should be calculated for a pressure increment of 100 kPa in excess of the effective overburden pressure on the soil at the depth in question. Some typical values of m_v are given in Table 9.7b.

Table 9.7 Range of compressibility of fine grained soils.

a) Compressibility index

Soil type	Range C_c	Degree of compressibility
Soft clay	over 0.3	Very high
Clay	0.3–0.15	High
Silt	0.15–0.075	Medium
Sandy clay	less than 0.075	Low

b) Coefficient of volume compressibility. Some typical values of coefficient of volume compressibility.

Coefficient of volume compressibility (m^2 MN^{-1})	Degree of compressibility	Soil types
above 1.5	Very high	Organic alluvial clays and peats
0.3–1.5	High	Normally consolidated alluvial clays
0.1–0.3	Medium	Varved and laminated clays. Firm to stiff clays
0.05–0.1	Low	Very stiff or hard clays. Tills
below 0.05	Very low	Heavily overconsolidated tills

Table 9.8 Some values of gravels, sands and silts.

	Gravels	Sands	Silts
Specific gravity	2.5–2.8	2.6–2.7	2.64–2.66
Bulk density (Mg m^{-3})	1.45–2.3	1.4–2.15	1.82–2.15
Dry density (Mg m^{-3})	1.4–2.1	1.35–1.9	1.45–1.95
Porosity (%)	20–50	23–35	–
Void ratio	–	–	0.35–0.85
Liquid limit (%)	–	–	24–35
Plastic limit (%)	–	–	14–25
Coefficient of consolidation (m^2 yr^{-1})	–	–	12.2
Cohesion (kPa)	–	–	75
Angle of friction (deg)	35–45	32–42	32–36

The compressibility of a soil is dependent on the average rate of compression and the soil structure has a substantial time-dependent resistance to compression. As time proceeds, the excess pore water pressure is dissipated gradually and finally disappears. The ratio between the decrease of the void ratio at a given time and the ultimate decrease represents the degree of consolidation. With a given layer thickness of clay soil the degree of consolidation at a given time depends exclusively on the coefficient of consolidation (c_v). The coefficient of consolidation, which determines the rate at which settlement takes place, is calculated for each load increment and either a mean value or that value appropriate to the pressure range in question is used.

9.5 Coarse-grained soils

The micro-structure of a sand or gravel refers to its particle arrangement, which involves its packing. The void ratio of a well sorted and perfectly cohesionless aggregate of equidimensional grains can range between about 0.35 and 1.00. If the void ratio is more than unity the micro-structure is potentially collapsible or metastable. Generally, the larger the particles of coarse-grained soils, the higher the strength, and deposits consisting of a mixture of different sized particles usually are stronger than those that are uniformly graded (Table 9.8).

There are two basic mechanisms that contribute towards the deformation of coarse-grained soil, namely, distortion of the particles and the relative motion between them. These two mechanisms usually are interdependent. At any instant during the deformation process different mechanisms may be acting in different parts of the soil and these may change as deformation continues. Interparticle sliding can occur at all stress levels, the stress required for its initiation increasing with initial stress and decreasing void ratio. Crushing and fracturing of particles begins in a minor way at small stresses, becoming increasingly important when some critical stress is reached. This critical stress is smallest when the soil is loosely packed and uniformly graded, and consists of large angular particles with low strength.

The internal shearing resistance of an ideal coarse-grained soil is generated by friction when the grains in the zone of shearing slide, roll and rotate against each other, and the angle of shearing resistance is influenced by the grain size distribution and grain shape (Table 9.9). Figure 9.3 shows that dense sand has a high peak strength and that when subjected to shear stress it expands up to the point of failure, after which a slight decrease in volume may occur. By contrast, loose sand compacts under shearing stress and its residual strength may be similar to that of dense sand. Both curves in Fig. 9.3 exhibit strains that are approximately proportional to stress at low stress levels, suggesting a large component of elastic distortion. If the stress is reduced the unloading stress–strain curve indicates that not all the strain is

Table 9.9 Effect of grain shape and grading on the peak friction angle of sand.

Shape and grading	Loose	Dense
Rounded, uniform	30°	37°
Rounded, well graded	34°	40°
Angular, uniform	35°	43°
Angular, well graded	39°	45°

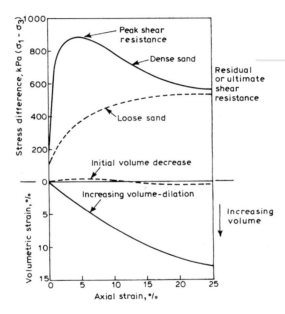

Fig. 9.3 Stress–strain and volumetric strain curves for dense and loose sand.

recovered on unloading. This hysteresis loss represents the energy lost in crushing and repositioning of grains. At higher shear stresses the strains are proportionally greater indicating greater crushing. The presence of water in the voids of a coarse-grained soil usually does not produce significant changes in the value of the angle of shearing resistance. However, if stresses develop in the pore water, then they may bring about changes in the effective stresses between the particles at which point the shear strength and the stress–strain relationships may be altered radically.

As water flows through sands and silts its energy is transferred to the particles past which it is moving, which creates a drag effect on the particles. If the drag effect is in the same direction as the force of gravity, then the effective pressure is increased and the soil is stable. Indeed, the soil tends to become denser. Conversely, if water flows towards the surface, then the drag effect is counter to gravity thereby reducing the effective pressure between particles. Moreover, if the velocity of upward flow is sufficient it can buoy up the particles so that the effective pressure is reduced to zero. This represents a critical condition where the weight of the submerged soil is balanced by the upward acting seepage force. A quick condition develops if the upward velocity of flow increases beyond the critical hydraulic gradient. Quicksands, if subjected to deformation or disturbance, can undergo a spontaneous loss of strength. This loss of strength causes them to flow like viscous liquids. Certain conditions should exist for quick conditions to develop. First, the sand or silt concerned must be saturated and loosely packed. Secondly, on disturbance the constituent grains become more closely packed. This leads to an increase in pore water pressure, reducing the forces acting between the grains and bringing about a reduction in strength. If the pore water can escape very rapidly the loss in strength is momentary. Hence, the third condition requires that pore water cannot escape readily. This is fulfilled if the sand or silt has a low permeability and/or the seepage path is long. As the velocity of the upward seepage force increases further from the critical gradient the soil begins to boil more violently. At such a point structures fail by sinking into the quicksand. Liquefaction of potential quicksands may be brought about by sudden shocks caused by the action of heavy machinery (notably pile driving), blasting and earthquakes. Such shocks increase the stress carried by the water, the neutral stress, and give rise to a decrease in the effective stress and shear strength of the soil.

9.6   Silts and loess

The grains in a deposit of silt frequently are rounded, which influences their degree of packing. The level of packing, however, is more dependent on the grain size distribution within a silt deposit – uniformly sorted deposits not being able to achieve such close packing as those in which there is a range of grain size. This, in turn, influences the values of porosity, void ratio, and bulk and dry densities (Table 9.8).

Dilatancy is characteristic of fine sands and silts. The environment is all important for the development of dilatancy since conditions must be such that expansion can take place. What is more, the soil particles must be well wetted and it appears that certain electrolytes exercise a dispersing effect. The moisture content at which fine sands and silts often become dilatant usually varies between 16 and 35%.

Consolidation of silt is influenced by grain size, particularly the size of the clay fraction, porosity and natural moisture content. Primary consolidation may account for over 75% of total consolidation. In addition, settlement may continue for several months after construction is completed because the rate at which water can drain from the voids under the influence of applied stress is slow. The angle of shearing resistance decreases with increasing void ratio. It also is dependent upon the plasticity index, grain interlocking and density.

Loess owes it engineering characteristics largely to the way in which it was deposited since this gave it a metastable structure, in that initially the particles were loosely packed. The porosity of the structure is enhanced by the presence of fossil root-holes. The latter are lined with carbonate cement, which helps bind the grains together. This means that the initial, loosely packed structure is preserved and the carbonate cement helps provide the bonding strength of loess. However, the chief binder is usually the clay matrix. On wetting, the clay bond in many loess soils becomes soft, which leads to the collapse of the metastable structure. The breakdown and collapse of the soil structure can occur under its own weight.

Loess deposits generally consist of 50 to 90% particles of silt size. In fact, sandy, silty and clayey loess can be distinguished. The undistorted densities of loess range from around 1.2 to 1.36 Mg m^{-3}. If this material is wetted, consolidated or reworked, then the density increases, sometimes to as high as 1.6 Mg m^{-3}. The liquid limit of loess averages about 30% (exceptionally liquid limits as high as 45% have been recorded), and their plasticity index ranges from about 4 to 9%, but averages 6%. As far as their angle of shearing resistance is concerned, this usually varies from 30 to 34°. In the unweathered state above the water table the unconfined compressive strength of loess may amount to several hundred kilopascals. On the other hand, if some loess soils are permanently submerged the metastable structure breaks down so that loess then becomes a slurry. Loess deposits are better drained (their permeability ranges from 10^{-7} to 10^{-5} m s^{-1}) than are true silts because of the fossil root-holes. As would be expected their permeability is appreciably higher in the vertical than in the horizontal direction.

Normally, loess can carry high loadings without significant settlement when natural moisture contents are low (e.g. around 10%). However, the density of loess is the most important factor controlling its shear strength and settlement. On wetting, large settlements due to collapse of the metastable structure (in some cases in the order of metres), and low shearing resistance are encountered when the density of loess is below 1.30 Mg m^{-3}, whereas if the density exceeds 1.45 Mg m^{-3} settlement is small and shearing resistance fairly high. The collapse process represents a rearrangement of soil particles into a denser state of packing. Collapse on saturation normally only takes a short period of time, although the more clay a collapsible loess contains, the longer the period tends to be. Collapse can lead to foundation failure.

Unlike silt, loess is not frost susceptible, this being due to its more permeable character, but like silt it can exhibit quick conditions and it is difficult, if not impossible, to compact. Because of its porous structure a 'shrinkage' factor must be taken into account when estimating earthwork.

Another problem that may be associated with some loess soils is the development of pipe systems. Extensive pipe systems, which have been referred to as loess karst, may run sub-parallel to a slope surface and the pipes may have diameters up to 2.0 m. Pipes tend to develop by weathering and widening taking place along the joint systems in loess. The depths to which pipes develop may be inhibited by changes in permeability associated with the occurrence of palaeosols.

9.7 Clay deposits

The principal minerals in a deposit of clay tend to influence its engineering behaviour. For example, the plasticity of a clay soil is influenced by the amount of its clay fraction and the type of clay minerals present since clay minerals greatly influence the amount of attracted water held in a soil. Usually, active clays have a relatively high water holding capacity and a high cation exchange capacity. They also are highly thixotropic and have a low permeability. Furthermore, the undrained shear strength is related to the amount and type of clay minerals present in a clay deposit, together with the presence of cementing agents. In particular, strength is reduced with increasing content of mixed-layer clay and montmorillonite in the clay fraction. The increasing presence of cementing agents, especially calcite, enhances the strength of the clay.

One of the most notable characteristics of some clay soils, from the engineering point of view, is their susceptibility to slow volume changes that can occur independently of loading due to swelling or shrinkage. Differences in the period and magnitude of precipitation and evapotranspiration are the major factors influencing the swell–shrink response of a clay soil. There are two modes of swelling in clay soils, namely, intercrystalline and intracrystalline swelling. Intercrystalline swelling takes place when the uptake of moisture is restricted to the external clay particle surfaces and the void spaces between them. Intracrystalline swelling, on the other hand, is characteristic of the smectite family of clay minerals, of montmorillonite in particular. The individual molecular layers that make up a crystal of montmorillonite are weakly bonded so that on wetting water enters not only between the crystals but also between the unit layers that comprise the crystals. Generally, kaolinite has the smallest swelling capacity of the clay minerals and nearly all of its swelling is of the intercrystalline type. Illite may swell by up to 15% but intermixed illite and montmorillonite may swell some 60 to 100%.

The volume change that occurs due to evapotranspiration from a clay soil can be conservatively predicted by assuming the lower limit of the soil moisture content to be the shrinkage limit. Desiccation beyond this value cannot bring about further volume change. The effect of desiccation on clay soils is similar to that of heavy overconsolidation. Desiccation cracks develop and may extend to depths of 2 m in expansive clays, and gape up to 150 mm. Transpiration from vegetative cover is a major cause of water loss from soils in semi-arid regions. Changes in soil suction may be expected over a depth of some 2.0 m between the wet and dry seasons. However, the complete depth of active clay profiles usually does not become fully saturated during the wet season in semi-arid regions. When vegetation is cleared from a site, its desiccating effect is removed. Hence, the subsequent regain of moisture by clay soils leads to them swelling. Swelling movements of over 350 mm have been reported for expansive clays in South Africa and Texas.

Volume changes in clay soils also occur as a result of loading (see Section 9.4) or unloading. When an excavation is made in a clay soil with weak diagenetic bonds, elastic rebound causes immediate dissipation of some stored strain energy in the soil. However, part of the strain energy is retained due to the restriction on lateral straining in the plane parallel to the ground surface. The lateral effective stresses either remain constant or decrease as a result of plastic deformation of the soil as time passes. This plastic deformation can result in significant time-dependent vertical heaving. However, creep of weakly bonded soils is not a common cause of heaving.

Fissures in clay soils play an extremely important role in the failure mechanism of overconsolidated clays. For example, the strength along fissures in clay soils is only slightly higher than the residual strength of the intact clay. Hence, the upper limit of the strength of fissured clay soil is represented by its intact strength whilst the lower limit corresponds to the strength along the fissures. The operational strength, which is somewhere between the two, often is significantly higher than the fissure strength. Fissures allow concentrations of shear stress to develop that locally exceed the peak strength, thereby giving rise to progressive failure. Under stress, the fissures in clay soil propagate and coalesce in a complex manner. The ingress of water into fissures means that the pore water pressure in the soil increases, which means that its strength is reduced. Fissures in normally consolidated clays have no significant practical consequences.

The greatest variation in the engineering properties of clay soils can be attributed to the degree of weathering that they have undergone. For instance, consolidation of a clay deposit gives rise to an anisotropic texture due to the rotation of the platey minerals. Secondly, diagenesis bonds particles together either by the development of cement, the intergrowth of adjacent grains or the action of van der Waals charges that are operative at very small grain separations. Weathering reverses these processes altering the anisotropic structure and destroying or weakening interparticle bonds. Ultimately, weathering through the destruction of interparticle bonds leads to a clay deposit reverting to a normally consolidated sensibly remoulded condition. Higher moisture contents are found in more weathered clay. This progressive degrading and softening also is accompanied by reductions in strength and deformation moduli with a general increase in plasticity.

9.8 Laterite and lateritic soils

Ferruginous and aluminous clay soils are frequent products of weathering in subhumid tropical and monsoon regions. They are characterized by the presence of iron and aluminium oxides and hydroxides. These compounds, especially those of iron, are responsible for the red, brown and yellow colours of the soils. The soils may be fine-grained or they may contain nodules or concretions. Concretions occur in the matrix where there are higher concentrations of oxides in the soil. Extensive accumulations of oxides give rise to laterite.

Laterite is a residual clay-like deposit that generally occurs below a hardened ferruginous crust or hardpan. The ratios of silica (SiO_2) to sesquioxides (Fe_2O_3, Al_2O_3) in laterites usually are less than 1.33 – ratios between 1.33 and 2.0 are indicative of lateritic soils, and those greater than 2.0 are indicative of non-lateritic types. Laterite tends to develop in regions where drier periods mean that the water table is lowered. The small amount of iron that has been mobilized in the ferrous state by the groundwater is then oxidized to form hematite (Fe_2O_3), or if hydrated to form goethite ($FeO(OH)$). The movement of the water table leads to the gradual accumulation of iron oxides at a given horizon in the soil profile. A cemented layer of laterite is formed that may be a continuous or honeycombed mass, or nodules may be formed, as in laterite gravel. Concretionary layers often are developed near the surface in lowland areas because of the high water table. Laterite hardens on exposure to air.

Laterite commonly contains all size fractions from clay to gravel and sometimes even larger material. Usually, at or near the surface the liquid limits of laterites do not exceed 60% and the plasticity indices are less than 30%. Consequently, laterites are of low to medium plasticity. The activity of laterites may vary between 0.5 and 1.75. Values of some common properties of laterite are given in Table 9.10.

Lateritic soils, particularly where they are mature, furnish a good bearing stratum. The hardened

Table 9.10 Changes in engineering properties of laterite with depth.

Depth(m)	Dry density(Mg m^{-3})	Moisture content(%)	Plastic limit(%)	Liquid limit(%)	Strength ϕ'	Strength c' (kPa)
3.0–3.6	1.63	28	31	44	6°	84
4.2–4.9	1.51	20	26	54	15°	138
7.9–8.5	1.55	26	49	65	5°	63
11.0–11.6	1.71	18	NP	–	23°	216

NP = non-plastic

crust has a low compressibility and therefore settlement is likely to be negligible. In such instances, however, the strength of laterite may decrease with increasing depth.

Red earths or latosols are residual ferruginous soils in which oxidation readily occurs. Such soils tend to develop in undulating country and most of them appear to have been derived from the first cycle of weathering of the parent material. They differ from laterite in that they behave as a clay soil and do not possess strong concretions. They do, however, grade into laterite.

9.9 Cretes and sabkhas

A common feature of arid regions is the cementation of soils by the precipitation of salts in the upper layers from the groundwater. The types of salt held in solution, and also those precipitated, depend on the character of the groundwater, as well as the prevailing temperature and humidity conditions. The process may lead to the development of various crusts or cretes in which unconsolidated deposits are cemented. Cretes and crusts may form continuous sheet or isolated patch-like masses at the ground surface when the groundwater table is at, or near, this level, or at some other position within the ground profile depending on the position of the water table. Therefore, duricrusts or pedocretes are soils that have been to a greater or lesser extent cemented and/or replaced by calcite (calcrete), dolomite (dolocrete), iron oxides (ferricrete, plinthite or laterite; these require a subhumid climate that allows chemical weathering to release the iron from parental material for their formation), gypsum (gypcrete), magnesite (magnesicrete), phosphate (phoscrete) or silica (silcrete). These materials take three forms, namely, indurated (e.g. hardpans and nodules), non-indurated (soft or powder forms), and mixtures of the two (e.g. nodular pedocretes). The most commonly precipitated material is calcium carbonate. As the carbonate content increases it first occurs as scattered concentrations of flaky habit, then as hard concretions. Once it exceeds 60% the concentration becomes continuous. The calcium carbonate in calcrete profiles decreases from top to base, as generally does the hardness. One type of pedocrete can replace another, for instance, calcretes may be silicified. The development of calcrete is inhibited beyond a certain aridity since the low precipitation is unable to dissolve and drain calcium carbonate towards the water table. Consequently, in more arid regions gypcrete may take the place of calcrete.

In arid regions sabkha conditions commonly develop in low lying coastal zones and inland plains with shallow water tables. These are extensive saline flats that are underlain by sand, silt or clay and often are encrusted with salt (Fig. 9.4). Highly developed sabkhas tend to retain a greater proportion of soil moisture than moderately developed sabkhas. In addition, the higher the salinity of the groundwater, the greater the amount of water retained by the sabkha at a particular drying temperature. The height to which water may rise from the water table is a function of the size and continuity of the pore spaces in the soil.

Minerals that are precipitated from groundwater in arid deposits also have high solution rates so that flowing groundwater may lead to the development of solution features. Problems such as increased

Fig. 9.4 Salt tepees on the Devil's Golf Course, Death Valley, California.

permeability, reduced density and settlement are liable to be associated with engineering works or natural processes that result in a decrease in the salt concentration of groundwaters. Changes in the hydration state of minerals, such as swelling clays and calcium sulphate, also causes significant volume changes in soils. In particular, low density sands that are cemented with soluble salts such as sodium chloride are vulnerable to salt removal by dissolution by freshwater, leading to settlement. Hence, rainstorms and burst water mains present a hazard, as does watering of grassed areas and flower beds. The latter should be controlled and major structures should be protected by drainage measures to reduce the risks associated with rainstorms or burst water pipes. Within coastal sabkhas the dominant minerals are calcite, dolomite and gypsum with lesser amounts of anhydrite, magnesite ($MgCO_3$), halite ($NaCl$) and carnalite ($KMgCl_3.6H_2O$), together with various other sulphates and chlorides. Highly saline groundwater may contain up to 23% $NaCl$, and occur close to ground level. In fact, the sodium chloride content of the water can be high enough to represent a corrosion hazard. In the case of inland sabkhas, the minerals precipitated within the soil are much more variable than those of coastal sabkhas since they depend on the composition of the local groundwater.

Sabkha soils frequently are characterized by low strength. Furthermore, some surface clay soils that are normally consolidated or lightly overconsolidated may be sensitive to highly sensitive. The low strength is attributable to the concentrated salt solutions in sabkha brines; the severe climatic conditions under which sabkha deposits are formed (e.g. large variations in temperature and excessive wetting–drying cycles) that can give rise to instability in sabkha soils; and the ready solubility of some of minerals that act as cements in these soils. As a consequence, the bearing capacity of sabkha soils and their compressibility frequently do not meet routine design requirements.

9.10 Dispersive soils

Dispersive soils deflocculate in the presence of relatively pure water forming colloidal suspensions. Consequently, such soils are highly susceptible to erosion and piping. They occur in semi-arid regions where the annual rainfall is less than 860 mm. Such soils contain a higher content of dissolved sodium in their pore water than ordinary soils (they may contain up to 12% sodium). There are no significant differences in the clay contents of dispersive and non-dispersive soils, except that soils with less than 10% clay particles may not have enough colloids to support dispersive piping. Usually, dispersive soils contain a moderate to high content of clay.

Dispersive erosion depends on the mineralogy and chemistry of the soil on the one hand, and the dissolved salts in the pore and eroding water on the other. The presence of exchangeable sodium is the main chemical factor contributing towards dispersive soil behaviour. This is expressed in terms of the exchangeable sodium percentage (ESP):

$$ESP = (exchangeable\ sodium\ /\ cation\ exchange\ capacity) \times 100 \qquad (9.5)$$

where the units are given in meq/100 g of dry soil. Above a threshold ESP value of 10%, soils have their free salts leached by seepage of relatively pure water and are prone to dispersion. Soils with ESP values above 15% are highly dispersive. On the other hand, those soils with low cation exchange values (15 meq/100 g of soil) are non-dispersive at ESP values of 6% or below. Similarly, soils with high cation exchange capacity (CEC) values and a plasticity index greater than 35% swell to such an extent that dispersion is not significant. High ESP values and piping potential generally exist in soils in which the clay fraction is composed largely of smectitic clays. Some illitic soils are highly dispersive. High values of ESP and high dispersibility generally are not common in clays composed largely of kaolinites. Another property governing the susceptibility of clayey soils to dispersion is the total content of dissolved solids (TDS) in the pore water. In other words, the lower the content of dissolved solids in the pore water the greater the susceptibility of sodium saturated clays to dispersion. The total dissolved solids for this specific purpose is the total content of calcium, magnesium, sodium and potassium in milli-equivalents per litre (Fig. 9.5a). The dispersive potential also may be determined from the relationship between ESP and CEC (Fig. 9.5b).

Damage due to internal erosion of dispersive soil leads to the formation of pipes and internal cavities on slopes. Piping is initiated by dispersion of clay particles along desiccation cracks, fissures and rootholes. Colloidal erosion damage also has led to the failure of earth dams. The pipes become enlarged rapidly and this can lead to failure of a dam. Dispersive erosion may be caused by initial seepage through an earth dam in areas of higher soil permeability, especially areas where compaction may not be so effective such as around conduits, against concrete structures and at the foundation interface; through desiccation cracks; cracks due to differential settlement or those due to hydraulic fracturing. Indications of piping take the form of small leakages of muddy coloured water after initial filling of a reservoir. In addition, severe erosion damage, forming deep gullies, occurs on embankments after rainfall. Fortunately, when dispersive soils are treated with lime they are transformed to a non-dispersive state.

9.11 Glacial soils
9.11.1 Tills
Till usually is regarded as being synonymous with boulder clay. It is deposited directly by ice. The character of till deposits varies appreciably and depends on the lithology of the material from which it was derived, on the position in which it was transported in the glacier, and on the mode of deposition. For instance, argillaceous rocks, such as shales and mudstones, are more easily abraded and so produce fine-grained tills that are richer in clay minerals and therefore are more plastic than other tills. Mineral composition also influences the natural moisture content, it being slightly higher in tills containing appreciable quantities of clay minerals or mica.

Lodgement till is plastered on to the ground beneath a moving glacier in increments as the basal ice melts. Because of the overlying weight of ice such deposits are overconsolidated. Due to abrasion and grinding the proportion of silt and clay size material is relatively high in lodgement till (e.g. the clay fraction varies from 15–40%). Lodgement till is commonly stiff, dense and relatively incompressible. Hence, it is practically impermeable. Fissures frequently are present in lodgement till, especially if it is clay matrix dominated.

Ablation till accumulates on the surface of the ice when englacial debris melts out, and as a glacier decays the ablation till is lowered to the ground. It is therefore normally consolidated. Because it has not been subjected to much abrasion, ablation till is characterized by abundant large stones that are angular, the proportion of sand and gravel is high, and clay is present only in small amounts (usually less than

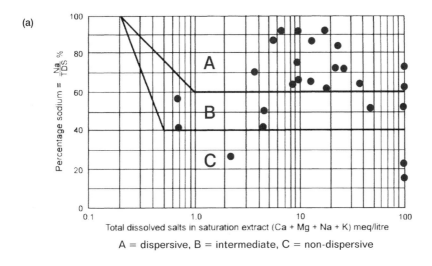

A = dispersive, B = intermediate, C = non-dispersive

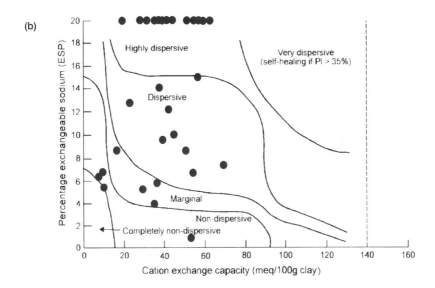

Fig. 9.5 (a) Potential dispersivity chart. (b) Chart for classification of the degree of dispersivity of soils. Showing some soils from Kwazulu-Natal, South Africa.

10%). Since the texture is loose, ablation till can have a low in situ density. Ablation till usually forms a thinner deposit than lodgement till as it consists of the load carried at the time of ablation. In addition, ablation tills have a high proportion of far-travelled material and may not contain any of the local bedrock.

The percentage of 'fines' in tills can be used to distinguish granular, well-graded and matrix-dominated tills, the boundaries being placed at 15 and 45% respectively. Tills are frequently gap graded, the gap generally occurring in the sand fraction (Fig. 9.6). Large, often very local, variations can occur in the gradings of till that reflect local variations in the formation processes, particularly the comminution processes. The range in the proportions of coarse and fine fractions in tills dictates the degree to which the properties of the fine fraction influence the properties of the composite soil. The variation in the engineering properties of the fine soil fraction is greater than that of the coarse fraction, and this often tends to dominate the engineering behaviour of the till. The specific gravity of till deposits is often remarkably uniform, varying from 2.77 to 2.78. These values suggest the presence of fresh minerals in the fine fraction, that is, rock flour rather than clay minerals. Rock flour behaves more like granular material than cohesive and has a low plasticity. The consistency limits of tills are dependent upon water content, grain size distribution and the properties of the fine-grained fraction. Generally, however, the plasticity index is small and the liquid limit of tills decreases with increasing grain size. The compressibility and consolidation of tills are principally determined by the clay content. For example, the value of compressibility index tends to increase linearly with increasing clay content. As the degree of weathering of till increases, so does the clay fraction and moisture content. This, in turn, leads to changes in the liquid and plastic limits and in the shear strength.

Fissures can have a preferred orientation in till and tend to be variable in character, spacing, orientation and areal extent. Opening and softening along fissures leads to a rapid reduction in undrained shear strength along the fissures. In fact, the undrained shear strength along fissures in till may be as little as one-sixth that of the intact soil. The nature of the various fissure coatings (sand, silt or clay-size material) is of critical importance in determining the shear strength behaviour of fissured tills. Deformation and permeability also are controlled by the nature of the fissure surfaces and coatings.

9.11.2 Fluvio-glacial deposits

Fluvio-glacial deposits can be subdivided into two categories, namely, those which develop in contact with the ice (the ice contact deposits) and those that accumulate beyond the limits of the ice, forming in streams, lakes or seas (i.e. the proglacial deposits). Examples of ice contact deposits include outwash fans, kames and eskers. Outwash fans range in particle size from coarse sands to boulders. When they

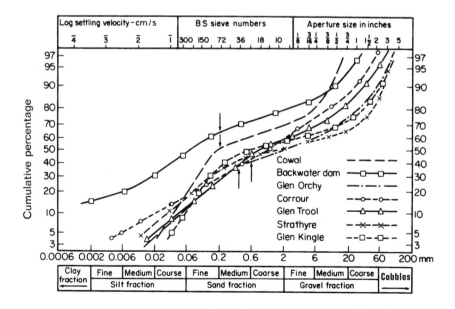

Fig. 9.6 Typical grading curves of tills from Scotland (courtesy of Geological Society).

are first deposited their porosity may be anything from 25 to 50% and they tend to be very permeable. The finer silt-clay fraction is transported further downstream. Kames, kame terraces and eskers usually consist of sands and gravels.

The most familiar proglacial deposits are varved clays. A varve is a couplet that consists of a coarser layer of silt size and a finer layer of clay size. The thickness of the individual varve frequently is less than 2 mm, although much thicker layers have been noted in a few deposits. Varved clays tend to be normally consolidated or lightly overconsolidated, however, it usually is difficult to make the distinction. In many cases the precompression may have been due to subsequent ice loading.

The range of liquid limits for varved clays tends to vary between 30 and 80% whilst that of the plastic limit often varies between 15 and 30%. These limits allow the material to be classified as inorganic silty clay of medium to high plasticity or compressibility. In some varved clays the natural moisture content is near the liquid limit. Consequently, they are soft and have sensitivities generally around 4. The varves are responsible for the anisotropic behaviour of these clays. For example, the angle of friction varies with the angle of orientation, the minimum occurring when the varves are orientated at 45° to the maximum principal stress. Similarly, the values of the coefficients of volume consolidation and of permeability are higher parallel to the varves.

9.11.3 Quick clay

The material of which quick clays are composed is predominantly smaller than 0.002 mm. Many deposits, however, seem to be very poor in clay minerals, containing high proportions of ground-down fine quartz. It has been suggested that this is of glacial origin and quick clays do occur in certain parts of the northern hemisphere that were subjected to glacial action during Pleistocene times. The fabric of these soils contains aggregations. Granular particles, whether aggregations or primary minerals, are rarely in direct contact, being linked generally by bridges of fine particles. Clay minerals usually are non-oriented and clay coatings on primary minerals tend to be uncommon, as are cemented junctions. Networks of platelets occur in some soils.

Quick clays generally exhibit little plasticity, their plasticity index generally varying from 8 to 12%. Their liquidity index normally exceeds 1, and their liquid limit is often less than 40%. Quick clays usually are inactive, their activity frequently being less than 0.5. The most extraordinary property possessed by quick clays is their very high sensitivity. In other words, a large proportion of their undisturbed strength is lost permanently following shear. The small fraction of the original strength regained after remoulding may be attributable to the development of some different form of interparticle bonding.

9.12 Peat

Peat is an accumulation of partially decomposed and disintegrated plant remains that have been fossilized under conditions of incomplete aeration and high water content. Physico-chemical and biochemical processes cause this organic material to remain in a state of preservation over a long period of time. Macroscopically, peaty material can be divided into three basic groups, namely, amorphous granular, coarse fibrous and fine fibrous peat. The amorphous granular peats have a high colloidal fraction, holding most of their water in an adsorbed rather than free state. In the other two types the peat is composed of fibres, these usually being woody. The mineral content of peat may be as low as 2% in some highly organic peats, although very occasionally it may be as high as 50%. It usually consists of quartz sand and silt, and in many deposits it increases with depth. The mineral content influences the engineering properties of peat.

The void ratio of peat ranges between 9 (for dense amorphous granular peat) and 25 (for fibrous types with a high content of sphagnum). Generally, it tends to decrease with depth within a peat deposit. Such high void ratios give rise to phenomenally high water contents. The latter varies according to the type of peat and may be as low as 500% in some amorphous granular varieties, whilst values exceeding 3000% have been recorded from coarse fibrous varieties. The volumetric shrinkage of peat increases up to a maximum and then remains constant, the volume being reduced almost to the point of complete

Fig. 9.7 The Holme Posts at Whittlesey Mere, Cambridgeshire, England, where peat has undergone drainage for more than 200 years. The original post was installed in 1851 through 7 m of peat into clay beneath, the top of the post then being flush with the ground surface. Since then the ground surface has been lowered by some 4 m.

dehydration. The amount of shrinkage generally ranges between 10 and 75% of the original volume of peat and it can involve reductions in void ratio from over 12 down to about 2. Because of the very high moisture content of peat, it undergoes notable subsidence when drained. The Fenlands of England provide a classic example, the peat in places having been drained for over 400 years so that in some areas the thickness of peat has been halved (Fig. 9.7). Amorphous granular peat has a higher bulk density than the fibrous types. For instance, in the former it can range up to 1.2 Mg m^{-3} whilst in woody fibrous peat it may be half this figure. However, the dry density is a more important engineering property of peat, influencing its behaviour under load. Dry densities of drained peat fall within the range 65 to 120 kg m^{-3}. The dry density is influenced by the mineral content and higher values than those quoted occur when peat possesses high mineral residues. Due to its extremely low submerged density, which may be between 15 and 35 kg m^{-3}, peat is especially prone to rotational failure or failure by spreading.

Although the unconfined compressive strength of undrained peat is negligible, it is increased by drainage to values between 20 and 30 kPa and the modulus of elasticity to between 100 and 140 kPa. When loaded, peat deposits undergo high deformations but their modulus of deformation tends to increase with increasing load. If peat is very fibrous it appears to suffer indefinite deformation without planes of failure developing. On the other hand, failure planes nearly always form in dense amorphous granular peats.

If the organic content of a soil exceeds 20% by weight, consolidation becomes increasingly dominated by the behaviour of the organic material. In the case of peat, it undergoes a decrease in permeability of several orders of magnitude on loading. Moreover, residual pore water pressure affects primary consolidation and considerable secondary consolidation further complicates settlement prediction. Differential and excessive settlement are the principal engineering problems of peaty soil. When a load is applied to peat, settlement occurs because of the low lateral resistance offered by the adjacent unloaded peat. Serious shearing stresses are induced even by moderate loads. Should the load exceed a given minimum, then settlement may be accompanied by creep, lateral spreading, or in extreme cases by rotational slip and upheaval of adjacent ground. At any given time the total settlement of peat due to loading involves settlement with and without volume change. Settlement without volume change is the more serious for it can give rise to the types of failure mentioned. What is more, it does not enhance the strength of peat.

9.13   Description of rocks and rock masses

Description is the initial step in an engineering assessment of rocks and rock masses. Intact rock may be described from a geological or engineering point of view. As far as geology is concerned, the origin and mineral content of a rock are of prime importance, as is its texture and any change that has occurred since its formation. In this respect the type of a rock provides an indication of its origin, mineralogical composition and texture (Table 1.1). Only a basic petrographical description of a rock is required when describing a rock mass. The micro-petrographic description of rocks for engineering purposes includes the determination of all parameters that cannot be obtained from a macroscopic examination of a rock sample, such as mineral content, grain size, texture, and amount of alteration, that have a bearing on the mechanical behaviour of the rock or rock mass. For example, grain size exerts some influence on the physical properties of a rock in that finer grained rocks usually are stronger than coarser grained varieties. Rock material tends to deteriorate in quality as a result of weathering and or alteration (see Chapter 1).

As far as engineering properties are concerned the dry density and porosity of rocks can be described as shown in Table 9.11, and the unconfined compressive strength as shown in Table 9.12. The strength of a rock can be estimated as shown in Table 9.13 if it is not determined in the laboratory, although such an estimate can only be very approximate. The point load strength provides an indirect measure of the tensile strength of a rock by causing it to fail between two loaded platens that are conical in shape. A

Table 9.11   Description of dry density and porosity of intact rock.

Class	Dry density (Mg m^{-3})	Description	Porosity (%)	Description
1	less than 1.8	Very low	over 30	Very high
2	1.8–2.2	Low	30–15	High
3	2.2–2.55	Moderate	15–5	Medium
4	2.55–2.75	High	5–1	Low
5	over 2.75	Very high	less than 1	Very low

Table 9.12   Grades of unconfined compressive strength.

Term	Strength (MPa)
Very weak	less than 1.25
Weak	1.25–5.00
Moderately weak	5.00–12.50
Moderately strong	12.50–50
Strong	50–100
Very strong	100–200
Extremely strong	over 200

Table 9.13 Estimation of the strength of intact rock.

Description	Approximate unconfined compressive strength(MPa)	Field estimation
Very strong	Over 100	Very hard rock – more than one blow of geological hammer required to break specimen
Strong	50–100	Hard rock – hand-held specimen can be broken with a single blow of hammer
Moderately strong	12.5–50	Soft rock – 5 mm indentations with sharp end of pick
Moderately weak	5.0–12.5	Too hard to cut by hand
Weak	1.25–5.0	Very soft rock – material crumbles under firm blows with the sharp end of a geological hammer

Table 9.14 Point-load strength classification.

	Point load strength index(MPa)	Equivalent uniaxial compressive strength(MPa)
Extremely high strength	over 10	over 160
Very high strength	3–10	50–160
High strength	1–3	15–60
Medium strength	0.3–1	5–16
Low strength	0.1–0.3	1.6–5
Very low strength	0.03–0.1	0.5–1.6
Extremely low strength	less than 0.03	less than 0.5

description of the point load strength is given in Table 9.14. The description of the permeability of rock masses is referred to in Chapter 4 and of discontinuities in rock masses in Chapter 1.

Classifications of intact rock are based upon some selected mechanical properties. The specific purpose for which a classification is developed obviously plays an important role in determining which mechanical properties of intact rock are chosen. The object of the classification is to provide a reliable basis for assessing rock quality. Any classification of intact rock for engineering purposes should be relatively simple, being based on significant mechanical properties (e.g. the unconfined compressive strength and Young's modulus) so that it has a wide application.

9.14 Engineering aspects of igneous and metamorphic rocks

The plutonic igneous rocks are characterized by granular texture and relatively homogeneous composition. In their unaltered state they are essentially sound and durable with adequate strength for any engineering requirement (Table 9.15). In some instances, however, these rocks may be altered by weathering or hydrothermal action thereby reducing their strength. Fissure zones are not uncommon in granites. The rock mass may be very much fragmented along such zones, indeed it may be reduced to sand size material, and it may have undergone varying degrees of kaolinization. Generally, the weathered products of plutonic rocks have large clay contents.

Table 9.15 Geomechanical properties of some British igneous and metamorphic rocks.

	Specific gravity	Unconfined compressive strength(MPa)	Point load strength(MPa)	Young's modulus(GPa)
Mount Sorrel Granite	2.68	176.4 (VS)[1]	11.3 (EHS)[2]	60.6 (VL)[3]
Eskdale Granite	2.65	198.3 (VS)	12.0 (EHS)	56.6 (L)
Dalbeattie Granite	2.67	147.8 (VS)	10.3 (EHS)	41.1 (L)
Markfieldite	2.68	185.2 (VS)	11.3 (EHS)	56.2 (L)
Granophyre (Cumbria)	2.65	204.7 (ES)	14.0 (EHS)	84.3 (VL)
Andesite (Somerset)	2.79	204.3 (ES)	14.8 (EHS)	77.0 (VL)
Basalt (Derbyshire)	2.91	321.0 (ES)	16.9 (EHS)	93.6 (VL)
Slate* (North Wales)	2.67	96.4 (S)	7.9 (VHS)	31.2 (L)
Slate+ (North Wales)		72.3 (S)	4.2 (VHS)	
Schist* (Aberdeenshire)	2.66	82.7 (S)	7.2 (VHS)	35.5 (L)
Schist+		71.9 (S)	5.7 (VHS)	
Gneiss	2.66	162.0 (VS)	12.7 (EHS)	46.0 (L)
Hornfels (Cumbria)	2.68	303.1 (ES)	20.8 (EHS)	109.3 (VL)

1. Classification of strength: ES = extremely strong, over 200 MPa; VS = very strong, 100 to 200 MPa; S = strong, 50 to 100 MPa.
2. Classification of point load strength: EHS = extremely high strength, over 10 MPa; VHS = very high strength, 3 to 10 MPa.
3. Classification of deformability (Young's modulus): VL = very low, over 60 GPa; L = low, 30 to 60 GPa.
* Tested normal to cleavage or schistosity + Tested parallel to cleavage or schistosity

Older volcanic deposits frequently do not prove a problem in engineering, ancient lavas having strengths commonly in excess of 200 MPa. On the other hand, volcanic deposits of geologically recent age at times prove treacherous, particularly if they have to carry heavy loads. This is because they often represent markedly anisotropic sequences in which lavas, pyroclasts and mudflows are interbedded. In addition, weathering during periods of volcanic inactivity may have produced fossil soils, these being of much lower strength. The individual laval flows may be thin and transected by a polygonal pattern of cooling joints. They also may be vesicular or contain pipes, cavities or even tunnels (Fig. 9.8). Certain basalts, known as slaking basalts, that have been subject to deuteric action (e.g. of hot solutions or gases) can breakdown rapidly on exposure.

Pyroclasts usually give rise to extremely variable ground conditions due to wide variations in strength, durability and permeability. Their behaviour very much depends upon their degree of induration, for example, many agglomerates have sufficient strength to support heavy loads and also have low permeability. By contrast, ashes are invariably weak and often highly permeable. One particular hazard concerns ashes, not previously wetted, which are metastable, that is, they exhibit a significant decrease in their void ratio on saturation. Ashes frequently are prone to sliding. Montmorillonite is not an uncommon constituent in the weathered products of basic ashes.

Slates, phyllites and schists are characterized by cleavage and schistosity that means that they are fissile and as such appreciably stronger across than along the lineation (Table 9.15). Normally, slates, phyllites and schists weather slowly but these rocks may be fractured and deformed in regions that have suffered extensive folding. Schists, slates and phyllites are variable in quality, some forming excellent foundations for heavy structures; others are so poor as to be wholly undesirable. For instance, talc,

Fig.9.8   Collapse of a tunnel in basalt lavas, Hawaii.

chlorite and sericite schist are weak rocks containing planes of schistosity only a millimetre or so apart. Some schists become slippery upon weathering and therefore fail under a moderately light loading. The engineering performance of gneiss is usually similar to that of granite. Fresh thermally metamorphosed rocks such as quartzite and hornfels are very strong and afford good ground conditions. Marble has the same advantages and disadvantages as other carbonate rocks.

9.15   Engineering aspects of sedimentary rocks

9.15.1   Mudrocks

Consolidation and the parallel orientation of platey minerals, notably micas, give rise to the fissility of shales. An increasing content of siliceous or calcareous material gives a less fissile shale whilst carbonaceous or organic shales are exceptionally fissile. Moderate weathering increases the fissility of shale by partially removing the cementing agents along the laminations or by expansion due to the hydration of clay particles. Intense weathering produces a soft clay-like soil.

The porosity of shale may range from slightly under 5% to just over 50%, with the natural moisture content varying from less than 5%, to as high as 35% in some clayey shales. When the natural moisture content of shales exceeds 20% they tend to develop high pore water pressures. Desiccation of shale following exposure, leads to the creation of negative pore pressures and consequent tensile failure of weak intercrystalline bonds. This leads to the production of shale particles of coarse sand/fine gravel size. Alternate wetting and drying causes a rapid breakdown of compaction shales, for example, some undergo complete disintegration after several cycles of drying and wetting. Mudstones tend to break down along irregular fracture patterns, which when well developed, can mean that these rocks also disintegrate within one or two cycles of wetting and drying. The swelling properties of certain clay shales have proved extremely detrimental to the integrity of many engineering structures. Swelling is attributable to the absorption of free water by certain clay minerals, notably, montmorillonite, in the clay fraction of shale. Highly fissured overconsolidated shales have greater swelling tendencies than poorly fissured clayey shales, the fissures providing access for water.

The unconfined compressive strengths of shales vary significantly, for instance, from 200 kPa to 20 MPa (Table 9.16a). Those samples that exhibit the high strengths generally are the cemented types. In weak compaction shales cohesion may be lower than 15 kPa and the angle of friction as low as 5°. As noted, cemented shales are stronger and more durable than compacted shales, for instance, certain dolomitic shales of Ordovician age have values of cohesion and angle of friction of 750 kPa and 56° respectively. Generally, shales with values of cohesion less than 20 kPa and angles of friction of less than 20° are likely to present problems. The elastic moduli of compaction shales range between 140 and 1400 MPa, whilst well cemented shales have elastic moduli in excess of 14 000 MPa. Values of strength and Young's modulus can be up to five times greater when shale is tested normal as opposed to parallel to the direction of lamination.

Severe settlements may take place in low grade compaction shales. Conversely, uplift frequently occurs in excavations in shales and is attributable to swelling and heave. Rebound on unloading of shales during excavation is attributed to heave due to the release of stored strain energy. The greatest amount of rebound occurs in heavily overconsolidated compaction shales. Sulphur compounds frequently are present in shales, clays, and mudstones. An expansion in volume large enough to cause structural damage can occur when sulphide minerals such as pyrite and marcasite suffer oxidation to give hydrous sulphates.

Table 9.16 Some geomechanical properties of sedimentary rocks.

a. Mudrocks

	Tow Law*	Wrexham* (weathered)	Bearpaw Shale+	Bearpaw Shale+ (weathered)
Natural moisture content (%)				
Range	5.9–10.5	6.4–14.3	19–27	25–36
Mean	7.3		19	35
Bulk density (Mg m^{-3})				
Range	2.43–2.65	2.04–2.12		
Mean	2.51			
Dry density (Mg m^{-3})				
Range	2.2–2.4	1.84–1.92		
Mean	2.34			
Unconfined compressive strength (MPa)				
Range	25.7–45.4		1–28	0.5–1.0
Mean	35.5			
Point load index (MPa)				
Diametral	0.9–0.37			
Axial	1.22–2.67			
Liquid limit (%)				
Range		31–41	50–150	80–150
Mean			120	120
Plasticity index (%)				
Range		15–16	30–130	30–130
Mean			95	95

*From Coal Measures in Britain   +Clay shales from Canada

Table 9.16 Cont'd

b. Sandstones

	Fell Sandstone, Northumberland, UK	Sherwood Sandstone, Nottinghamshire, UK	Natal Group Sandstone, South Africa	Clarens Sandstone, South Africa
Dry density (Mg m⁻³)				
Range	2.14–2.4	1.77–1.96	2.40–2.54	1.93–2.58
Mean	2.32	1.83	2.48	2.27
Effective porosity (%)				
Range	6.5–20.5	24.0–28.8	2.4–9.3	1.5–19.2
Mean	9.8	26.2	6.3	15.1
Unconfined compressive strength (MPa)				
Range	33.2–112.4	6.3–20.8	77–214	13.6–159
Mean	74.1	11.8	136.6	61.7
Point load strength (MPa)				
Range	0.2–9.5		6–13	1.2–8.7
Mean	4.4		9.3	3.5
Young's modulus (GPa)				
Range	27.3–46.2	3.24–9.83	19.5–99.9	3.8–14.3
Mean	32.7	6.16	50.8	6.8

Clay shales usually have permeabilities of the order 1×10^{-8} m s⁻¹ to 10^{-12} m s⁻¹. However, sandy and silty shales and closely jointed cemented shales may have permeabilities as high as 1×10^{-6} m s⁻¹.

9.15.2 Sandstones
Sandstones may vary from thinly laminated micaceous types to very thickly bedded varieties. Moreover, they may be cross-bedded and are invariably jointed. With the exception of shaley sandstone, sandstone is not subject to rapid surface deterioration on exposure.

The dry density and especially the porosity of sandstone are influenced by the amount of cement and/or matrix material occupying the pores. The compressive strength of sandstone is influenced by its porosity, amount and type of cement and/or matrix material, as well as the composition of the individual grains. If cement binds the grains together a stronger rock is produced than one in which a similar amount of detrital matrix performs the same function. However, the amount of cementing material is more important than the type of cement, although if two sandstones are equally well cemented, one having a siliceous, the other a calcareous cement, then the former is the stronger. Pore water plays a very significant role as far as the compressive strength and deformation characteristics of a sandstone are concerned. For example, it can reduce the unconfined compressive strengths by 30 to 60%.

9.15.3 Carbonate rocks
The engineering properties of carbonate rocks are influenced by grain size and those post-depositional changes that bring about induration. Because induration can take place at the same time as deposition is occurring, this means that carbonate sediments can sustain high overburden pressures, which means that they can retain high porosities to considerable depths. Indeed, a layer of cemented grains may

Table 9.16 Cont'd

c. Limestones, England

	Chee Tor, Limestone, Carboniferous, (Buxton, Derbyshire)	Magnesium Limestone, Permian, (Whitwell, Yorkshire)	Lincolnshire Limestone, Jurassic (Ancaster, Lincs.)	Great Oolite, Jurassic (Corsham, Wiltshire)
Dry density (Mg m^{-3})				
Range	2.55–2.61	2.46–2.58	2.09–2.38	1.91–2.21
Mean	2.58	2.51	2.27	1.98
Effective porosity (%)				
Range	2.4–3.60	8.5–12.0	11.1–19.9	13.8–23.7
Mean	2.9	10.4	14.1	17.7
Unconfined compressive strength (MPa)				
Range	65.2–170.9	34.6–69.6	17.6–38.7	8.9–20.1
Mean	106.2	54.6	28.4	15.6
Point load strength (MPa)				
Range	1.9–3.5	2.2–3.1	1.2–2.9	0.6–1.2
Mean	2.8	2.7	1.9	0.9
Young's modulus (GPa)				
Range	53.9–79.7	22.3–53.0	14.1–35.0	9.7–27.8
Mean	68.9	41.3	19.5	16.1
Permeability ($\times 10^{-8}$ m s^{-1})				
Range	0.01–0.3	1.8–8.4	6.9–41.9	17.2–45.2
Mean	0.07	4.1	17.5	26.6

overlie one that is poorly cemented. Eventually, however, high overburden pressures, creep and recrystallization produces crystalline limestone with very low porosity.

Representative values of some physical properties of carbonate rocks are given in Table 9.16c. Generally, the density of carbonate rocks increases with age, whilst the porosity is reduced. Diagenetic processes mainly account for the lower porosities of the older limestones. Limestone when dolomitized undergoes an increase in porosity of a few per cent and therefore tends to possess a lower compressive strength than limestone that has not been dolomitized. For example, the Great Limestone of the north of England has a compressive strength ranging from 110 to 210 MPa with an average porosity of 4%. When dolomitized its average porosity is 7.5% with a compressive strength of between 70 and 165 MPa.

Joints in limestone generally have been subjected to various degrees of dissolution so that some may gape. Sinkholes may develop where joints intersect and these may lead to an integrated system of subterranean galleries and caverns (see Chapter 3). The progressive opening of discontinuities by dissolution leads to an increase in mass permeability. An important effect of dissolution in limestone is enlargement of the pores, which enhances water circulation thereby encouraging further solution. This brings about an increase in stress within the remaining rock framework that reduces the strength of the rock mass and leads to increasing stress corrosion. On loading, the volume of the voids is reduced by fracture of the weakened cement between the particles and by the reorientation of the intact aggregations of rock that become separated by loss of bonding. Settlement in such conditions takes place rapidly within a few days of the application of load.

Table 9.16 Cont'd

d. Evaporites, England

| | Anhydrite | | Gypsum | |
	Sandwich	Newbiggin	Kirby Thore	Sherburn-in-Elmet
Dry density (Mg m^{-3})				
Range	2.77–2.82	2.74–2.84	2.16–2.33	2.16–2.32
Mean	2.79	2.78	2.29	2.21
Effective porosity (%)				
Range	3.1–3.7	3.0–3.5	3.6–4.6	3.4–9.1
Mean	3.3	2.9	4.0	5.1
Permeability (m s^{-1})				
Range	0.4–3.0 $\times 10^{-8}$	1.2–3.6 $\times 10^{-8}$	3.1–8.0 $\times 10^{-7}$	4.0 $\times 10^{-7}$–12.6 $\times 10^{-7}$
Mean	0.6 $\times 10^{-8}$	3.0 $\times 10^{-8}$	6.2 $\times 10^{-7}$	9.6 $\times 10^{-7}$
Unconfined compressive strength (MPa)				
Range	77.9–126.8	66.1–120.8	28.1–42.4	19.0–40.8
Mean	102.9	97.5	34.8	27.5
Point load strength (MPa)				
Range	3.4–4.9	3.0–4.6	1.8–2.5	0.9–2.4
Mean	4.0	3.4	2.0	1.9
Young's modulus (GPa)				
Range	57.0–86.4	48.8–83.0	18.1–46.8	15.6–36.0
Mean	78.7	69.4	35.3	24.8

The dry density of chalk tends to range from 1.35 to 2.0 Mg m^{-3} and the porosity from 30 and 50%. Although the unconfined compressive strength of chalk varies from moderately weak to moderately strong, it undergoes a marked reduction when it is saturated. For instance, moderately weak chalk may suffer a loss on saturation amounting to approximately 70%. Chalk compresses elastically up to a critical pressure, the apparent preconsolidation pressure. Marked breakdown and substantial consolidation occurs at higher pressures. Moderately weak chalk is deformable, a typical value of Young's modulus being 5 × 10^3 MPa. Such chalk may exhibit elastic–plastic deformation, with perhaps incipient creep prior to failure. The deformation properties of chalk in the field depend upon its hardness, and the spacing, tightness and orientation of its discontinuities, as well as the amount of weathering it has undergone. Discontinuities are the fundamental factors governing the mass permeability of chalk. Chalk also is subject to dissolution along discontinuities, and solution pipes and sinkholes can be present.

9.15.4 Evaporitic rocks

Evaporitic rocks are formed by precipitation from saline waters. Representative dry densities of gypsum and anhydrite are given in Table 9.16d, as are porosity values. Anhydrite is a strong rock, gypsum is moderately strong, whilst rock salt is moderately weak. Evaporitic rocks exhibit varying degrees of plastic deformation before failing. For example, in rock salt the yield strength (the boundary between elastic and plastic deformation) may be as little as one-tenth the ultimate compressive strength, whereas anhydrite undergoes comparatively little plastic deformation prior to rupture. Creep (i.e. deformation under constant same load) may account for anything between 20 and 60% of the strain at failure when these evaporitic rocks are subjected to creep tests. Rock salt is most prone to creep.

Gypsum is more readily soluble than limestone, 2100 mg l⁻¹ can be dissolved in non-saline waters as compared with 400 mg l⁻¹. Sinkholes and caverns therefore can develop in thick beds of gypsum more rapidly than they can in limestone (see Chapter 3). As gypsum is weaker than limestone it therefore collapses more readily. Cavern collapse can lead to extensive cracking and subsidence at the ground surface. Massive anhydrite can be dissolved to produce uncontrollable runaway situations in which seepage flow rates increase in a rapidly accelerating manner. Heave is another problem associated with anhydrite. This takes place when anhydrite is hydrated to form gypsum, in so doing there is a volume increase of between 30 and 58% that exerts pressures that have been variously estimated between 2 and 12 MPa. It is thought that no great length of time is required to bring about such hydration. When it occurs at shallow depths it causes expansion but the process is gradual and usually is accompanied by the removal of gypsum in solution. At greater depths anhydrite is confined during the process, resulting in a gradual build-up of pressure with the stress finally being liberated in a rapid manner.

Rock salt or halite is highly soluble, that is, the solubility of $NaCl$ in water is 35.5% by weight at 25°C, and it is 7500 times more soluble than limestone. Hence, because halite is so soluble it only survives at the surface in arid areas. In fact, karst features similar to those developed in carbonate rock masses are developed in rock salt but there are only a few places in the world with extensive outcrops of salt where such karst features exist.

10

Geology and Construction Materials

10.1 Gravels and sands

Gravels and sands may occur as river deposits, river terrace deposits, beach deposits, raised beach deposits and fluvio-glacial deposits. Sand also may accumulate as wind blown deposits. Gravels and sands represent sources of coarse and fine aggregate respectively.

10.1.1 Gravel

Gravel deposits frequently represent local accumulations, for example, channel fillings. In such instances, they are restricted in width and thickness but may have considerable length. Fan-shaped deposits of gravels or aprons may accumulate at the snouts of ice masses, or blanket deposits may develop on transgressive beaches. The latter type of deposit is usually thin and patchy whilst the former is frequently wedge-shaped. A gravel deposit consists of a framework of pebbles between which there are voids. The voids are rarely empty, being occupied by sand, silt or clay material. River and fluvio-glacial gravels are notably bimodal, the principal mode being in the gravel grade, the secondary in the sand grade. Marine gravels, however, are often unimodal and tend to be more uniformly sorted than fluvial types of similar grade size.

The composition of a gravel deposit reflects not only the type of rocks in the source area but it is also influenced by the agent(s) responsible for its formation and the climatic regime in which it was or is being deposited. Furthermore, relief influences the character of a gravel deposit, for example, under low relief gravel production is small and the pebbles tend to be chemically inert residues such as vein quartz, quartzite, chert and flint. By contrast, high relief and rapid erosion yield coarse immature gravels. All the same, a gravel achieves maturity much more rapidly than does a sand under the same conditions. Gravels that consist primarily of one type of rock fragment are called oligomictic. Such deposits are usually thin and well sorted. Polymictic gravels usually consist of a varied assortment of rock fragments and occur as thick, poorly sorted deposits.

The shape and surface texture of the pebbles in a gravel deposit are influenced by the agent responsible for transportation and the length of time taken in transport, although shape also is dependent on the initial shape of the fragment, which is controlled by the fracture pattern within the parental rock. As far as the shape of gravel particles are concerned, they can be classified as rounded, irregular, angular, flaky and elongated. The surface texture of pebbles can be described as glassy, smooth, granular, rough, crystalline and honeycombed. Particles of gravel often possess surface coatings that may be the result of weathering or may represent mineral precipitates derived from circulating groundwaters. The latter type of coating may be calcareous, ferruginous, siliceous or occasionally gypsiferous. Clay also may form a coating about pebbles. Surface coatings generally reduce the value of gravels for use as concrete aggregate, thick and/or soft and loosely adhering surface coatings are particularly suspect. Clay and gypsum coatings, however, often can be removed by screening and washing. Siliceous coatings tend to react with the alkalis in high alkali cements and are therefore detrimental to concrete (see below).

In a typical gravel pit the material is dug from the face by a mechanical excavator. This loads the material into trucks or onto a conveyor that transport it to the primary screening and crushing plant

Fig. 10.1 Screening sand from gravel at a pit in Derbyshire, England.

(Fig. 10.1). After crushing the material is further screened and washed. This sorts the gravel into various grades and separates it from the sand fraction. The latter usually is sorted into coarser and finer grades, the coarser is used for concrete and the finer is preferred for mortar. Because gravel deposits are highly permeable, if the water table is high, then the gravel pit will flood. The gravels then have to be worked by dredging. Sea dredged aggregates are becoming increasingly important.

10.1.2 Sand

The textural maturity of sand varies appreciably. A high degree of sorting coupled with a high degree of rounding characterizes a mature sand. The shape of sand grains, however, is not greatly influenced by length of transport. Maturity also is reflected in the mineralogical composition of sand, many mature sands being concentrates of quartz, the less stable minerals having disappeared due to mechanical and/ or chemical breakdown during erosion, transportation, or after the sand has been deposited.

Sands are used for building purposes to give bulk to concrete, mortars, plasters and renderings. For example, sand is used in concrete to lessen the void space created by the coarse aggregate. A sand consisting of a range of grade sizes gives a lower proportion of voids than one in which the grains are of uniform size. Indeed, grading is probably the most important property as far as the suitability of sand for concrete is concerned. Poorly graded sands can be improved by adding the lacking/missing grade sizes to them, so that a high quality material can be produced with correct blending. Generally, sand with rounded particles produces a slightly more workable concrete than one consisting of irregularly shaped particles. Sands used for building purposes often are siliceous in composition and should be as free from impurities as possible. They should contain less than 3% by weight of silt or clay, since these particles need a high water content to produce a workable concrete mix. This, in turn, leads to shrinkage and cracking on drying. The presence of clay or shaley material tends to retard setting and hardening of cement, and particles coated with clay form a poor bond with cement so producing a weaker, less durable concrete. Organic material adversely affects the setting and hardening properties of cement by retarding hydration and thereby reduces its strength and durability. Organic matter also causes popping, pitting and blowing. If iron pyrite occurs in sand, then it gives rise to unsightly rust stains when used in concrete. The salt content of marine sands is unlikely to produce any serious adverse effects in good quality concrete although it will probably give rise to efflorescence. Salt can be removed by washing sand.

Glass sands must have a silica content of over 95% (over 96% for plate glass). The amount of iron oxides present in glass sands must be very low – under 0.05% in the case of sand used for clear glass. Uniformity of grain size is another important property as this means that the individual grains melt in the furnace at approximately the same temperature. High grade quartz sands are also important for making silica bricks used for refractory purposes.

10.2   Mudrocks and brick manufacture

The suitability of a mudrock for brick making is determined by its physical, chemical and mineralogical character and the changes that occur when the material is fired. The unfired properties such as plasticity, workability (i.e. the ability of the clay to be moulded into shape without fracturing, and to maintain its shape when the moulding action ceases), dry strength, dry shrinkage and vitrification range are dependent upon the raw material. On the other hand, the fired properties such as colour, strength, total shrinkage on firing, porosity, water absorption, bulk density and tendency to bloat are controlled by the nature of the firing process. The ideal raw material should possess moderate plasticity, good workability, high dry strength, total shrinkage on firing of less than 10%, and a long vitrification range.

10.2.1   Raw material

Although bricks can be made from most mudrocks, the varying proportions of the different clay minerals have a profound influence on the processing and character of a fired brick. Those mudrocks that contain a single predominant clay mineral have a shorter temperature interval between the onset of vitrification and complete fusion, than those consisting of a mixture of clay minerals. This is more true of montmorillonitic and illitic clay materials than those composed mainly of kaolinite. Moreover, clays that consist of a mixture of clay minerals tend not to shrink as much when fired as those composed chiefly of one type of clay mineral. Mudrocks that contain significant amounts of disordered kaolinite tend to have moderate to high plasticity and are easily workable. They undergo little shrinkage during brick manufacture and possess a long vitrification range thereby yielding a fairly refractory product. Conversely, mudrocks containing appreciable quantities of well-ordered kaolinite are poorly plastic and less workable. Illitic mudrocks are more plastic and less refractory than those in which disordered kaolinite is dominant, and fire at somewhat lower temperatures. Smectites are the most plastic and least refractory of the clay minerals. They show high shrinkage on drying since they require high proportions of added water to make them workable. As far as the unfired properties of the raw materials are concerned the non-clay minerals present act mainly as a diluent, but they may be of considerable importance in relation to the fired properties. The non-clay material also may enhance the working properties, for instance, colloidal silica improves the workability by increasing plasticity.

The presence of quartz, in significant amounts, affords strength and durability to a brick. This is because during the vitrification period quartz combines with the basic oxides of the fluxes released from the clay minerals on firing to form glass, which improves the strength. However, as the proportion increases, the plasticity of the raw material decreases.

The accessory minerals in mudrocks play a significant role in brick making. The presence of carbonates is particularly important and can influence the character of the bricks produced. When heated above 900°C carbonates break down yielding carbon dioxide and leaving behind reactive basic oxides, especially of calcium and magnesium. The escape of carbon dioxide can cause lime popping or bursting if large pieces of carbonate such as shell fragments are present, thereby pitting the surface of a brick. To avoid lime popping the material must be finely ground to pass a 20 mesh sieve. The residual lime and magnesia form fluxes that give rise to low viscosity silicate melts. The reaction lowers the temperature of the brick during firing and hence, unless additional heat is supplied, lowers the firing temperature and shortens the range over which vitrification occurs. The reduction in temperature can result in inadequately fired bricks. If excess oxides remain in the brick they will hydrate on exposure to moisture, thereby destroying the brick. The expulsion of significant quantities of carbon dioxide can increase the porosity of bricks, so reducing their strength. Accordingly, engineering bricks must be made from a raw material that has a low carbonate content.

Sulphate minerals in mudrocks are detrimental to brick making. For instance, calcium sulphate does not decompose within the range of firing temperature of bricks. It is soluble and, if present in trace amounts in the fired brick, causes efflorescence when the brick is exposed to the atmosphere. Soluble sulphates dissolve in the water used to mix the raw material. During drying and firing they often form a white scum on the surface of a brick. Barium carbonate may be added to render such salts insoluble and so prevent scumming. Iron sulphides, such as pyrite and marcasite, frequently occur in mudrocks. When heated in oxidizing conditions, the sulphides decompose to produce ferric oxide and sulphur dioxide. In the presence of organic matter oxidation is incomplete yielding ferrous compounds that combine with silica and basic oxides, if present, to form black glassy spots. This may lead to a black vitreous core being present in some bricks that can reduce strength significantly. If the vitrified material forms an envelope around the ferrous compounds and heating continues until this decomposes, then the gases liberated cannot escape causing bricks to bloat and distort. Under such circumstances the rate of firing should be controlled in order to allow gases to be liberated prior to the onset of vitrification. Too high a percentage of pyrite or other iron bearing minerals gives rise to rapid melting that can lead to difficulties on firing.

Organic matter commonly occurs in mudrock. It may be concentrated in lenses or seams or be finely disseminated throughout a mudrock. Incomplete oxidation of the carbon upon firing may result in black coring or bloating. Even minute amounts of carbonaceous material can give black coring in dense bricks if it is not burned out. Black coring can be prevented by ensuring that all carbonaceous material is burnt out below the vitrification temperature. This means that if a raw material contains much carbonaceous material it may be necessary to admit cool air into the firing chamber to prevent the temperature rising too quickly.

The first process in brick manufacture is to dig the raw material from the pit (Fig. 10.2). It then usually is stored in heaps for a period of time to weather before being crushed, sieved, mixed with water and pressed to shape. Then, the green bricks are dried before being placed in the kiln. Three stages can be recognized in brick burning. During the water smoking stage, which takes place up to approximately 600°C, water is given off. In other words, pore water and the water with which the clay material was mixed is driven off at about 110°C, whilst water of hydration disappears between 400°C and 600°C. The next stage is that of oxidation during which the combustion of carbonaceous matter and the decomposition of pyrite takes place, and carbon dioxide and sulphur dioxide are given off. The final stage is that of vitrification. Above 800°C the centre of the brick gradually develops into a highly viscous mass, the fluidity of which increases as the temperature rises. The components are now bonded together by the formation of glass. Bricks are fired at temperatures around 1000°C to 1100°C for about 3 days. The degree of firing depends on the fluxing oxides, principally H_2O, Na_2O, CaO and Fe_2O_3. Mica is one of the chief sources of alkalis in mudrocks. Because illites are more intimately associated with micas than kaolinites, illitic clays usually contain a higher proportion of fluxes and so are less refractory than kaolinitic clays.

10.2.2 Bricks

The strength of the brick depends largely on the degree of vitrification. Theoretically, the strength of bricks made from mudrocks containing clay minerals such as illite should be higher than those containing the coarser kaolinite. Illitic clays, however, vitrify more easily and there is a tendency to underfire, particularly if the material contains fine-grained calcite or dolomite. Kaolinitic clays are much more refractory and can stand harder firing, therefore greater vitrification is achieved. Permeability also depends on the degree of vitrification. Mudrocks containing a high proportion of clay minerals produce less permeable products than those with a high proportion of quartz, but the former types may give a high drying shrinkage and high moisture absorption.

Fig. 10.2   Digging clay for brickmaking, near Bletchley, England.

The colour of the mudrock prior to burning gives no indication of the colour it will have after leaving the kiln. The iron content, however, is important in this respect. For instance, as there is less scope for iron substitution in kaolinite than in illite, this often means that kaolinitic clays give a whitish or pale yellow colour on firing whilst illitic clays generally produce red or brown bricks. More particularly a clay material possessing about 1% of iron oxides when burnt tends to produce a cream or light yellow colour, 2 to 3% gives buff, and 4 to 5% red. Under reducing conditions, however, ferrous silicate develops and the clay product has a blackish colour. Reducing conditions are produced if carbonaceous material is present in the clay material (or they may be brought about by the addition of coal dust or sawdust to the clay before it is burnt). Blue bricks also are produced under reducing conditions. The clay material should contain about 5% iron together with lime and alkalis. The presence of other constituents, notably calcium, magnesium or aluminium oxides, tends to reduce the colouring effect of iron oxide, whereas the presence of titanium oxide enhances it. High original carbonate content tends to produce yellow bricks. The presence of manganese in a clay material may impart a purplish shade to the burnt product.

10.3   Some special types of clay deposits

The principal clay minerals belong to the kandite, illite, smectite, vermiculite and palygorskite families. The kandites, of which kaolinite is the chief member, are the most abundant clay minerals. Deposits of kaolin or china clay are associated with granite masses that have undergone kaolinization. The soft china clay is excavated by strong jets of water under high pressure, the material being washed to the base of the quarry. This process helps separate the lighter kaolin fraction from the quartz. The lighter material is pumped to the surface of the quarry where it is fed into a series of settling tanks. These separate mica, which is itself removed for commercial use, from china clay. Washed china clay has a comparatively coarse size, only approximately 20% of the constituent particles being below 0.01 mm, accordingly the material is non-plastic. Kaolin is used for the manufacture of white earthenwares and stonewares, in white Portland cement and for special refractories.

Ball clays are composed almost entirely of kaolinite. These clays have a high plasticity because 70 to 90% of the individual particles are below 0.01 mm in size. Their plasticity at times is enhanced by the presence of montmorillonite. Ball clays contain a low percentage of iron oxide and consequently when burnt give a light cream colour. They are used for the manufacture of sanitary ware and refractories.

A clay or shale that can be used to manufacture refractory bricks is termed a fireclay. Such material should fuse above 1600°C and should be capable of taking a glaze. In fact, ball clays and china clays are fireclays, fusing at 1650°C and 1750°C respectively. Most fireclays are highly plastic and contain kaolinite as their predominant material. Some of the best fireclays in Britain are found beneath coal seams, indeed these fireclays are restricted almost entirely to strata of Coal Measures age. The material in a bed of fireclay that lies immediately beneath a coal seam is often of better quality than that found at the base of the bed. As fireclays represent fossil soils that have undergone severe leaching they consist chiefly of silica and alumina, and contain only minor amounts of alkalis, lime and iron compounds. This accounts for their refractoriness (alkalis, lime, magnesia and iron oxides in a clay deposit tend to lower its temperature of fusion and act as fluxes). Very occasionally a deposit contains an excess of alumina and in such cases it possesses a very high refractoriness. After a fireclay has been quarried or mined it usually is left to weather for an appreciable period of time to allow it to breakdown before it is crushed. The crushed fireclay is mixed with water and moulded. Bricks, tiles and sanitary ware are made from fireclay.

Bentonite is formed by the alteration of volcanic ash, the principal clay mineral present being either montmorillonite or beidellite. When water is added to bentonite it swells to many times its original volume to produce a soft gel. Bentonite is markedly thixotropic and this, together with its plastic properties, has given the clay a wide range of uses. For example, it is added to poorly plastic clays to make them more workable and to cement mortars for the same purpose. It is used for clay grouting, for drilling mud, and for slurry trenches and diaphragm walls in the construction industry.

10.4 Building or dimension stone

Stone for use as construction material often is available locally and requires little energy for extraction and processing. Indeed, it is used more or less as it is found except for the seasoning, shaping and dressing that is necessary before it is used for building purposes. A number of factors determine whether a rock mass will be worked as a building stone. These include the volume of material that can be quarried, the ease with which it can be quarried, the wastage consequent upon quarrying and the cost of transportation, as well as its appearance and physical properties. As far as volume is concerned, the life of the quarry should be at least 20 years. The amount of overburden that has to be removed also affects the economics of quarrying in that the greater the quantity that has to be stripped, the less economic does the operation become. As weathered rock represents waste, the ratio of fresh to weathered rock is another factor of economic significance. The ease with which a rock can be quarried depends to a large extent upon geological structures, notably the dip, and the geometry of joints and bedding planes, where present. Ideally, rock for building stone should be massive, certainly it must be free from closely spaced joints or other discontinuities as these control block size. The stone should be free of fractures and other flaws. In the case of sedimentary rocks, where beds dip steeply, quarrying has to take place along the strike. Steeply dipping rocks also can give rise to problems of slope stability when excavated. On the other hand, if beds of rock dip gently it is advantageous to develop the quarry floor along the bedding planes. The massive nature of igneous rocks such as granite, means that a quarry can be developed in any direction, within the constraints of planning permission.

10.4.1 Properties of stone

The appearance of a stone largely depends upon its colour, which is determined by its mineral composition, and in the case of sedimentary rocks like sandstone by the type of cement. Texture also affects the appearance of a stone, as does the way in which it weathers. For example, some minerals, such as pyrite, may weather to produce rusty stains. Generally, rocks of light colour are used as building stone.

An unconfined compressive strength of 35 MPa is satisfactory for most building purposes and the strength of most rocks used for building stone is well in excess of this figure. Tensile strength is important in some situations, for example, tensile stresses may be generated in a stone subjected to flexuring as a result of ground movements. Hardness is a factor of minor consequence except where a stone is subjected to continual wear such as in steps or pavings. The texture and porosity of a rock affect its ease of dressing, and the amount of expansion, freezing and dissolution it may undergo. For instance, fine-grained rocks are more easily dressed than coarse varieties. The retentivity of water in a rock with small pores is greater than in one with large pores, and so they are more prone to frost attack.

The durability of a stone is a measure of its ability to resist weathering and so to retain its size, shape, strength and appearance over an extensive period of time. It is one of the most important factors that determines whether or not a rock will be worked for building stone. The amount of weathering undergone by a rock in field exposures or quarries affords some indication of its qualities of resistance. However, there is no guarantee that the durability is the same throughout a rock mass. Damage can occur to stone of low tensile strength due to alternate wetting and drying as water in the pores can expand enough when warmed to cause its disruption. For instance, when the temperature of water is raised from 0°C to 60°C it expands some 1.5% and this can exert a pressure of up to 52 MPa in the pores of a rock.

Frost damage is a major factor causing deterioration in building stone in certain climatic regimes. Fortunately, most igneous rocks, and the better quality sandstones and limestones, are immune. Porosity, pore size and degree of saturation all play important roles in relation to frost susceptibility. As water turns to ice it increases in volume thus giving rise to an increase in pressure within the pores, and ice pressures rapidly increase with decreasing temperature. Usually, coarse-grained rocks withstand freezing better than the fine-grained types. Indeed, the critical pore size for freeze–thaw durability appears to be around 0.005 mm, that is, larger mean pore diameters allow outward drainage and escape of fluid from the frontal advance of the ice line and are therefore less frost susceptible. Fine-grained rocks that have over 5% sorped water often are susceptible to frost damage whilst those containing less than 1% are durable.

Deleterious salts when present in a building stone may cause efflorescence by crystallizing on the surface of a stone. In subflorescence crystallization takes place just below the surface and may be responsible for surface scabbing. The pressures produced by crystallization of salts in small pores are appreciable and are often sufficient to cause disruption. Crystallization caused by freely soluble salts such as sodium chloride, calcium sulphate or sodium hydroxide can lead to the surface of a stone crumbling or powdering. Deep cavities may be formed in magnesian limestone when it is attacked by magnesium sulphate. Salt action can give rise to honeycomb weathering in some sandstones and porous limestones (Fig. 10.3). Conversely, surface induration of a stone by the precipitation of salts may give rise to a protective hard crust referred to as case hardening. If the stone is the sole supplier of these salts, then the interior is correspondingly weaker.

The rate of weathering of most coarse-grained igneous rocks used for building stone is slow and so they generally suffer negligible decay. Building stones derived from sedimentary rocks may undergo varying amounts of decay, especially in urban atmospheres where weathering is accelerated due to the presence of aggressive impurities such as SO_2, SO_3, NO_3, Cl_2 and CO_2 in the air, which produce corrosive acids. Limestones are most suspect. For instance, weak sulphuric acid reacts with the calcium carbonate of limestones to produce calcium sulphate. The latter often forms just below the surface of a stone and the expansion that takes place upon crystallization causes slight disruption. If this reaction continues, then the outer surface of the limestone begins to flake off.

10.4.2 Working stone

In some cases stone may be obtained by splitting along bedding and/or joint surfaces by using a wedge and feathers. Another method of quarrying rock for building stone consists of drilling a series of closely spaced holes (often with as little as 25 mm spacing between them) in line in order to split a large block from the face. Stone also may be cut from the quarry face by using a diamond impregnated wire saw (Fig. 10.4). Flame cutting has been used primarily for winning granitic rocks. It is claimed that this

Fig. 10.3 Honeycomb weathering worn through a sandstone tombstone, Tynemouth, England. The North Sea is some 300 m distant.

Fig. 10.4 Working marble using diamond impregnated wire saws, Estremoz, Portugal.

technique is the only way of cutting stone in areas of high stress relief. If explosive is used to work building stone, then the blast should only weaken the rock along joint and/or bedding planes, and not fracture the material. The object is to obtain blocks of large dimension that can be sawn to size. Hence, the blasting pattern and amount of charge (black powder) are very important and every effort should be made to keep rock wastage and hair cracking to a minimum.

When stone is won from a quarry it contains a certain amount of pore water referred to as quarry sap. As this dries out it causes the stone to harden. Consequently, it is wise to shape the material as soon as possible after it has been quarried. Blocks then are sawn to the required size. After this some material may be planed or turned, before final finishing (e.g. polished, flame textured). Careless operation of dressing machines or tooling of stone may produce bruising. Subsequently, scaling may develop at points where the stone was bruised, so spoiling its appearance.

10.4.3 Types of stone

Granite is ideally suited for building, engineering and monumental purposes. Its crushing strength varies between 160 and 240 MPa. It has exceptional weathering properties and most granites are virtually indestructible under normal climatic conditions. For instance, the polish on granite is such that it is only after exposure to very heavily polluted atmospheres, for a considerable length of time, that any sign of deterioration is apparent. The maintenance cost of granite compared with other materials therefore is very much less and in most cases there is no maintenance cost at all for an appreciable number of years.

Limestones show a variation in their colour, texture and porosity, and those that are fossiliferous are highly attractive when cut and polished. However, carbonate stone can undergo dissolution by acidified water. This results in dulling of polish, surface discolouration and structural weakening. Carvings and decoration are subdued and eventually may disappear, natural features such as grain, fossils, etc. are emboldened (Fig. 10.5).

The colour and strength of sandstone are largely attributable to the type and amount of cement binding the constituent grains. The cement content also influences the porosity and therefore water absorption. Sandstones that are used for building purposes are found in most of the geological systems,

Fig. 10.5 Highly weathered limestone statuary, Wells Cathedral, Somerset, England.

the exception tending to be those of the Cainozoic era. The sandstones of this age generally are too soft and friable to be of value. The properties of some sandstones and limestones that are used in Britain for building purposes are given in Table 10.1.

10.5 Roofing and facing materials

Rocks used for roofing purposes must possess a sufficient degree of fissility to allow them to split into thin slabs, as well as being durable and impermeable. Consequently, slate is one of the best roofing materials available and has been extensively used. Slates are derived from argillaceous rocks that, because they were involved in major earth movements, were metamorphosed. They are characterized by their cleavage, which allows the rock to break into thin slabs (Fig. 1.15a). Some slates, however, may possess a grain that runs at an angle to the cleavage planes and may tend to fracture along it. Therefore, in slate used for roofing purposes any such grain should run along its length. Slates are differently coloured, they may be grey, blue, purple, red or mottled. Red slates contain more than twice as much ferric as ferrous oxide. A slate may be greenish coloured if the reverse is the case. Manganese is responsible for the purplish colour of some slates. Blue and grey slates contain little ferric oxide. Green coloured slates also may be obtained from some cleaved volcanic tuffs.

The specific gravity of a slate is about 2.7 to 2.9 with an approximate density of 2.59 Mg m^{-3}. The maximum permissible water absorption of a slate is 0.37%. Calcium carbonate may be present in some slates of inferior quality, which may result in them flaking and eventually crumbling upon weathering. Accordingly, a sulphuric acid test is used to test their quality. Top quality slates, which can be used under moderate to severe atmospheric pollution conditions, reveal no signs of flaking, lamination or swelling after the test.

There is a large amount of wastage when explosives are used to quarry slate. Consequently, they sometimes are quarried by using a wire saw threaded through drill holes, the cutting actually being done by sand fed onto the wire saw. Once won, the slate is sawn into blocks, and then into slabs about 75 mm thick. These slabs are split into slate tiles by hand.

Stone is commonly applied as facings to buildings to enhance their appearance. Hence, an important property of facing stone is that it should be attractive. Facing stone also provides a protective covering. Various thicknesses are used from 20 mm to 40 mm in the case of granite and slate. If granite is used as a facing stone, then it should not be over dried, but should retain some quarry sap, otherwise it becomes too tough and hard to fabricate. Slabs of limestone, marble and sandstone are somewhat thicker, that is, they may vary between 50 and 100 mm. Because of their comparative thinness facing stones should not be fixed too rigidly otherwise differential expansion, due to changing temperatures, can cause them to crack. Rocks used for facing stones have a high tensile strength in order to resist cracking. The high tensile strength also means that thermal expansion is not a great problem when slabs are spread over large faces.

When fissile stones are used as facing stone and are given a riven or honed finish they are extremely attractive. A flame textured finish also produces an attractive facing stone. Usually, however, facing stones have a polished finish and are even more striking. Such stones are more or less self-cleansing.

10.6 Crushed rock: concrete aggregate

Crushed rock is produced for a number of purposes, the chief of which is for aggregate for concrete and roadstone (Fig. 10.6). Approximately 75% of the volume of concrete consists of aggregate, therefore its properties have a significant influence on the engineering behaviour of the material. Aggregate is divided into coarse and fine varieties, the former usually consists of rock material that passes the 37 mm sieve and is retained on the 4.8 mm mesh sieve. The latter passes through this sieve and is caught on the 100 mesh sieve. Fines passing through the 200 mesh sieve should not exceed 10% by weight of an aggregate.

High explosives such as gelignite, dynamite or trimonite are used in drill holes when quarrying crushed rocks. ANFO, a mixture of diesel oil and ammonium nitrate, is also used frequently. The holes are

Table 10.1 Some physical properties of British sandstones and limestones used for building purposes.

a. Sandstones

Stone	Colour	Dry density (Mg m⁻³)	Porosity (%)	Unconfined compressive strength (MPa)	Young's modulus (GPa)	Acid immersion test	Crystal-lization test	Durability classification
Red St. Bees Trias St. Bees	Red	2.15	19.6	12.5	8.9	failed	-	E,F
Lazonby Permian Near Penrith	Dark pink to red	2.38	9.3	40	21.8	passed	37	B,C
Delph Coal Measures Wingerworth	Grey to deep buff	2.33	13.5	62	36.8	passed	33	B,C
Shipley Coal Measures Barnard Castle	Grey to buff	2.41	10.5	75	40.4	passed	16	A,B
Birchover Namurian Near Matlock	Buff to pink	2.34	12.4	48	25.6	passed	40	B,C
Dunhouse Namurian Staindrop	White to buff	2.36	12.3	69	40.6	passed	17	A,B
Blaxter Lr Carboniferous Elsdon	White to buff	2.24	16.6	50	35.4	passed	56	B,C
Monmouth Old Red Sandstone	Red to pinkish brown	2.43	8.8	22	17.4	failed	-	E,F

b. Limestones

Stone	Hopton Wood	Anstone	Doulting	Ancaster	Bath	Portland	Purbeck
Age	Lr. Carboniferous	Magnesian Limestone	Inferior Oolite	Lincolnshire Limestone	Great Oolite	Portland	Purbeck
Location	Middleton-by Wirksworth	KivetonPark	Shepton Mallet	Ancaster	Monks Park	Isle of Portland	Swanage
Property							
Specific gravity	2.72	2.83	2.7	2.7	2.71	2.7	2.7
Dry density (Mg m^{-3})	2.57	2.51	2.34	2.27	2.3	2.25	2.21
Porosity (%)	7.6	10.4	12.8	19.3	18.3	22.4	9.6
Microporosity (% saturation)	29	23	30	60	77	43	62
Saturation coefficient	0.61	0.64	0.69	0.84	0.94	0.58	0.62
Unconfined compressive strength (MPa)	80.5	54.6	35.6	28.4	15.6	20.2	24.1
Young's modulus (GPa)	53.9	41.3	24.1	19.5	16.1	17.0	17.4
Crystallization test (% wt loss)	4	5	8	20	52	13	3
Durability classification	B	B	C	D	E	C	B

Classification of porosity: 1–5% = low; 5–15% = medium; 15–30% = high.

Classification of unconfined compressive strength: 12.5–50 MPa = moderately strong; 50–100 MPa: strong.

Durability classification: A = best i.e. passes acid immersion and crystallization tests or the latter test in the case of limestone; E = poorest i.e. should not be used in areas of high pollution or exposed coastal areas.

Fig. 10.6   Quarrying granite for aggregate in Hong Kong.

drilled at an angle of about 10° to 20° from vertical for safety reasons, and are usually located 3 to 6 m from the working face and a similar distance apart. Usually, one but sometimes two or more rows of holes are drilled. It is common practice to have millisecond delay intervals between firing individual holes, in this way the explosions are complementary. The object of blasting is to produce stone of a size that can be loaded onto dump trucks without difficulty, for conveyance to the primary crusher. Large stone must be further reduced in size by using a drop-ball or by secondary blasting. After crushing the material is screened to separate it into different grade sizes (Fig. 10.7).

10.6.1   Properties of concrete aggregate

The crushing strength of rock used for aggregate generally ranges between 70 and 300 MPa. Aggregates that are physically unsound lead to the deterioration of concrete, inducing cracking, popping or spalling. Cement shrinks on drying but the amount of shrinkage is minimized if the aggregate is strong and the cement–aggregate bond is good.

It usually is assumed that shrinkage in concrete does not exceed 0.045%, this taking place in the cement. However, basalt, gabbro, dolerite, mudstone and greywacke have been shown to be shrinkable, that is, they have large wetting and drying movements of their own, so much so that they affect the total shrinkage of concrete. Clay and shale absorb water so are likely to expand if they are incorporated in concrete, and on drying they shrink causing injury to the cement. Consequently, the proportion of clay material in aggregate should not exceed 3%. Granite, limestone, quartzite and felsite are unaffected.

The shape of aggregate particles is an important property and is governed mainly by the fracture pattern within a rock mass. Rocks like basalts, dolerites, andesites, granites, quartzites and limestones tend to produce angular fragments when crushed. However, argillaceous limestones produce an excessive amount of fines when crushed. The crushing characteristics of sandstone depend upon the closeness of its texture and how well the grains are cemented together. Angular fragments may produce a mix that is difficult to work, that is, it can be placed less easily and offers less resistance to segregation. Nevertheless, angular particles produce denser concrete. Rounded, smooth fragments produce workable mixes. The less workable the mix, the more sand, water and cement must be added to produce a satisfactory concrete.

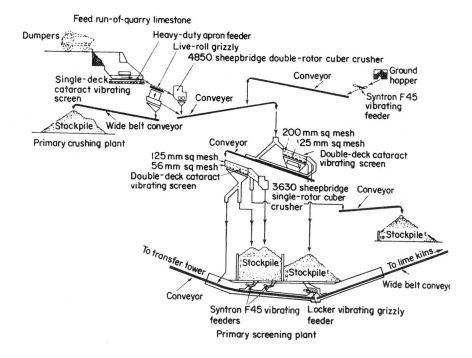

Fig. 10.7 Flow sheet of the primary crushing and primary screening plant at a typical aggregate quarry.

Fissile rocks such as those that are strongly cleaved, schistose, foliated or laminated have a tendency to split and, unless crushed to a fine size, give rise to tabular or planar shaped particles. Planar and tabular fragments not only make concrete more difficult to work but they also pack poorly and so reduce its compressive strength and bulk weight. Furthermore, they tend to lie horizontally in the cement so allowing water to collect beneath them, which inhibits the development of a strong bond on their under surfaces.

The surface texture of aggregate particles largely determines the strength of the bond between the cement and themselves. A rough surface creates a good bond whereas a smooth surface does not.

10.6.2 Aggregate reactivity

As concrete sets, hydration takes place and alkalis (Na_2O and K_2O) are released which react with siliceous material. Table 10.2 lists some of the reactive rock types. If any of these types of rock are used as aggregate in concrete made with high alkali cement, then the concrete is liable to expand and crack, thereby losing strength. When concrete is wet the alkalis that are released are dissolved by its water content and as the water is used up during hydration so the alkalis are concentrated in the remaining liquid. This caustic solution attacks reactive aggregates to produce alkali-silica gels. The osmotic pressures developed by these gels as they absorb more water eventually may rupture the cement around reacting aggregate particles. The gels gradually occupy the cracks so produced and these eventually may extend to the surface of the concrete. If alkali reaction is severe a polygonal pattern of cracking develops at the surface (Fig. 10.8). These troubles can be avoided if a preliminary petrological examination is made of the aggregate. In other words, material that contains over 0.25% opal, over 5% chalcedony, or over 3% glass or crypto-crystalline acidic to intermediate volcanic rock, by weight, will produce an alkali reaction in concrete unless low alkali cement (i.e. containing less than 0.6% of Na_2O and K_2O) is used. A deleterious

269

Table 10.2   Rock types that react with high-alkali cements.

Reactive rocks	Reactive component
Siliceous rocks	
Opaline cherts	Opal
Chalcedonic cherts	Chalcedony
Siliceous limestones	Chalcedony and/or opal
Volcanic rocks	
Rhyolites and rhyolitic tuffs	
Dacites and dacitic tuffs	Glass, devitrified glass and tridymite.
Andesites	
Metamorphic rocks	
Phyllites	Hydromica (illite)
Miscellaneous rocks	
Any rocks containing veinlets, inclusions, coatings, or detrital grains of opal, chalcedony or tridymite. Quartz highly fractured by natural processes.	

Fig. 10.8   Alkali aggregate reaction in concrete at a multi-storey car park, Plymouth, England.

reaction may be avoided if aggregate containing reactive material is mixed with inert matter. The deleterious effect of alkali aggregate reaction also can be avoided if a pozzolan is added to the mix, the reaction taking place between it and the alkalis.

Reactivity may be related not just to composition but also to the percentage of strained quartz that a rock contains. For instance, aggregates containing 40% or more of strongly undulatory or granulated quartz may be highly reactive whilst those with between 30 and 35% are moderately reactive. Basaltic rocks with 5% or more secondary chalcedony or opal exhibit deleterious reactions when mixed with high alkali cements. Sandstones and quartzites containing 5% or more chert behave in a similar manner.

Certain argillaceous dolostones may expand when used as aggregates in high alkali cement, thereby causing failure in concrete. This phenomenon has been referred to as alkali carbonate rock reaction. The expansion is due to the uptake of moisture by the clay minerals present in the argillaceous dolostones. This is made possible by dedolomitization caused by alkalis that leads to an increase in porosity thereby providing access for moisture. However, expansion usually only occurs when the dolomite crystals are less than 75 microns.

10.7 Road aggregate

Aggregate constitutes the basic material for road construction and forms the greater part of a road surface. As a consequence, it has to carry the main stresses imposed by traffic and has to resist wear. The rock material used therefore should have a high resistance to impact and abrasion, polishing and skidding, and frost action, as well as possessing a high strength. It also must be impermeable, chemically inert and have a low coefficient of expansion. The principal tests carried out in order to assess the value of a roadstone are the aggregate crushing test, the aggregate impact test, the aggregate abrasion test and the test for the assessment of the polished stone value. Some typical values of materials commonly used as roadstone are given in Table 10.3. Other tests of consequence are those for water absorption, specific gravity and density, and the aggregate shape tests.

The way in which alteration develops can strongly influence roadstone durability. Weathering may reduce the bonding strength between grains to such an extent that they are easily plucked out. Chemical alteration is not always detrimental to the mechanical properties, indeed a small amount of alteration may improve the resistance of a rock to polishing (see below). On the other hand, resistance to abrasion decreases progressively with increasing content of altered minerals, as does the crushing strength. The combined hardness of the minerals in a rock together with the degree to which they are cleaved, as well as the texture of the rock, also influence its rate of abrasion. The crushing strength is related to porosity and grain size, the higher the porosity and the larger the grain size, the lower the crushing strength.

One of the most important parameters of road aggregate is the polished stone value, which influences skid resistance. A skid-resistant surface is one that is able to retain a high degree of roughness whilst in service. The rate of polish initially is proportional to the volume of the traffic carried by a road. Straight stretches of road are less subject to polishing than bends. The latter may polish up to seven times more rapidly. Stones are polished when fine detrital powder is introduced between tyre and road surface. Detrital powder on a road surface tends to be coarser during wet than dry periods, which suggests that

Table 10.3 Some representative values of the roadstone properties of some common aggregates.

Rock type	Water absorption	Specific gravity	Aggregate crushing value	Aggregate impact value	Aggregate abrasion value	Polished stone value
Basalt	0.9	2.91	14	13	14	58
Dolerite	04	2.95	10	9	6	55
Granite	08	2.64	17	20	15	56
Micro-granite	05	2.65	12	14	13	57
Hornfels	0.5	2.81	13	11	4	59
Quartzite	1.8	2.63	20	18	15	63
Limestone	0.5	2.69	14	20	16	54
Greywacke	0.5	2.72	10	12	7	62

polishing is better developed when the road surface is dry than wet, the coarser detritus more readily roughening the surface of the roadstone.

Rocks within the same major petrological group may differ appreciably in their polished stone characteristics. The best resistance to polish occurs in rocks that contain a proportion of softer alteration materials. Coarser grain size and the presence of cracks in individual grains also tends to improve resistance to polishing. In the case of sedimentary rocks, the presence of hard grains set in a softer matrix produces a good resistance to polish. Sandstones, greywackes and gritty limestones offer a good resistance to polishing, but unfortunately not all of them possess sufficient resistance to crushing and abrasion to render them useful in the wearing course of a road. Purer limestones show a significant tendency to polish. A good resistance to polish in igneous and contact metamorphic rocks is due to variation in hardness between the minerals present. An improvement in skid resistance can be brought about by blending aggregates.

The petrology of an aggregate determines the nature of the surface to be coated, the adhesion attainable depending on the affinity between the individual minerals and the binder, as well as the surface texture of the individual aggregate particles. If the adhesion between the aggregate and binder is less than the cohesion of the binder, then stripping may occur. Insufficient drying and the non-removal of dust before coating are, however, the principal causes of stripping. Acid igneous rocks generally do not mix well with bitumen as they have a poor ability to absorb it. By contrast, basic igneous rocks such as basalts possess a high affinity for bitumen, as does limestone.

Igneous rocks are commonly used for roadstone. Dolerite has been used extensively. It has a high strength and resists abrasion and impact, but its polished stone value usually does not meet motorway specification in Britain, although it is suitable for other roads. Felsite, basalt and andesite are also much used. The coarse-grained igneous rocks such as granite are not generally so suitable as the fine-grained types, as they crush more easily. Moreover, the very fine-grained and glassy volcanics often are unsuitable since they produce chips with sharp edges when crushed, and tend to develop a high polish. Igneous rocks with a high silica content resist abrasion better than those in which the proportion of ferromagnesian minerals is high, that is, acid rocks like rhyolites are harder wearing than basic rocks such as basalts. Unfortunately, some basalts and dolerites have a tendency to slake and therefore are unsuitable for use as roadstone (see Chapter 9). Some rocks that are the products of thermal metamorphism such as hornfels and quartzite, because of their high strength and resistance to wear make good roadstones. Conversely, many rocks of regional metamorphic origin are either cleaved or schistose and therefore are unsuitable because they tend to produce flaky particles when crushed. Such particles do not achieve good interlock and so impair the development of dense mixtures for surface dressing. Coarse-grained gneisses offer a similar performance to granites. The amount and type of cement and/or matrix material that bind grains together in a sedimentary rock influence roadstone performance. Limestone and greywacke frequently are used as roadstone. Greywacke, in particular, has a high strength, resists wear and develops a good skid resistance. Some quartz arenites are used, as are gravels.

10.8   Armourstone

Armourstone refers to large blocks of rock that are used to protect civil engineering structures. Large blocks of rock, which may be single-size or more frequently widely graded (rip-rap), are used to protect the upstream face of dams against wave action. They also are used in the construction of river training schemes, in river bank and bed protection and stabilization, as well as in the prevention of scour around bridge piers. Armourstone is used in coastal engineering for the construction of rubble mound breakwaters, for revetment covering embankments, for the protection of sea walls, and for rubble rock groynes (Fig. 10.9). Indeed, breakwaters and sea defences represent a major use of armourstone. As the marine environment is one of the most aggressive in which construction occurs, armourstone must afford stability against wave action, and accordingly block size and density are all-important. Shape also is important since this affects how blocks interlock together. In addition, armourstone must be able to withstand rapid and severe changes in hydraulic pressure, alternate wetting and drying, thermal changes,

Fig. 10.9 left: Revetment of armourstone used to protect an embankment, Freemantle, Australia. right: The Dawlish Warren revetment in S W England, with PermoTrias Red Sandstone arch stack in the background (Courtesy Dr J-P Latham, Imperial College).

wave and sand/gravel impact and abrasion, as well as salt and solution damage. Consequently, the size, grading, shape, density, water absorption, abrasion resistance, impact resistance, strength and durability of the rock material used for armourstone must be considered during the design stage of a particular project.

Usually, armourstone is specified by weight, a median weight of between 1 and 10 tonnes normally being required. Blocks up to 20 tonnes, however, may be required for breakwaters that will be subjected to large waves. The median weight of secondary armourstone and underlayer rock material may range upwards from 0.1 tonne. In the case of rip-rap used for revetment and river bank protection, the weight of the blocks required usually is less than 1.0 tonne and may grade down to 0.05 tonne. The size of blocks that can be produced at a quarry depends on the incidence of discontinuities and to a lesser extent the method of extraction. Detailed discontinuity surveys can provide the data required for prediction of in situ block size and shape.

The location of armourstone on a breakwater is an important factor that should be considered when making an assessment of rock durability. Rock durability concerns its resistance to chemical decay and mechanical disintegration, including reduction in size and change of shape, during its working life. The intrinsic properties of a rock such as its mineralogy, fabric, grain size, grain interlock, porosity, and in the case of sedimentary rocks, type and amount of cementation, all affect its resistance to breakdown. In addition, the amount of damage that armourstone undergoes is influenced by the action to which it is submitted. For instance, the damage suffered by armourstone used on breakwaters depends on the type of waves (plunging or breaking), their height, period and duration, notably during storm conditions, on the one hand, and the slope and permeability of the structure on the other. Abrasion due to wave action is the principal reason for the reduction in size of blocks of armourstone used in breakwaters, as well as rounding of their shape.

In summary, the parameters that should be included in any assessment and specification of armourstone quality are rock type, its density, water absorption, degree of weathering and potential for further weathering, the potential for development of planes of weakness, and the abrasion resistance.

Hence, the selection of a suitable source of rock for armourstone requires an inspection and evaluation of the quarry or quarries concerned, as well as an assessment of the quality of the intact and processed stone.

10.9 Cement, lime and plaster

Portland cement is manufactured by burning pure limestone or chalk with suitable argillaceous material (clay, mud or shale) in the proportions 3 to 1. First, the raw materials are crushed and ground to a powder, and then blended. Next they are fed into a rotary kiln and heated to a temperature of over 1800°C (Fig. 10.10). Carbon dioxide and water vapour are driven off and the lime fuses with the aluminium silicate in the clay to form a clinker. This is ground to a fine powder and less than 3% gypsum is added to retard setting. Lime is the principal constituent of Portland cement but too much lime produces a weak cement. Silica constitutes approximately 20% and alumina 5%, both being responsible for the strength of the cement. However, a high content of the former produces a slow-setting cement whilst a high content of the latter gives a quick-setting cement. The percentage of iron oxides is low and in white Portland cement it is kept to a minimum. The proportion of magnesia (MgO) should not exceed 4% otherwise the cement is unsound. Similarly, sulphate (SO_4) must not exceed 2.75%. Sulphate-resisting cement is made by the addition of a very small quantity of tricalcium aluminate to normal Portland cement.

Lime is made by heating limestone, including chalk, to a temperature of between 1100 and 1200°C in a current of air, at which point carbon dioxide is driven off to produce quicklime (CaO). Approximately 56 kg of lime can be obtained from 100 kg of pure limestone. Slaking and hydration of quicklime take place when water is added, giving calcium hydroxide. Carbonate rocks vary from place to place both in chemical composition and physical properties so that lime produced in different districts can vary somewhat in its behaviour. For example, dolostones also produce lime, however, the resultant product slakes more slowly than does that derived from limestones.

When gypsum ($CaSO_4.nH_2O$) is heated to a temperature of 170°C it loses three-quarters of its water of crystallization, becoming calcium sulphate hemi-hydrate or plaster of Paris. Anhydrous calcium sulphate forms at higher temperatures. These two substances are the chief materials used in plasters. Gypsum plasters have now more or less replaced lime plasters.

Fig. 10.10 Rotary tunnel kiln that calcines chalk for the manufacture of Portland cement, Gravesend, Kent, England.

11
Geology and Construction

Geology obviously is one of the most important factors in construction since construction takes place either at the ground surface or within the ground. Hence, geology has a key influence on most construction operations since it helps determine their nature, form and cost. For example, route design and tunnel construction are governed to a notable extent by geological considerations. Indeed, a tunnel can be an uncertain and sometimes hazardous undertaking because information on geological conditions along its alignment is never complete, no matter how good the site investigation.

11.1   Open excavation

Open excavation refers to the removal of material at the surface, within certain specified limits, for construction purposes. In order to accomplish this economically and without hazard the character of the rocks and soils involved, and their geological setting must be investigated. Indeed, the method of excavation and the rate of progress are very much influenced by the geology on site. Furthermore, the position of the water table in relation to the base level of an excavation is of prime importance, as are any possible effects of construction operations on surrounding ground and/or buildings.

11.1.1   Methods of excavation

Drillability in rock masses is influenced by their hardness, abrasiveness and grain size, and the discontinuities present. The harder the rock, the stronger the bit that is required for drilling since higher pressures need to be exerted. Abrasiveness refers to the ability of a rock to wear away drill bits. Bit wear is a more significant problem in rotary than percussive drilling. The size of the fragments produced during drilling operations influences abrasiveness. For example, large fragments may cause scratching but comparatively little wear whereas the production of dust causes polishing. Even diamonds lose their cutting ability on polishing. Generally, coarse-grained rocks can be drilled more quickly than can fine-grained varieties or those in which the grain size is variable. The ease of drilling in rocks in which there are many discontinuities is influenced by their orientation in relation to the drill hole. Drilling across the dip is generally less difficult than drilling with it. If a drill hole crosses discontinuities at a low angle, then the bit may stick or the hole may go off line. Drilling over an open discontinuity means that part of the energy controlling drill penetration is lost. Where discontinuities are filled with clay, this may penetrate the flush holes in the bit, causing it to bind or deviate from alignment. The drill hole may require casing if the ground is badly broken.

Spacing of blast holes is determined by the strength, density and fracture pattern within a rock mass, as well as the size of the charge. As a rule, spacing varies between 0.75 and 1.25 times the burden (i.e. the width of the strip blown from the face). Generally, 1 kg of high explosive brings down about 8 to 12 tonnes of rock. Good fragmentation reduces or eliminates the amount of secondary blasting, while minimizing wear and tear on loading machinery. Rocks characterized by high specific gravity and high intergranular cohesion with no preferred orientation of mineral grains resist crack initiation and propagation on blasting. Blasting in rocks that are relatively brittle with a low resistance to dynamic stresses, may give rise to extensive pulverization immediately around the blast holes, leaving the area between largely unfractured. Those rocks that possess marked preferred orientation present difficulties on blasting due to their mechanical anisotropy, splitting easily along the lineation but crack propagation across it is limited.

In many excavations it is important to keep overbreak to a minimum, as damage to the walls or floor of an excavation may lower their strength and necessitate further excavation. Also, smooth faces are more stable. Line drilling is the method most commonly used to improve the peripheral shaping of excavations. It consists of drilling alternate holes between the pattern blastholes forming the edge of the excavation. The quantity of explosive placed in each line hole is significantly smaller and if the holes are closely spaced, from 150 to 250 mm, then explosive may be placed only in every second or third hole. These holes are timed to fire ahead, with or after the nearest normally charged holes of the blasting pattern. In pre-splitting a line of trimming holes is charged and fired to produce a shear plane. This acts as a limiting plane for the blast proper and is carried out prior to blasting the main round inside the proposed break lines. Once pre-split, the rock excavation can be blasted with a normal pattern of holes.

The major objective of ripping is to break the rock just enough to enable economic loading to take place (Fig. 11.1). Rippability depends on intact strength, fracture index and abrasiveness; that is, strong, massive and abrasive rocks do not lend themselves to ripping. By contrast, if sedimentary rocks are well bedded and jointed or if strong and weak rocks are thinly interbedded, then they can be excavated by ripping rather than by blasting. The run direction during ripping should be normal to any vertical joint planes, down-dip to any inclined strata, and downhill on sloping ground. Ripping runs of 70 to 90 m usually give the best results. Where possible the ripping depth should be adjusted so that a forward speed of 3 km h^{-1} can be maintained since this generally is found to be the most productive. Adequate breakage depends on the spacing between ripper runs, which is governed by the fracture pattern in the rock mass.

Diggability depends principally upon the intact strength of the ground, its bulk density, bulking factor and natural water content. When material is excavated it increases in bulk, this being brought about by the decrease that occurs in density per unit volume. Some examples of typical bulking in soils are given in Table 11.1. The bulking factor is important in relation to the loading and removal of material from the working face. The natural moisture content influences the adhesion or stickiness of soils, especially clay soils.

Fig. 11.1 A ripper at work at the surge pond, Dinorwic Pumped Storage Scheme, Llanberis, North Wales.

Table 11.1 Density, bulking factor and diggability of some common soils.

Soil type	Density (Mg m^{-3})	Bulking factor	Diggability
Gravel, dry	1.8	1.25	E
Sand, dry	1.7	1.15	E
Sand and gravel, dry	1.95	1.15	E
Clay, light	1.65	1.3	M
Clay, heavy	2.1	1.35	M-H
Clay, gravel and sand, dry	1.6	1.3	M

E = easy digging, loose, free-running material such as sand and small gravel; M = medium digging, partially consolidated materials such as clayey sand and clay; M-H = medium hard digging, materials such as heavy wet clay, gravels and large boulders.

11.1.2 Stability of slopes

The stability of slopes is a critical factor in open excavation. This is particularly the case in cuttings, as for instance for roads, canals and railways, where slopes should be designed to resist disturbing forces over long periods. Instability in a soil or rock mass occurs when slip surfaces develop and movements are initiated within it. Undesirable properties in a soil such as low shearing strength, development of fissures and high pore water pressures aid instability, and are likely to lead to deterioration of slopes. Generally, the most important factors relating to the stability of a rock mass are the incidence, geometry and nature of discontinuities, and the water content. In the case of open excavation, removal of material can give rise to the dissipation of residual stress, which can further aid instability.

There are several methods available for analysis of the stability of slopes in soils. Most of these may be classed as limit equilibrium methods in which the basic assumption is that the failure criterion is satisfied along the assumed path of failure. Starting from known or assumed values of the forces acting on a free mass within a slope, calculation is made of the shear resistance required for equilibrium of the soil. This shearing resistance then is compared with the estimated or available shear strength of the soil to give an indication of the factor of safety, assessed in two dimensions. The analysis gives a conservative result. The types of failure modes of slopes in soil masses have been discussed in Chapter 3.

The design of a slope excavated in a rock mass requires as much information as possible on the discontinuities present. Information relating to their spatial relationships affords some indication of the modes of failure that may occur and information relating to the shear strength of the rock mass, or more particularly the shear strength along discontinuities, is required for use in stability analysis. The joint inclination is always the most important parameter for slopes of medium and large height, whereas density is more important for small slopes than friction. Cohesion becomes less significant with increasing slope height, while the converse is true as far as the effects of pore water pressure are concerned.

The principal types of failure that occur in rock slopes are rotational, translational and toppling failures (Fig. 3.10). Rotational failures normally only occur in structureless overburden, highly weathered material or very high slopes in closely jointed rock. Relict jointing may persist in highly weathered materials, along which sliding may take place. These failure surfaces are often intermediate in geometry between planar and circular slides. Translational or planar failures take place in inclined layered sequences of rock, the movement occurring along a planar surface such as a bedding plane. Wedge failure is a type of translational failure in which two planar discontinuities intersect, the wedge so formed daylighting into the face, that is, failure may occur if the line of intersection of both planes dips into the slope at an angle less than that of the slope. Toppling failure generally is associated with steep slopes in which the discontinuities are near vertical. It involves the overturning of individual blocks and therefore is governed by discontinuity spacing as well as orientation. The likelihood of toppling increases with increasing inclination of the discontinuities.

Faces excavated in fresh massive plutonic igneous rocks such as granite and gabbro can be left more or less vertical after removal of loose fragments. On the other hand, volcanic rocks such as basalts and andesites are generally bedded and jointed, and may contain layers of ash or tuff, which are usually softer and weather more rapidly. Thus, slope angles have to be reduced accordingly.

Gneiss, quartzite and hornfels are highly weather resistant and slopes in them may be left almost vertical. Schists vary in character and some of the softer schists may be weathered and tend to slide along their planes of schistosity. Slate generally resists weathering although slips may occur where the cleavage dips into a cut face.

If strata are horizontal, then excavation is relatively straightforward and slopes can be determined with some degree of certainty. Near vertical slopes can be excavated in massive limestones and sandstones that are horizontally bedded. In brittle cemented shales slopes of 60° and 75° usually are safe but increasing fissility and decreasing strength necessitate flatter slopes. Even in weak shales slopes are seldom flatter than 45°. However, excavated slopes may have to be modified in accordance with the dip and strike directions in inclined strata. The most stable excavation in dipping strata is one in which the face is orientated normal to the strike, since in such situations there is a low tendency for rocks to slide along their bedding planes. Conversely, if the strike is parallel to the face, then the strata dip into one slope. This is most critical where the rocks dip at angles varying between 30° and 70°. If the dip exceeds 70° and there is no alternative to working against the dip, then the face should be developed parallel to the bedding planes for safety reasons.

Sedimentary sequences in which thin layers of shale or clay are present may have to be treated with caution, especially if the bedding planes are dipping at a critical angle. Weathering may reduce such material to an unstable state within a short period of time, which can lead to slope failure.

A slope of 1:1.5 generally is used when excavating dry sand, this more or less corresponding to the angle of repose, that is, 30° to 40°. This means that a cutting in granular soil will be stable, irrespective of its height, as long as the slope is equal to the lower limit of the angle of internal friction, provided that the slope is suitably drained. Slope failure in coarse-grained soils is a surface phenomenon, which is caused by the particles rolling over each other down the slope. As far as sands are concerned, their packing density is important. For example, densely packed sands that are very slightly cemented may have excavated faces with high angles that remain stable. The moisture content is of paramount importance in loosely packed sands, for if these are saturated they are likely to flow on excavation.

The most frequently used gradients in many clay soils vary between 30° and 45°. In some clay soils, however, the slope angle may have to be less than 20° in order to achieve stability. The stability of slopes in clay soil depends not only on its strength and the angle of the slope but also on the depth to which the excavation is taken and on the depth of a firm stratum, if one exists, not far below the base level of the excavation. Slope failure in uniform clay soil takes place along a near circular surface of slippage. In stiff fissured clays the fissures appreciably reduce the strength below that of intact samples. Thus, reliable estimation of slope stability in stiff fissured clays is difficult. Generally, steep slopes can be excavated in such clay soils initially but their excavation means that fissures open due to the relief of residual stress and there is a change from negative to positive pore water pressures along the fissures (the former tend to hold the fissures together). This change can occur within a matter of days or hours. Not only does this weaken the clay soil but it also permits a more significant ingress of water, which means that the soil is softened. Irregular shaped blocks may begin to fall from the face and slippage may occur along well-defined fissure surfaces, which are by no means circular. If there are no risks to property above the crests of slopes in stiff fissured clay soils, then they can be excavated at about 35°. Although this will not prevent slips, those that occur are likely to be small.

The stability of the floor of a large excavation may be affected by ground heave. The amount of heave and the rate at which it occurs depends on the degree of reduction in vertical stress during construction operations, on the type and succession of underlying strata, and on the surface and groundwater conditions. It generally is greater in the centre of a level excavation in relatively homogeneous ground, as for example, clay soils and shales. Long-term swelling involves absorption of water from the ground surface or is due to water migrating from below. Where the excavation is in overconsolidated clay soils or shales, then swelling and softening are quite rapid. In the case of clay soils with a low degree of saturation, swelling and softening take place very rapidly if surface water gains access to the excavation area.

11.1.3 Treatment of slopes (see also 3.3.6)

It is rarely economical to design a rock slope so that no subsequent rock falls occur, indeed many roads in rough terrain could not be constructed with the finance available without accepting some degree of risk. Therefore, except where absolute security is essential, slopes should be designed to allow small falls of rock under controlled conditions. The design of systems to prevent rockfall require data concerning trajectory (height of bounce), velocity, impact energy and total volume of accumulation. Rock traps in the form of a ditch and/or barrier can be installed at the foot of a slope. Benches also may act as traps to retain rock fall, especially if a barrier is placed at their edge. Wire mesh fixed to the face provides yet another method for controlling rock fall.

Excavation involving the removal of material from the head of an unstable slope, flattening of a slope, benching of a slope, or complete removal of unstable material, are all ways to stabilize a slope. If some form of reinforcement is required to provide support for a rock slope, then it is advisable to install it as quickly as possible after excavation. Dentition is masonry or concrete infill placed in fissures or cavities in a rock slope. Thin to medium bedded rocks dipping parallel to a slope can be held in place by steel dowels grouted into drilled holes, which are up to 2 m in length. Holes are drilled beneath the slip surface and are normal to the bedding. Rock bolts provide additional strength on critical planes of weakness within the rock mass. They may be up to 8 m in length with tensile working loads of up to 100 kN and are anchored in stable rock. They are put in tension so that the compression induced in the rock mass improves shearing resistance on potential failure planes. Light steel sections or steel mesh may be used between bolts to support the rock face. Rock anchors are used for major stabilization works, especially in conjunction with retaining structures. They may exceed 30 m in length. Anchors can be installed at different inclinations in order to dissipate the load within a rock mass. Gunite or shotcrete are frequently used to preserve the integrity of a rock face by sealing the surface and inhibiting the action of weathering (Fig. 11.2). They are pneumatically applied mortar and concrete respectively, and adapt to the surface configuration. In addition, they can be coloured to match the colour of the surrounding rocks. Coatings may be reinforced with wire mesh and used in combination with rock bolts. It generally is considered that such surface treatment offers negligible support to the overall slope structure. Fibrecrete also is pneumatically sprayed and contains short fibres of either steel or glass fibre. Heavily fractured rocks may be grouted in order to stabilize them. Soil nailing has been used to retain slopes, the nails consisting of steel bars, metal rods or metal tubes that are driven into in situ soil or soft rock, or grouted into bored holes. The ground surface between the nails is covered with a layer of shotcrete reinforced with wire mesh.

Restraining structures control sliding by increasing the resistance to movement. They include retaining walls, cribs, gabions and buttresses. The ability of a retaining wall to resist shearing action, overturning and sliding on or below its base must be considered before a retaining wall is used for slope control. They often are used where there is a lack of space for the full development of a slope, such as along many roads and railways. As retaining walls are subjected to unfavourable loading, a large wall width is necessary to increase slope stability. Pre-stressed anchor walls can be used to oppose the movement of a soil or rock mass. Walls also can be formed of contiguous piles. Reinforced earth can be used for retaining earth slopes. Such a structure is flexible and so can accommodate some settlement. Thus, reinforced earth can be used on poor ground. Reinforced earth walls are constructed by erecting a thin front skin at the face of the wall at the same time as the earth is placed. Strips of steel or geogrid are fixed to the

Fig. 11.2   Shotcrete used to stabilize a slope in Goteburg, Sweden.

Fig. 11.3   A crib wall in Hamilton, New Zealand.

facing skin at regular intervals. Cribs are constructed of precast reinforced concrete or steel units set up in cells that are filled with gravel or stone (Fig. 11.3). Gabions consist of strong wire mesh surrounding placed stones (Fig. 3.14). Concrete buttresses occasionally have been used to support large blocks of rock, usually where they overhang.

Drainage reduces the effectiveness of one of the principal causes of instability, namely, excess pore water pressure. The most likely zone of failure must be determined so that the extent of the slope mass that requires drainage treatment can be defined. Surface run-off should be prevented from flowing

Fig. 11.4 Surface drainage ditches filled with coarse granular material leading into a lined interceptor drain that, in turn, leads into a sink. The material being drained is till near Loch Lomond, Scotland.

unrestrained over a slope, which usually is done by the installation of a drainage ditch at the top of an excavated slope to collect drainage from above. The ditch, especially in soils, should be lined to prevent erosion, otherwise it will act as a tension crack. It may be filled with cobble aggregate. Infiltration can be lowered by sealing the cracks in a slope by regrading or filling with cement, bitumen or clay. A surface covering such as geotextile has a similar purpose and function. Herringbone ditch drainage normally is employed to convey water from the surfaces of slopes. These drainage ditches lead into an interceptor drain at the foot of the slope (Fig. 11.4). Trench drains are filled with free draining materials and may be lined with geotextiles. Deep wells are used to drain slopes where the depths involved are too deep for the construction of trench drains. Usually, the collected water flows from the base of the drainage wells under gravity but occasionally pumps may be installed in the bottom of wells to remove the water. Successful use of subsurface drainage depends on tapping the source of water; the presence of permeable material that aids free drainage; the location of the drain on relatively unyielding material to ensure continuous operation (flexible perforated PVC drains frequently are used); and the installation of a filter to minimize silting in the drainage channel. Drainage galleries are costly to construct and in areas subject to slip may experience caving. They should be backfilled with stone to ensure their drainage capacity if they are likely to be partially deformed by subsequent ground movements. Drill holes may be made about the perimeter of a gallery to enhance drainage.

11.1.4 Treatment of groundwater

Groundwater frequently provides one of the most difficult problems during excavation and its removal can prove costly. Not only does water make working conditions difficult, but piping, uplift pressures and flow of water into an excavation can lead to erosion and failure of the sides. Then, collapsed material has to be removed and the damage made good. Subsurface water normally is under pressure, which increases with increasing depth below the water table. Under high pressure gradients, weakly cemented rock can disintegrate. High piezometric pressures may cause the floor of an excavation to heave or, worse still, cause a blowout (i.e. the ground is ruptured by water under high pressure that enters the excavation). Hence, data relating to the groundwater conditions should be obtained prior to the commencement of operations.

281

Fig. 11.5 A wellpoint system dewatering an excavation.

Some of the worst conditions are met in excavations that have to be taken below the water table. In such cases the groundwater level can be lowered by some method of dewatering. The method adopted depends upon the permeability of the ground and its variation within the stratal sequence, the depth of base level below the water table and the piezometric conditions in underlying horizons. Pumping from sumps within an excavation, bored wells or wellpoints are the dewatering methods most frequently used. Pumping from a sump within an excavation generally can be achieved when the rate of inflow does not lead to instability of the sides or base. Ditches are dug in the floor of the excavation that convey the water to a sump located at a lower level.

In waterlogged silts and sands inflow of water may be high enough to cause the sides of an excavation to slump or the floor to boil. Therefore, predrainage is called for. Predrainage of a site can be accomplished by installing wellpoints (Fig. 11.5), an eductor system or bored wells about the perimeter. Such groundwater lowering techniques depend on excessive pumping that lowers the water table and thereby develops a cone of depression. In wellpointing the radii of influence of the individual wellpoints overlap and they are laid out so as to lower the water table by approximately a metre below the base level of the excavation. Wellpoint pumping is most useful where the required lowering is not more than about 4 to 5.5 m. In order to achieve greater lowering another tier of wellpoints, header main and pumps must be installed. Wellpoints are used for dewatering sands, however, those using vacuum assisted drainage are effective in silty sands. An eductor is a jet pump and is most effective when used to dewater deep excavations made in stratified soils. They are used where it is impossible or inconvenient to install a multi-stage wellpoint system. A bored filter well consists of a perforated tube placed in a borehole and surrounded by an annulus of filter media. Their operational depth, in theory, may be unlimited. An electric submersible pump is connected by a riser pipe to a common discharge main. Bored wells are preferable to wellpointing for deep excavations where the area of the excavation is small in relation to the depth. They also are preferable in ground containing cobbles and boulders where wellpoint installation is difficult. Deep wells are suited to variable and multi-layer soils, as well as to the control of groundwater under artesian or subartesian conditions.

However, whenever the phreatic or piezometric surface is lowered, the effective load on the soil is increased, causing consolidation and consequent settlement. Usually, settlement due to the abstraction

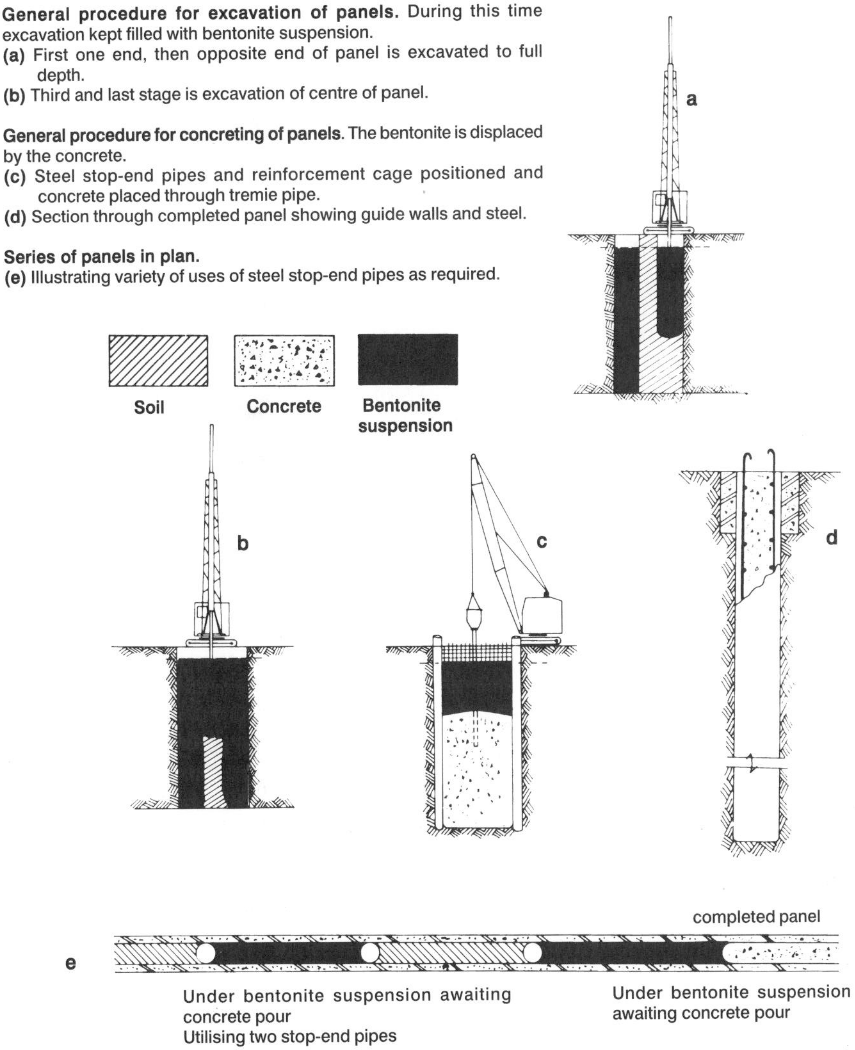

General procedure for excavation of panels. During this time excavation kept filled with bentonite suspension.
(a) First one end, then opposite end of panel is excavated to full depth.
(b) Third and last stage is excavation of centre of panel.

General procedure for concreting of panels. The bentonite is displaced by the concrete.
(c) Steel stop-end pipes and reinforcement cage positioned and concrete placed through tremie pipe.
(d) Section through completed panel showing guide walls and steel.

Series of panels in plan.
(e) Illustrating variety of uses of steel stop-end pipes as required.

Soil   Concrete   Bentonite suspension

completed panel

Under bentonite suspension awaiting concrete pour
Utilising two stop-end pipes

Under bentonite suspension awaiting concrete pour

Fig. 11.6   Construction of a reinforced diaphragm wall.

of water from clean sands is insignificant unless the sand was initially very loosely packed. On the other hand, pumping from ground containing layers of soft clay, peat or other compressible soils, or from a confined aquifer overlain by compressible soils, may cause significant settlements. The amount of settlement undergone depends on the thickness of the compressible layers and their compressibility, as well as on the amount of groundwater lowering. The permeability of the soil and the length of the pumping period influence the rate of settlement. As a result it may be necessary to limit the radius of influence by the use of groundwater recharge methods.

Where surrounding property has to be safeguarded it usually is more appropriate to provide a barrier about an excavation in order to prevent inflow of water, whilst maintaining the surrounding water table at its level, rather than to adopt a dewatering technique. Methods of forming such a barrier, include steel sheet piling, diaphragm walls, secant pile walls, cement or clay-cement grout curtains and frozen soil barriers. Ideally, these structures should be keyed into an impermeable horizon beneath the excavation. Steel sheet piling cannot be used in ground containing numerous boulders or other obstructions. In these conditions a diaphragm wall can be constructed since the boulders can be dealt with by grabbing or chiselling (Fig. 11.6). Secant piles, which are piles keyed into each other, have been used in gravels, sands or silts. If for some reason groundwater control in sand or silt, especially if they are loosely packed, is not effective, then a quick condition may develop in the floor of the excavation. Relief wells can be used in such instances to reduce the hydrostatic head. Jet grouting offers a means of forming an impermeable barrier both vertically and horizontally, and can be used in all types of soils. In its simplest form jet grouting involves inserting an injection pipe into the soil to the required depth. The soil is then subjected to a horizontally rotating jet of water and at the same time is mixed with grout (cement or cement-bentonite) to form plastic soil-cement. The injection pipe is raised gradually. Replacement jet grouting involves the removal of soil from the zone to be treated by a high-energy erosive jet of water and air, the operation proceeding upwards from the base of a borehole. The grout is emplaced simultaneously. The soil that is removed is brought to the surface by air lift pressure. Frozen ground is dealt with in Section 11.3.

11.2 Tunnels and tunnelling

Geology is the most important factor determining the nature, form and cost of a tunnel. For example, the route, design and construction of a tunnel are largely dependent upon geological considerations. Accordingly, tunnelling can be an uncertain and sometimes hazardous undertaking because information on ground conditions along the alignment is never complete, no matter how good the site investigation. Estimating the cost of tunnel construction, particularly in areas of geological complexity, is uncertain.

A pilot tunnel is probably the best method of exploring tunnel location and should be used if a major sized tunnel is to be constructed in ground that is known to have critical geological conditions. It also drains the rock ahead of the main excavation. If the inflow of water is excessive, the rock can be grouted from the pilot tunnel before the main excavation reaches the water bearing zone.

Reliable information relating to the ground conditions ahead of the advancing face obviously is desirable during tunnel construction. This can be achieved with a varying degree of success by drilling long horizontal holes between shafts, or by direct drilling from the tunnel face at regular intervals. In extremely poor ground conditions tunnelling progresses behind an array of probe holes that fan outwards some 10 to 30 m ahead of the tunnel face. Although this slows progress it ensures completion. Holes drilled upwards from the crown of the tunnel and forwards from the side-walls help locate any abnormal features such as faults, buried channels, weak seams or solution cavities. Drilling equipment for drilling in a forward direction can be incorporated into a tunnelling machine. The penetration rate of a probe drill must exceed that of the tunnel boring machine, ideally it should be about three times faster. Maintaining the position of the hole, however, presents a major problem when horizontal drilling is undertaken. In particular, variations in hardness of the ground oblique to the direction of drilling can cause radical deviations. The inclination of a hole therefore must be surveyed.

11.2.1 Tunnels and problem ground

As far as tunnelling in soft ground is concerned, the difficulties and costs of construction depend almost exclusively on the stand-up time of the ground and this is greatly influenced by the position of the water table in relation to the tunnel. Above the water table the stand-up time depends principally on the shearing and tensile strength of the ground whereas below it, it also is influenced by the permeability of the material involved.

A number of types of soft ground have been distinguished. Firm ground has sufficient shearing and tensile strength to allow a tunnel heading to be advanced without support, typical representatives are stiff clays of low plasticity and loess above the water table. In ravelling ground, blocks fall from the roof and sides of the tunnel some time after the ground has been exposed. The strength of the ground usually decreases with increasing duration of load. It also may decrease due to excess pore water pressures induced by ground movements in clay soil, or due to evaporation of water with subsequent loss of apparent cohesion in silt and fine sand. If ravelling begins within a few minutes of exposure it is described as fast ravelling, otherwise it is referred to as slow ravelling. Fast ravelling may take place in residual soils and sands with a clay binder below the water table. These materials above the water table are slow ravelling.

In running ground the removal of support from a surface inclined at more than 34° gives rise to a run, the latter occurring until the angle of rest of the material involved is attained. Runs take place in clean loosely packed gravel, and clean coarse-to-medium grained sand, in both cases when they are above the water table. In clean fine-grained moist sand a run usually is preceded by ravelling, such behaviour being termed cohesive running.

Flowing ground moves like a viscous liquid. It can invade a tunnel from any angle and, if not stopped, ultimately will fill the excavation. Flowing conditions occur in sands and silts below the water table.

Squeezing ground advances slowly and imperceptibly into a tunnel. There are no signs of fracturing of the sides. Ultimately, the roof may give and this can produce a subsidence trough at the surface. The two most common reasons why ground squeezes are, firstly, excessive overburden pressure and, secondly, the dissipation of residual stress, both eventually leading to failure. Soft clay soils display squeezing behaviour. Other materials in which squeezing conditions may occur include shales and highly weathered granites, gneisses and schists.

Swelling ground also expands into a tunnel but the movement is associated with a considerable volume increase in the ground immediately surrounding the tunnel. Swelling occurs as a result of water migrating from the surrounding strata into the material at the tunnel perimeter. These conditions develop in overconsolidated clays of high plasticity and in certain shales and mudstones, especially those containing montmorillonite. Swelling pressures are of unpredictable magnitude and may be extremely large. The development period may take a few weeks or several months. Immediately after excavation the pressure is insignificant but then it increases at a higher rate. In the final stages the increase slows down. This condition is usually dealt with by imposing no restriction on swelling until it has attained a certain limit and by constructing the permanent lining at a later date. Swelling also has occurred in evaporite formations as a consequence of anhydrite being hydrated to form gypsum. Swelling pressures up to 12 MPa have been recorded.

Boulders within a soft ground matrix may prove difficult to remove, while if boulders are embedded in a hard cohesive matrix they may impede progress of and may render a mechanical excavator of almost any type impotent. Large boulders may be difficult to handle unless they are broken apart by jackhammer or by blasting.

11.2.2 Tunnels and ground stress

Rocks, especially those at depth, are affected by the mass of overburden and the stresses so developed cause the rocks to be strained. In certain areas, particularly orogenic belts, the state of stress also is influenced by tectonic factors. However, because the rocks at depth are confined they suffer partial strain. The stress that does not give rise to strain, that is, which is not dissipated, remains in the rocks as residual stress. While the rocks remain in a confined condition the stresses accumulate and may reach high values, sometimes in excess of the yield point (i.e. the point or loading at which elastic deformation gives way to plastic deformation) of the rocks. If the confining condition is removed, as in tunnelling, then the residual stress can cause displacement. The amount of movement depends upon the magnitude of the residual stress. The pressure relief, which represents a decrease in residual stress, may be instantaneous or slow in character, and is accompanied by movement of the rock mass with variable degrees of violence. Rock may suddenly break from the sides of the excavation in tunnels driven at great depths. This phenomenon is referred to as rock bursting and it may release hundreds of tonnes of rock with explosive force. Most rock bursts occur at depths in excess of 600 m and the most explosive failures occur in rocks that have unconfined compressive strengths and values of Young's modulus greater than 140 MPa and 34.5 GPa respectively.

Popping is a similar but less violent form of failure, where the sides of the excavation bulge before exfoliating. Spalling tends to occur in jointed or cleaved rocks. To a certain extent such a rock mass can bulge as a sheet, collapse occurring when a key block either fails or is detached from the mass. In fissile rock such as shale, the beds may bend slowly into the tunnel. In this case the rock is not necessarily detached from the main mass, but the deformation may cause fissures and hollows in the rock surrounding the tunnel. Another pressure relief phenomenon is bumping ground. Bumps are sudden and somewhat violent earth tremors, which at times dislodge rock from the sides of a tunnel.

11.2.3   Tunnels, discontinuities and faults

As far as conventional tunnelling is concerned (i.e. excavation by drilling and blasting), the roof of a tunnel will be flat where horizontally lying rocks are thickly bedded and contain few joints. Large planar surfaces also form most of the roof in a formation that is not inclined at a high angle and strikes more or less parallel to the axis of a tunnel (Fig. 11.7a). Conversely, a peaked roof is formed if the rocks are thinly bedded and intersected by many joints (Fig. 11.7b). This type of stratification is more dangerous where the beds dip at 5° to 10° since this may lead to the roof spalling, as the tunnel is driven forward. In tunnels where jointed strata dip into the side at 30° or more, the up-dip side may be unstable. Discontinuities that are parallel to the axis of a tunnel and that dip at more than 45° may prove especially treacherous, leading to slabbing of the walls and fall-outs from the roof. When the tunnel alignment is normal to the strike of jointed rocks and the dips are less than 15°, then large blocks again are likely to fall from the roof. The sides, however, tend to be reasonably stable. If the axis of a tunnel runs parallel to the strike of vertically dipping rocks, then the mass of rock above the roof is held by friction along the bedding planes. When the joint spacing in horizontally layered rocks is greater than the width of a tunnel, then the beds bridge the tunnel as a solid slab and are only subject to bending under their own weight.

Faults generally mean non-uniform rock pressures on a tunnel and therefore at times necessitate special treatment such as the construction of box sections. Generally, problems increase as the strike of a fault becomes more parallel to a tunnel opening. However, even if the strike is across a tunnel, faults with low dips can represent a hazard. If a tunnel is driven from the hanging wall, the fault first appears at the invert and it generally is possible to provide adequate support or reinforcement when driving through

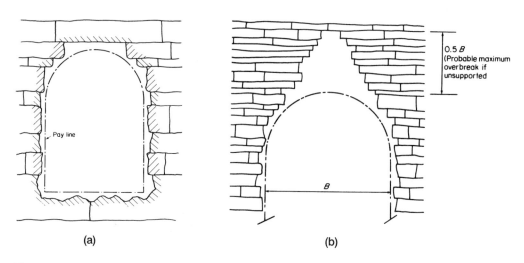

(a)   (b)

Fig. 11.7   (a) Bridge action in strong, horizontally bedded, rocks with few joints. (b) Overbreak in thinly bedded horizontal strata with vertical joints during tunnelling by blasting. Ultimate overbreak occurs if no support is installed.

the rest of the zone. Conversely, when a tunnel is driven from the footwall side, the fault first appears in the crown, and there is a possibility that a wedge-shaped block, formed by the fault and the tunnel, will fall from the roof without warning. Major faults are usually associated with a number of minor faults and the dislocation zone may occur over many metres. What is more, rock material within a faulted zone may be shattered and unstable. Problems tend to increase with increasing width of a fault zone. Sometimes a fault zone is filled with sand-sized crushed rock that has a tendency to flow into a tunnel. If, in addition, the tunnel is located beneath the water table, a sandy suspension may rush into the excavation. When a fault zone is occupied by clay gouge and a section of a tunnel follows the gouge zone, then if this material swells it can cause displacement or breakage of tunnel supports during construction. Large quantities of water in a permeable rock mass may be impounded by a fault zone occupied by impervious gouge and are released when tunnelling operations penetrate through the fault zone.

Movements along major active faults in certain parts of the world can disrupt a tunnel lining and even lead to a tunnel being offset. Consequently, a tunnel alignment should avoid active faults or, if possible, use open cut within fault zones. The earthquake risk to an underground structure is influenced by the material in which it occurs. For instance, a tunnel at shallow depth in alluvial deposits will be seriously affected by the large relative displacements of the ground surrounding it. On the other hand, a deep tunnel in solid rock will be subjected to displacements that are considerably less than those that occur at the surface. The main causes of stresses in shallow underground structures arise from the interaction between the structure and the displacements of the ground. If the structure is sufficiently flexible it will follow the displacements and deformations to which the ground is subjected.

11.2.4   Tunnels, groundwater and gas

Construction of a tunnel may alter the groundwater regime of a locality, as a tunnel generally acts as a drain. Generally, however, the amount of water flowing into a tunnel decreases as construction progresses. This is due to the gradual exhaustion of water at source and to the decrease in hydraulic gradient, and hence in flow velocity. On the other hand, there may be an increase in flow as construction progresses, if construction operations cause fissuring (e.g. blasting may open new water conduits around a tunnel, shift the direction of flow and in some cases may even cause partial flooding). The amount of water held in a rock mass depends on its reservoir storage properties that, in turn, influence the amount of water that can drain into a tunnel. Isolated heavy flows of water may occur in association with faults, solution pipes and cavities, abandoned mine workings, or even from pockets of gravel. Tunnels driven under lakes, rivers and other surface bodies of water may tap considerable volumes of flow. Flow also may take place from a perched water table to a tunnel beneath. One of the principal problems created by water entering a tunnel is that of face instability. Secondary problems include removal of excessively wet muck and the placement of a precision fitted primary lining or of ribs.

Correct estimation of water inflow into a projected tunnel is of vital importance, as inflow influences the construction programme. Not only is the value of the maximum inflow required but so is the distribution of inflow along the tunnel section and the changes of flow with time. The greatest groundwater hazard in underground work is the presence of unexpected water bearing zones, and therefore whenever possible the position of hydrogeological boundaries should be located. Obviously, the location of the water table and its possible fluctuations are of major importance. The techniques used to help control groundwater include drainage, compressed air, grouting or freezing. Water pressures are more predictable than water flows as they are nearly always a function of the head of water above the tunnel location. They can be very large, especially in confined aquifers.

Naturally occurring gas can occupy the pore spaces and voids in rock. This gas may be under pressure and so may burst into underground workings, causing the rock to fail, sometimes with explosive force. Wherever possible, the likelihood of gas hazards should be noted during the geological survey, but this is one of the most difficult tunnel hazards to predict. If the flow of gas appears to be fairly continuous, the entrance of the flow may be sealed with concrete. Often the supply of gas is quickly exhausted, but cases have been reported where it continued for up to three weeks.

Many gases are dangerous. For example, methane, CH_4, is not only toxic, it also is combustible and highly explosive when mixed with air. It is lighter than air. Carbon dioxide, CO_2, and carbon monoxide, CO, are both toxic. The former is heavier than air whereas carbon monoxide is slightly lighter than air. All three of these gases are found in coal bearing strata. Carbon dioxide also may be associated with volcanic deposits and limestones. Hydrogen sulphide, H_2S, is heavier than air, is highly toxic and is explosive when mixed with air. The gas may be generated by the decay of organic substances or by volcanic activity. Sulphur dioxide, SO_2, is a colourless, pungent, asphyxiating gas, which dissolves readily in water to form sulphuric acid. It usually is associated with volcanic emanations or it may be formed by the breakdown of pyrite.

11.2.5   Tunnels and ground temperatures

Temperatures in tunnels are not usually of concern unless the tunnel is more than 170 m below the surface. When rock is exposed by excavation the amount of heat liberated depends on the virgin rock temperature (VRT); the thermal properties of the rock; the length of time of exposure; the area, size and shape of exposed rock; the wetness of rock; the air flow rate; the dry bulb temperature; and humidity of the air. In deep tunnels high temperatures can make work more difficult. Indeed, high temperatures and rock pressures place limits on the depth of tunnelling. The moisture content of the air in tunnels is always high and in saturated air the efficiency of labour declines when the temperature exceeds 25°C, dropping to almost zero when the temperature reaches 35°C. Conditions can be improved by increased ventilation, by water spraying or by using refrigerated air. Air refrigeration is essential when the virgin rock temperature exceeds 40°C. The rate of increase in rock temperature with depth depends on the geothermal gradient. Although the geothermal gradient varies with locality, according to rock type and structure, on average it increases with depth at a rate of 1°C per 30 to 35 m.

11.2.6   Tunnelling methods

The choice of tunnelling method is influenced by a number of factors among which ground conditions are the most important. Tunnelling in soft ground usually employs a shield to provide safe working conditions for operatives, as well as for achieving efficient tunnel excavation. The strength of rock masses varies enormously, from less than 10 MPa to over 300 MPa in terms of unconfined compressive strength, and their behaviour is influenced by the presence of discontinuities. Strong massive rocks may preclude machine excavation but will require minimal temporary support. Mixed face conditions can provide problems for machine tunnelling, as well as with the provision of temporary support. The diameter of a tunnel may influence the choice of tunnelling method, as may location. For example, drill and blast excavation may be unacceptable in urban areas.

The conventional method of advancing a tunnel in hard rock is by full-face drivage in which the complete face is drilled and blasted as a unit. However, full-face drivage should be used with caution where the rocks are variable. The usual alternatives are the top heading and bench method or the top heading method whereby the tunnel is worked on an upper and lower section or heading. The sequence of operations in these three methods is illustrated in Fig. 11.8. In tunnel blasting a cut is opened up approximately in the centre of the face in order to provide a cavity into which subsequent shots can blast, the shots being detonated in a predetermined sequence.

Drilling and blasting can damage the rock structure, depending on the properties of the rock mass and the blasting technique. The stability of a tunnel roof in fissured rocks depends upon the formation of a natural arch and this is influenced by the extent of any disturbance, the irregularities in the profile, and the relationship between tunnel size and fracture pattern. The amount of overbreak (i.e. excavation beyond the payline) tends to increase with increased lengths of pull since drilling inaccuracies are magnified. In such situations not only does the degree of overbreak become very expensive in terms of grout and concrete backfill but it may give rise to support problems and subsidence over the crown of a tunnel. However, overbreak can be reduced by controlled blasting either by presplitting the face to the desired contour or by smooth blasting. In the presplitting method a series of holes is drilled around the

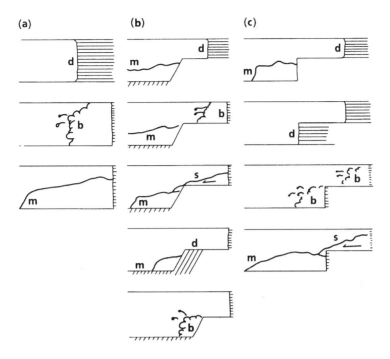

Fig. 11.8 Conventional tunnelling methods (i.e. drilling and blasting) (a) full-face (b) top heading and bench (c) top heading bench drilled horizontally. Phases: D = drilling; B = blasting; M = mucking; S = scraping.

perimeter of the tunnel, loaded with explosives that have a low charging density and detonated before the main blast, thereby developing a fracture that spreads between the holes and produces an accurate profile. The technique is not particularly suited to slates and schists because of their respective cleavage and schistosity. Also, in jointed rock masses, the tunnel profile is influenced by the pattern of the jointing. Again, explosives with a low charging density are used in closely spaced perimeter holes in smooth blasting so that crack formation is controlled between the drill holes and is concentrated within the final contour. The holes are fired after the main blast, their purpose being to break away the last fillet of rock between the main blast and the perimeter. Smooth blasting cannot be carried out without good drilling precision.

Machine tunnelling in rock uses either a roadheader machine or a tunnel boring machine (TBM). A roadheader generally moves on a tracked base and has a cutting head, usually equipped with drag picks, mounted on a boom. Twin-boom machines have been developed in order to increase the rate of excavation. Roadheaders can cut a range of tunnel shapes and are particularly suited to stratified formations. Some of the heavier roadheaders can excavate massive rocks with unconfined compressive strengths in excess of 100 MPa and even up to 200 MPa. Obviously, the cutting performance is influenced by the presence and character of discontinuities. Basically, excavation with a TBM is accomplished by a cutter head, equipped with an array of suitable cutters, which usually is rotated at a constant speed and thrust into the tunnel face by a hydraulic pushing system (Fig. 11.9). The stresses imposed on the surrounding rock by a TBM are much less than those produced during blasting and therefore damage to the perimeter is minimized and a sensibly smooth face usually is achieved. What is more, overbreak normally is less

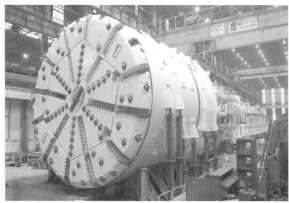

Fig. 11.9 Top left: Assembly and test of the Barcelona metro TBM in NFM Technologies factory. Left: The tunnel boring machine used for the Shiraz Metro, Iran, diameter 6.9m. Above: Cutterhead of the Barcelona TBM before excavation, 11.95 m diameter (all courtesy NFM Technologies, Lyon, France).

during TBM excavation than during drilling and blasting, on average 5% as compared with up to 25% for conventional methods. This means that less support is required.

The rate of tunnel drivage obviously is an important economic factor in tunnelling, especially in hard rock. Tunnel boring machines have provided increased rates of advance and so shortened the time taken to complete tunnelling projects. Indeed, they have achieved faster rates of drivage than conventional tunnelling methods in rocks with unconfined compressive strengths of up to 150 MPa. Consequently, tunnels now are excavated much more frequently by TBMs than by conventional drill and blast methods. The unconfined compressive strength commonly is one of the most important properties determining the rate of penetration of a TBM. The rate of penetration in low strength rocks is affected by problems of roof support and instability, as well as gripper problems. Problems associated with rocks of high strength are the increased cutter wear and larger thrust, and hence cost, required to induce rock fracture. The rate of penetration also is influenced by the necessity to replace cutters on the head of a TBM, which involves downtime. Cutter wear depends, in part, on the abrasive properties of the rock mass being bored. Whether a rock mass is massive, jointed, fractured, water bearing, weathered or folded, also affect cutter life. For instance, in hard blocky ground some cutters are broken by the tremendous impact loads generated during boring. Moreover, the performance of TBMs is more sensitive to changes in rock properties than conventional drilling and blasting methods, consequently their use in rock masses that have not been thoroughly investigated involves a high risk.

In soft ground support is vital and so tunnelling is carried out by using a shield. A shield is a cylindrical drum with a cutting edge around the circumference, the cut material being delivered onto a conveyor

Fig. 11.10 Installation of segmental lining, Delivery Tunnel North, Lesotho Highlands Water Scheme.

for removal. The limit of these machines usually is regarded as an unconfined compressive strength of around 20 MPa. Shield tunnelling means that construction can be carried out in one stage at the full tunnel dimension and that the permanent lining is installed immediately after excavation thereby providing support as the tunnel is advanced. Bentonite slurry is used to support the face in soft ground in a pressure bulkhead machine. The bentonite slurry counterbalances the hydrostatic head of groundwater in the soil and stability is further increased as the bentonite is forced into the pores of the soil, gelling once penetration occurs.

11.2.7 Tunnel support

The time a rock mass may remain unsupported in a tunnel is called its stand-up time or bridging capacity. This mainly depends on the magnitude of the stresses within the unsupported rock mass, which in their turn depend on its span, its strength and its discontinuity pattern. If the bridging capacity of the rock is high, the rock material next to the heading will stay in place for a considerable time. By contrast, if the bridging capacity is low, the rock will immediately start to fall at the heading so that supports have to be erected as soon as possible (Fig. 11.10).

The correct evaluation of the effect of the dip and strike of rock formations, their joint pattern and the direction of tunnel drive are of paramount importance as far as determining the type of support system to be used in a tunnel (see Section 11.2.3). The joint pattern proves one of the most difficult and crucial factors to appraise when determining the type of support system to employ. In addition to defining the dimensions and orientation of the joint pattern it is necessary to evaluate jointing with regard to the conditions of the joint surfaces, tunnel size, direction of drive and method of excavation. Rock pressures on the lining of a tunnel are influenced by the size and shape of the tunnel with respect to the intact strength of the rock mass(es) concerned and the nature of the discontinuities; the pre-excavation geostatic stress; the groundwater pressures; the method of excavation; the degree of overbreak; the length of time before placing the permanent lining; and the stiffness of the artificially injected infilling behind the lining and the rigidity of the lining itself. The redistributed ground stresses around circular tunnels in very competent rock are accepted by the rock as low radial and high tangential stresses. This gives a low loading on the tunnel lining. On the other hand, if a rock mass is weathered or highly jointed, redistribution of stress may weaken it and eventually lead to partial collapse of overlying strata. Partial collapse results in high stresses on the lining at the crown and redistributed invert and sidewall resisting radial stresses. The effect of groundwater on support requirements varies with respect to weathering, joint filler or condition of the joint surfaces and depth of cover. Probably the most difficult support situation experienced in tunnel driving occurs where heavy inflows of water under high pressures are encountered in conjunction with adverse rock properties. Many tunnels, however, have penetrated heavy inflow formations with little difficulty with respect to ground support.

Suitable support measures at times must be adopted to attain a stand-up time longer than that indicated by the class of the rock mass. These measures constitute the primary support, the purpose of which is to ensure tunnel stability until the permanent support system, for example, a concrete lining, is installed. The primary support for a tunnel in rock may be provided by rock bolts, gunite or steel arches. Rock bolts maintain the stability of an opening by suspending the dead weight of a slab from the rock above; by providing a normal stress on the rock surface to clamp discontinuities together and develop beam action; by providing a confining pressure to increase shearing resistance and develop arch action; and by preventing key blocks becoming loosened so that the strength and integrity of the rock mass is maintained. Gunite or shotcrete can be used for lining tunnels. For example, a layer 150 mm thick around a tunnel 10 m in diameter can safely carry a load of 500 kPa corresponding to a burden of approximately 23 m of rock, more than has ever been observed with rock falls. When combined with rock bolting and steel mesh, gunite has proved an excellent temporary support for all qualities of rock. In very bad cases steel arches can be used for reinforcement of weaker tunnel sections.

Rating systems or classifications have been developed to assess rock mass quality in relation to tunnel support. The two most notable are the Geomechanics system and the Q system. In the former system the uniaxial compressive strength of rock material; the rock quality designation (RQD, i.e. the percentage core stick in excess of 100 m lengths to the length drilled); the spacing, orientation, and condition of the discontinuities; and groundwater inflow are grouped into five categories and the categories are given a rating (Table 11.2). Once determined, the ratings of the individual parameters are summed to give the total rating or class of the rock mass. The higher the total rating is, the better the rock mass conditions. However, the value of the rock mass rating in certain situations may be open to question, for example, it does not take into account the effects of blasting on rock masses. Neither does it consider the influence of in situ stress on stand-up time nor the durability of the rock. The latter can be assessed in terms of the geodurability classification (see Chapter 1). In the Q system the rock mass quality is defined in terms of six parameters:

1. The RQD or an equivalent estimate of joint density.
2. The number of joint sets (J_n), which is an important indication of the degree of freedom of a rock mass. The RQD and the number of joint sets provide a crude measure of relative block size.
3. The roughness of the most unfavourable joint set (J_r). The joint roughness and the number of joint sets determine the dilatancy of the rock mass.
4 The degree of alternation or filling of the most unfavourable joint set (J_a). The roughness and degree of alteration of the joint walls or filling materials provides an approximation of the shear strength of the rock mass.
5. The degree of water seepage (J_w).
6. The stress reduction factor (SRF), which accounts for the loading on a tunnel caused either by loosening loads in the case of clay-bearing rock masses, or unfavourable stress-strength ratios in the case of massive rock. Squeezing and swelling are also taken account of in the SRF.

The ratings for each of the six parameters allow the rock mass quality (Q) to be derived from:

$$Q = \frac{RQD}{J_n} \cdot \frac{J_r}{J_a} \cdot \frac{J_w}{SRF} \tag{11.1}$$

The numerical value of Q ranges from 0.001 for exceptionally poor quality squeezing ground, to 1000 for

Table 11.2 The rock mass rating – geomechanics classification of rock masses (after Bieniawski, 1989).

Classification parameters and their ratings

Parameter		Range of values						
1. Strength of intact rock material	Point-load strength index (MPa)	>10	4–10	2–4	1–2	For this low range uniaxial compressive test preferred		
	Uniaxial compressive strength (MPa)	>250	100–250	50–100	25–50	5–25	1–5	<1
	Rating	15	12	7	4	2	1	0
2. Drill core quality RQD (%)		90–100	75–90	50–75	25–50	<25		
	Rating	20	17	13	8	3		
3. Spacing of discontinuities		>2 m	0.6–2 m	200–600 mm	60–200 mm	<60 mm		
	Rating	20	15	10	8	5		
4. Condition of discontinuities		Very rough surfaces Not continuous No separation Unweathered wall rock	Slightly rough surfaces Separation <1 mm Slightly weathered walls	Slightly rough surfaces Separation <1 mm Highly weathered walls	Slickensided surfaces or gouge <5 mm thick or Separation 1–5 mm Continuous	Soft gouge >5 mm thick or Separation >5 mm Continuous		
	Rating	30	25	20	10	0		
5. Groundwater	Inflow per 10 m tunnel length (litre/min) or	none	<10	10–25	25–125	>125		
	ratio of joint water pressure major principal stress or	0	<0.1	0.1–0.2	0.2–0.5	>0.5		
	General conditions	completely dry	damp	wet	dripping	flowing		
	Rating	15	10	7	4	0		

Table 11.2 cont'd.

Parameter	Range of values				
Rating adjustment for discontinuity orientations					
Strike and dip orientations of discontinuities	very favourable	favourable	fair	unfavourable	very unfavourable
Ratings tunnels and mines	0	−2	−5	−10	−12
foundations	0	−2	−7	−15	−25
slopes	0	−5	−25	−50	−60
Rock mass classes determined from total ratings					
Rating	100←81	80←61	60←41	40←21	<20
Class number	I	II	III	IV	V
Description	very good rock	good rock	fair rock	poor rock	very poor rock
Meaning of rock mass classes					
Class number	I	II	III	IV	V
Average stand-up time	20 yr for 15 m span	1 yr for 10 m span	1 wk for 5 m span	10 hr for 2.5 m span	30 min for 1 m span
Cohesion of the rock mass (kPa)	>400	300–400	200–300	100–200	<100
Friction angle of the rock mass (degrees)	>45	35–45	25–35	15–25	<15

exceptionally good quality rock that is practically unjointed. The Q value is related to the type and amount of support by deriving the equivalent dimensions of the excavation. The latter is related to the size and purpose of the excavation, and is obtained from:

$$\text{equivalent dimension} = \text{span or height of wall/ESR} \qquad (11.2)$$

where ESR is the excavation support ratio related to the use of the excavation and the degree of safety required.

11.3 Shafts and raises

A shaft is driven vertically downwards whereas a raise is driven either vertically or at a steep angle upwards. Excavation is either by drilling and blasting or by shaft boring machine. Drilling is usually easier in shaft sinking than in tunnelling but blasting is against gravity and mucking is slow and therefore expensive. By contrast, mucking and blasting are simpler in raising operations but drilling is difficult. As the cost of drilling generally is exceeded by that of blasting and mucking, raising is more economic than shaft sinking where both are practical. Since a raise is of small cross-sectional area, if the excavation is to have a large diameter, then enlargement is done from above, the primary raise excavation being used as a muck shute. Where a raise emerges at the surface through unconsolidated material, then this section is excavated from the surface.

Usually, a shaft is sunk through a series of different rock types and the two principal problems likely to be encountered are varying stability of the walls and ingress of water. Indeed, these two problems frequently occur together and they are likely to be met with in rock with a high fracture index, weak zones being particularly hazardous. They are most serious, however, in unconsolidated deposits, especially loosely packed gravels, sands and silts. The position of the water table is highly significant. The geological investigation prior to shaft sinking therefore should provide detailed information relating to the character of the soil/rock masses involved, noting where appropriate their porosity, permeability, fracture index and strength.

An effective method of dealing with groundwater in shaft sinking is to pump from a sump within the shaft. However, problems arise when the quantity pumped is so large that the rate of inflow under high head causes instability in the sides of the shaft or prevents fixing and back-grouting of the shaft lining. Surrounding a shallow shaft by a ring of bored wells may not achieve effective lowering of the water table in fissured rocks or variable water bearing soils as there is a tendency for the water to by-pass the wells and flow into the excavation. Where the stability of the wall and/or ingress of water are likely to present problems in shaft sinking one of the most frequent techniques resorted to is ground freezing. Freezing transforms weak waterlogged materials into ones that are self-supporting and impervious. Normally, the freeze probes are laid out in linear fashion so that an adequate boundary wall encloses the future excavation when the radial development of ice about each probe unites to form a continuous section of frozen ground. The classic coolant is brine, although cryogenic liquids, notably liquid nitrogen, are also used. The latter reduces the freezing time by about two-thirds. Freezing is a versatile technique and can deal with a variety of soil and rock types but a limitation is placed upon the process by unidirectional flow of groundwater. For example, a velocity exceeding 2 m per day will seriously affect and distort the growth of an ice wall in a brine freezing project. The tolerance is much wider when liquid nitrogen is used. Frozen soils undergo creep when subjected to uniform loading, in particular soils rich in silt–clay fractions may creep to a significant extent if excavations in them are to be open for an extended period of time. Frost heave in silty ground is uncommon when ground is frozen artificially. Grouting can be an economical method of eliminating or reducing the flow of groundwater into shafts if the soil or rock conditions are suitable for accepting cement or chemical grouts. This is because the perimeter of the grouted zone is relatively small in relation to the depth of the excavation.

11.4 Underground caverns

The site investigation for an underground cavern has to locate a sufficiently large mass of sound rock in which the cavern can be excavated. Because caverns usually are located at appreciable depth below ground surface, the rock mass often is beneath the influence of weathering and consequently the chief considerations are rock quality, geological structure and groundwater conditions. The orientation of an underground cavern generally is based on an analysis of the joint pattern, including the character of the different joint systems in the area and, where relevant, on the stress distribution. It normally is considered necessary to avoid an orientation whereby the long axis is parallel to steeply inclined major joint sets. Wherever possible caverns should be orientated so that fault zones are avoided.

Displacement data provide a direct means of evaluating cavern stability. Displacements that exceed the predicted elastic displacements by a factor of 5 or 10 generally result in decisions to modify support and excavation methods. In a creep sensitive material, such as may occur in a major shear zone or zone of soft altered rock, the natural stresses concentrated around an opening cause time-dependent displacements that, if restrained by support, result in a build-up of stress on the support. Conversely, if a rock mass is not sensitive to creep, stresses around an opening normally are relieved as blocks displace towards the opening. However, initial movements may be very much influenced by the natural stresses concentrated around an opening and under certain boundary conditions may continue to act even after large displacements have occurred.

The angle of friction for tight irregular joint surfaces commonly is greater than 45° and as a consequence the included angle of any wedge opening into the roof of a cavern has to be 90° or more, if the wedge is to move into the cavern. A tight rough joint system therefore only presents a problem when it intersects the surface of a cavern at relatively small angles or is parallel to the surface of the cavern. However, if the material occupying a thick shear zone has been reduced to its residual strength, then the angle of friction could be as low as 15° and in such an instance the included angle of the wedge would be 30°. Such a situation would give rise to a very deep wedge, which could move into a cavern. Displacement of wedges into a cavern is enhanced if the ratio of the intact unconfined compressive strength to the natural stresses concentrated around the cavern is low. Values of less than 5 are indicative of stress conditions in which new extension fractures develop about the cavern during its excavation. Wedge failures are facilitated by shearing and crushing of the asperities along discontinuities as wedges are displaced.

The walls of a cavern may be greatly influenced by the prevailing state of stress, especially if the tangential stresses concentrated around the cavern approach the intact compressive strength of the rock. In such cases, extension fractures develop near the surface of the cavern as it is excavated and

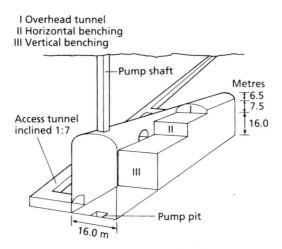

Fig. 11.11 Main stages in the excavation of underground chambers.

cracks produced by blast damage become more pronounced. The problem is accentuated if any lineation structures or discontinuities run parallel to the walls of the cavern. Indeed, popping of slabs of rock may take place from cavern walls.

Rock bursts have occurred in underground caverns at rather shallow depths, particularly where they were excavated in the sides of valleys and on the inside of faults when the individual fault passed through a cavern and dipped towards an adjacent valley. Bursting can take place at depths of 200 to 300 m when the tensile strength of the rocks varies between 3 and 4 MPa.

Three methods of blasting are normally used to excavate underground caverns (Fig. 11.11). First, in the overhead tunnel the entire profile is drilled and blasted together or in parts by horizontal holes. Secondly, benching with horizontal drilling commonly is used to excavate the central parts of a large cavern. Thirdly, the bottom of a cavern may be excavated by benching, the blastholes being drilled vertically. The central part of a cavern also can be excavated by vertical benching, provided that the upper part has been excavated to a sufficient height or that the walls of the cavern are inclined. Smooth blasting is used to minimize fragmentation in the surrounding rock. Once the crown of the cavern has been excavated, economical excavation of the walls calls for large deep bench cuts exposing substantial areas of wall in a single blast. Under these conditions, however, an unstable wedge can be exposed and fail before it is supported.

The support pressures required to maintain the stability of a cavern increase as its span increases so that for the larger caverns standard sized rock bolts arranged in normal patterns may not be sufficient to hold the rock in place. Most caverns have arched crowns with span-to-rise ratios of 2.5 to 5.0. In general, higher support pressures are required for flatter roofs. Frequently, the upper parts of the walls are bolted more heavily in order to help support the haunches and the roof arch while the lower walls may be either slightly bolted or even unbolted.

11.5 Highways

As highways are linear structures, they often traverse a wide variety of ground conditions along their length. In addition, the construction of a highway requires the excavation of soils and/or rocks, the provision of a stable foundation for the highway, and at times the construction of embankments, tunnels, cuttings and bridges, as well as construction materials. The ground beneath roads, and more particularly embankments, must have sufficient bearing capacity to prevent foundation failure and also be capable of preventing excess settlements due to the imposed load. Very weak and compressible ground may need to be entirely removed before construction takes place, although this depends on the quantity of material involved. In some cases, heave that occurs due to the removal of load may cause significant problems. In other cases, improvement of the ground by the use of lime or cement stabilization, compaction, surcharging, the use of drainage, the installation of piles, stone columns or mattresses may be carried out prior to embankment and road building. Usually, the steepest side slopes possible are used when constructing cuttings and embankments as this minimizes the amount of land required for the highway and the quantity of material to be moved. Where potential failure surfaces exist or where adverse groundwater conditions are present the slopes probably need to be at a shallower angle. Obviously, attention must be given to the stability of these slopes. Difficulties frequently occur in heterogeneous materials and those with properties that are marginal between soils and rocks. Slight variations in strength, spacing of discontinuities or the grade of weathering can have an effect on the rate of excavation, and may affect the suitability of material that has been removed for use in construction. Where the materials excavated are unsuitable for construction, then considerable extra expense is entailed in disposing of waste and importing fill. Geological features such as faults, crush zones and solution cavities, as well as man-made features such as abandoned mine workings, drains and areas of fill can cause considerable difficulties during construction.

Normally, a road consists of a number of layers, each of which has a particular function (Fig. 11.12). In addition, the type of pavement structure depends on the nature and number of the vehicles it has to carry, their wheel loads and the period of time over which it has to last. The wearing surface of a modern

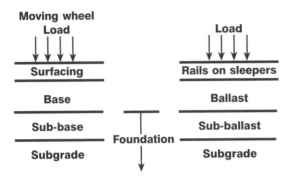

Fig. 11.12   Cross-sections of the pavement foundation of (left) a road (right) a rail track.

road consists either of 'black-top' (i.e. bituminous bound aggregate) or a concrete slab, although a bituminous surfacing may overlie a concrete base. A concrete slab distributes the load that the road has to carry, while in a bituminous road the load primarily is distributed by the base beneath. The base and sub-base below it generally consist of granular material, although in heavy duty roads the base may be treated with cement. The subgrade refers to the soil immediately beneath the sub-base. However much the load is distributed by the layers above, the subgrade has to carry the load of the road structure plus that of the traffic. Consequently, the top of the subgrade may have to be strengthened by compaction or stabilization. The strength of the subgrade, however, does not remain the same throughout its life. Changes in its strength are brought about by changes in its moisture content, by repeated wheel loading, and in some parts of the world by frost action. Although the soil in the subgrade exists above the water table and beneath a sealed surface, this does not stop the ingress of water. As a consequence, partially saturated or saturated conditions can exist in the soil. Also, road pavements are constructed at a level where the subgrade is affected by wetting and drying, which if expansive clay is present may lead to swelling and shrinkage respectively. Such volume changes are non-uniform and the associated movements may damage the pavement. Irrecoverable plastic and viscous strains can accumulate under repeated wheel loading. In a bituminous pavement, repeated wheel loading can lead to fatigue and cracking, and rutting occurs as a result of the accumulation of vertical permanent strains. Frost heave can cause serious damage to roads. Furthermore, the soil may become saturated when the ice melts, giving rise to thaw settlement and loss of bearing capacity. Repeated cycles of freezing and thawing change the structure of the soil, again reducing its bearing capacity. Rigid concrete pavements are more able to resist frost action than flexible bituminous pavements.

11.5.1   Road construction and geotextiles

The improvement in the performance of a pavement attributable to the inclusion of geotextiles comes mainly from their separation and reinforcing functions. This can be assessed in terms of either an improved system performance (e.g. reduction in deformation or increase in traffic passes before failure) or reduced aggregate thickness requirements (where reductions of the order 25–50% are feasible for low-strength subgrade conditions with suitable geotextiles).

The most frequent role of geotextiles in road construction is as a separator between the sub-base and subgrade. This prevents the subgrade material from intruding into the sub-base due to repeated traffic loading and so increases the bearing capacity of the system. The savings in sub-base materials, which would otherwise be lost due to mixing with the subgrade, can sometimes cover the cost of the geotextile. The range of gradings or materials that can be used as sub-bases with geotextiles normally is greater than when they are not used. Nevertheless, the sub-base materials preferably should be angular, compactible and sufficiently well graded to provide a good riding surface.

If a geotextile is to increase the bearing capacity of a subsoil or pavement significantly, then large deformations of the soil–geotextile system generally must be accepted as a geotextile has no bending stiffness, is relatively extensible, and usually is laid horizontally and restrained from extending laterally. Thus, considerable vertical movement is required to provide the necessary stretching to induce the tension that affords vertical load-carrying capacity to a geotextile. Therefore, geotextiles are likely to be of most use when included within low-density sands and very soft clays. Although large deformations may be acceptable for access and haul roads, they are not acceptable for most permanent pavements. In this case the geotextile at the sub-base/subgrade interface should not be subjected to mechanical stress or abrasion. When geotextile is used in temporary or permanent road construction it helps redistribute the load above any local soft spots that occur in a subgrade. In other words, the geotextile deforms locally and progressively redistributes load to the surrounding areas thereby limiting local deflections. As a result, the extent of local pavement failure and differential settlement is reduced.

Under wheel loading, the use of geogrid reinforcement in road construction helps restrain lateral expansive movements at the bottom of the aggregate forming the sub-base. This gives rise to improved load redistribution in the sub-base which, in turn, means a reduced pressure on the subgrade. In addition, the cyclic strains in the pavement are reduced. The stiff load bearing platform created by the interlocking of granular fill with geogrids is utilized effectively in the construction of roads over weak soil. Reduction in the required aggregate thickness of 30 to 50% may be achieved. Geogrids can be used within a granular capping layer when constructing roads over variable subgrades. They also have been used to construct access roads across peat bogs, the geogrid enabling the roads to be 'floated' over the surface.

In arid regions impermeable geomembranes can be used as capillary breaks to stop the upward movement of salts where they would destroy the road surface. Geomembranes also can be used to prevent the formation of ice lenses in permafrost and other frost prone regions. The geomembrane must be located below the frost line and above the water table.

Where there is a likelihood of uplift pressure disturbing a road constructed below the piezometric level, it is important to install a horizontal drainage blanket. This intercepts the rising water and conveys it laterally to drains at the side of the road. A geocomposite can be used for effective horizontal drainage. Problems can arise when sub-bases are used that are sensitive to moisture changes, that is, they swell, shrink or degrade. In such instances, it is best to envelop the sub-base, or excavate, replace and compact the upper layers of sub-base in an envelope of impermeable geomembrane.

11.5.2   Soil stabilization and road construction

The objectives of mixing additives with soil are to improve volume stability, strength and stress-strain properties, permeability, and durability. Swelling and shrinkage can be reduced in clay soils. Good mixing of stabilizers with soil is the most important factor affecting the quality of results. Cement and lime are the two most common additives.

The principal use of soil-cement is as a base material underlying pavements. One of the reasons soil-cement is used as a base is to prevent pumping of fine-grained subgrade soils into the pavement above. The thickness of the soil-cement base depends upon subgrade strength, pavement design, traffic and loading conditions, and thickness of the wearing surface. Frequently, however, soil-cement bases are around 150 to 200 mm in thickness. The principal use of the addition of lime to soil is for subgrade and sub-base stabilization, and as a construction expedient on wet sites where lime is used to dry out the soil. As far as lime stabilization for roadways is concerned, stabilization is brought about by the addition of between 3 and 6% lime (by dry weight of soil). Use of both cement or lime for subgrade stabilization involves stabilizing the soil in place, or stabilizing borrow materials that are used for sub-bases. After the soil, which is to be stabilized, has been brought to grade, the roadway is scarified to full depth and width, and then partly pulverized. A rooter, grader scarifier and/or disc-harrow are used for initial scarification followed by a rotary mixer for pulverization. After mixing in the cement or lime and any additional water needed to reach the optimum moisture content, compaction and grading to final level is carried out. Finally, the processed layer is covered with a waterproof membrane, commonly bitumen emulsion, to

Fig. 11.13 Pre-mixed soil-cement being laid as a sub-base for a road in Norfolk, England. Note the cover of bitumen in the centre ground that acts as a waterproof membrane.

prevent drying out and to ensure hydration (Fig. 11.13). The properties developed by compacted cement or lime stabilized soils are governed by the amount of cement/lime added on the one hand and compaction on the other. With increasing cement or lime content the strength and bearing capacity increase, as does the durability to wet-dry cycles. The permeability generally decreases but tends to increase in clayey soils. The swellability of clay soils generally is reduced.

11.6   Embankments

Embankments are mechanically compacted by laying and rolling soil in thin layers. The soil particles are packed more closely due to a reduction in the volume of the void space, resulting from the momentary application of loads such as rolling, tamping or vibration. Compaction involves the expulsion of air from the voids without the moisture content being changed significantly. Hence, the degree of saturation is increased. However, all the air cannot be expelled from soil by compaction so that complete saturation is not achievable. Nevertheless, compaction does lead to a reduced tendency for changes in the moisture content of soil to occur. The engineering properties of soils used for embankments, such as their shear strength and compressibility, are influenced by the amount of compaction they have undergone. Accordingly, the desired amount of compaction is established in relation to the engineering properties required for an embankment to perform its design function. A specification for compaction needs to indicate the type of compaction equipment to be used, its mass, speed and travel, and any other factors influencing performance such as frequency of vibration, thickness of layers to be compacted and number of passes of the compactor (Table 11.3).

11.6.1   Compaction

Most specifications for the compaction of fine-grained soils require that the dry density that is achieved represents a certain percentage of some laboratory standard. For example, the dry density required in the field can be specified in terms of relative compaction or final air void percentage achieved. The ratio between the maximum dry densities obtained in situ and those derived from a standard compaction test is referred to as relative compaction. If the dry density is given in terms of the final air percentage, a

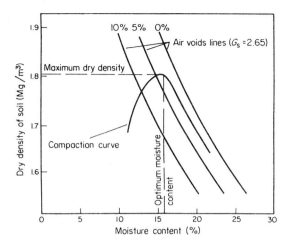

Fig. 11.14 Compaction curve showing relationship between dry density and moisture content.

value of 5 to 10% usually is stipulated, depending upon the maximum dry density determined from a standard compaction test (Fig. 11.14). In most cases of compaction of fine-grained soils, 5% variation in the value specified is allowable, provided that the average value attained is equal to or greater than that specified. The compaction characteristics of clay soil are governed largely by the moisture content. For instance, a greater compactive effort is necessary as the moisture content is lowered. Compaction of fine-grained soils should be carried out when the moisture content of the soil is not more than 2% above their plastic limit. If it exceeds this figure, then the soil should be allowed to dry. Overcompaction of soil on site, that is, compacting the soil beyond the optimum moisture content means that the soil becomes softer. In clay soils, it may be necessary to use thinner layers and more passes by heavier compaction plant than required for granular materials. The properties of fine-grained fills also depend to a much greater extent on the placement conditions than do those of a coarse-grained fill.

The shear strength of a given compacted fine-grained soil depends on its density and moisture content at the time of shear. The pore water pressures developed while the soil is being subjected to shear also are of great importance in determining the strength of such soils. For example, if a fine-grained soil is significantly drier than optimum moisture content, then it has a high strength due to high negative pore water pressures developed as a consequence of capillary action. The strength declines as the optimum moisture content is approached and continues to decrease on the wet side of optimum. Increases in pore water pressures produced by volume changes coincident with shearing tend to lower the strength of a compacted fine-grained soil. The compressibility of a compacted fine-grained soil also depends on its density and moisture content at the time of loading. However, its placement moisture content tends to affect compressibility more than does its dry density. If a soil is compacted significantly dry of optimum, and then is saturated, extra settlement occurs on loading. This does not occur when soils are compacted at optimum moisture content or on the wet-side of optimum. Fine-grained soil compacted on the dry-side of optimum moisture content swells more, at the same confining pressure, when given access to moisture than a sample compacted on the wet-side. This is because the former type has a greater moisture deficiency and a lower degree of saturation, as well as a more random arrangement of particles than the latter. Even at the same degrees of saturation, a soil compacted on the dry-side tends to swell more than one compacted wet of optimum. Conversely, a sample compacted wet of optimum shrinks more on drying than a soil, at the same density, which has been compacted dry of optimum. A minimum permeability occurs in a compacted fine-grained soil that is at or slightly above optimum moisture content, after which a slight increase occurs

Table 11.3 Typical compaction characteristics for soils used in earthwork construction (with permission of British Standards Institution).

Coarse soils – Major group: gravel sand and gravely soils

Subgroup	Suitable type of compaction plant	Minimum number of passes for satisfactory compaction	Maximum thickness of compacted layer (mm)	Remarks
Coarse soils – gravel sand and gravelly soils				
Well graded gravel and gravel/sand mixtures; little or no fines	Grid roller over 540 kg per 100 mm of roll	3–12 depending on type of plant	75–125 depending on type of plant	
Well graded gravel/sand mixtures with excellent clay binder	Pneumatic-tyred roller over 2000 kg per wheel Vibratory plate compactor over 1100 kg m^{-2} of baseplate			
Uniform gravel; little or no fines	Smooth wheeled roller Vibro-rammer			
Poorly graded gravel and gravel/sand mixtures; little or no fines	Self propelled tamping roller			
Gravel with excess fines, silty gravel, clayey gravel, poorly graded gravel/ sand/clay mixtures				
Coarse soils – sands and sandy soils				
Well graded sands and gravelly sands; little or no fines				
Well graded sands with excellent clay binder				

Subgroup	Suitable type of compaction plant	Minimum number of passes for satisfactory compaction	Maximum thickness of compacted layer (mm)	Remarks
Coarse soils – uniform sands and gravels				
Uniform gravels; little or no fines Uniform sands; little or no fines Poorly graded sands; little or no fines Sands with fines; silty sands, clayey sands poorly graded sand/clay mixtures	Smooth-wheeled roller below 500 kg per 100 mm of roll Grid roller below 540 kg per 100 mm Pneumatic-tyred roller below 1500 kg per wheel Vibrating roller Vibrating plate compactor Vibro-tamper	3–16 depending on type of plant	75–300 depending on type of plant	
Fine soils – soils having low plasticity				
Silts (inorganic) and very fine sands, silty or clayey fine sands with slight plasticity Clayey silts (inorganic) Organic silts of low plasticity	Sheepsfoot roller Smooth-wheeled roller Pneumatic-tyred roller Vibratory roller over 70 kg per 100 mm roll Vibratory plate compactor over 1400 kg m^{-2} of baseplate Vibro-tamper Power rammer	4–8 depending on type of plant	100–450 depending on type of plant	If moisture content is low it may be preferable to use a vibratory roller. Sheepsfoot rollers are best suited to soils at a moisture content below their plastic limit

Table 11.3 cont'd

Subgroup	Suitable type of compaction plant	Minimum number of passes for satisfactory compaction	Maximum thickness of compacted layer (mm)	Remarks
Fine soils – soils having medium plasticity				
Silty and sandy clays (inorganic) of medium plasticity				
Clays (inorganic) of medium plasticity				
Organic clays of medium plasticity				Generally unsuitable for earthworks
Fine soils – soils having high plasticity				
Fine sandy and silty soils, plastic silts				Should only be used when circumstances are favourable
Clay (inorganic) of high plasticity, fat clays				
Organic clays of high plasticity				Should not be used for earthworks

Specifications for the control of compaction of coarse-grained soils either require a stated relative density to be achieved or stipulate type of equipment, thickness of layer and number of passes required. Field compaction trials should be carried out in order to ascertain which method of compaction should be used, including the type of equipment. Such tests should take account of the variability in grading of the material used and its moisture content, the thickness of the individual layers to be compacted and the number of passes per layer. Because granular soils are relatively permeable, even when compacted, they are not affected significantly by their moisture content during the compaction process. In other words, for a given compactive effort, the dry density obtained is high when the soil is dry and high when the soil is saturated, with somewhat lower densities occurring when the soil contains intermediate amounts of moisture. Moisture can be forced from the pores of granular soils by compaction equipment and so a high standard of compaction can be obtained even if the material initially has a high moisture content. Normally, granular soils are easy to compact. However, if granular soils are uniformly graded, then a high degree of compaction near to the surface of the fill may prove difficult to obtain particularly when vibratory rollers are used. This problem usually is resolved when the succeeding layer is compacted in that the loose surface of the lower layer also is compacted. Improved compaction of uniformly graded granular material can be brought about by maintaining as high a moisture content as possible by intensive watering and by making the final passes at a higher speed using a non-vibratory smooth wheel roller or grid roller. When compacted, granular soils have a high load bearing capacity and are not compressible, and usually are not susceptible to frost action unless they contain a high proportion of fines. Unfortunately, however, if coarse-grained material contains a significant amount of fines, then high pore water pressures can develop during compaction if the moisture content of the soil is high. It is important to provide an adequate relative density in coarse-grained soil that may become saturated and subjected to static or, more particularly, to dynamic shear stresses. For example, a quick condition might develop in sandy soils with a relative density of less than 50% during ground accelerations of approximately 0.1 g. On the other hand, if the relative density is greater than 75%, liquefaction is unlikely to occur for most earthquake loadings. Consequently, in order to reduce the risk of liquefaction, sandy soils should be densified to a minimum relative density of 85% in the foundation area and to at least 70% within the zone of influence of the foundation.

11.6.2 Construction

The most critical period during the construction of an embankment is just before it is brought to grade or shortly thereafter. At this time pore water pressures, due to consolidation in the embankment and foundation, are at a maximum. The magnitude and distribution of pore water pressures developed during construction depend primarily on the construction water content, the properties of the soil, the height of the embankment and the rate at which dissipation by drainage can occur. Moisture contents above optimum can cause high construction pore water pressures in fine-grained soils, which increase the danger of rotational slips occurring in embankments.

Geogrids or geowebs can be used in the construction of embankments over poor ground without the need excavate the ground and substitute granular fill. They can allow acceleration of fill placement often in conjunction with vertical band drains. Layers of geogrids or geowebs can be used at the base of an embankment to intersect potential deep failure surfaces or to help construct an embankment over peat deposits. Geogrids also can be used to encapsulate a drainage layer of granular material at the base of an embankment. The use of band drains or a drainage layer help reduce and regulate differential settlement. A geocell mattress also can be constructed at the base of an embankment that is to be constructed on soft soil (Fig. 11.15). The cells are generally about 1 m high and are filled with granular material. This also acts as a drainage layer. The mattress intersects potential failure planes and its rigidity forces them deeper into firmer soil. The rough interface at the base of the mattress ensures mobilization of the maximum shear capacity of the foundation soil and significantly increases stability. Differential settlement and lateral spread are minimized.

If the soil beneath a proposed embankment, which is to carry a road, is likely to undergo appreciable

305

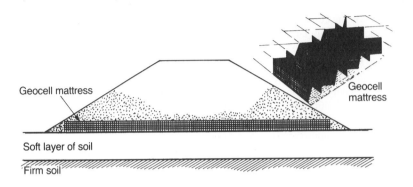

Fig. 11.15 A geocell mattress beneath an embankment.

settlement, then the soil can be treated. One of the commonest forms of treatment is precompression. Those soils that are best suited to improvement by precompression include compressible silts, saturated soft clays, organic clays and peats. The presence of thin layers of sand or silt in compressible soil may mean that rapid consolidation takes place on precompression. Unfortunately, this may be accompanied by the development of abnormally high pore water pressures in those layers beyond the edge of the precompression load. This lowers their shearing resistance ultimately to less than that of the surrounding weak soil. Such excess pore water pressures may have to be relieved by vertical drains.

Precompression normally is brought about by preloading, which involves the placement of a dead load. This compresses the foundation soils thereby inducing settlement prior to construction. If the load intensity from the dead weight exceeds the pressure that will be imposed by the final load, then this is referred to as surcharging. In other words, the surcharge is the excess load additional to the final load and is used to accelerate the process. The surcharge load is removed after a certain amount of settlement has taken place. Soil undergoes considerably more compression during the first phase of loading than during any subsequent reloading. Moreover, the amount of expansion following unloading is not significant. The installation of vertical drains (e.g. sand drains, sandwicks or band drains) beneath the precompression load helps shorten the time required to bring about primary consolidation. The water from the drains flows into a drainage blanket placed at the surface or to highly permeable layers deeper in the soil. Another method of bringing about precompression is by vacuum preloading by pumping from beneath an impervious membrane placed over the ground surface.

Reinforced earth is a composite material consisting of soil that is reinforced with metallic or geosynthetic strips. The effectiveness of the reinforcement is governed by its tensile strength and the bond it develops with the surrounding soil. It also is necessary to provide some form of barrier to contain the soil at the edge of a reinforced earth structure (Fig. 11.16). This facing can be either flexible or stiff but it must be strong enough to retain the soil and it must allow the reinforcement to be fixed to it. As reinforced earth is flexible, and the structural components are built at the same time as backfill is placed, it is particularly suited for use over compressible foundations where differential settlements may occur during or soon after construction. In addition, as a reinforced earth wall uses small prefabricated components and requires no formwork, it can be readily adapted to required variations in height or shape. Granular fill is the most suitable in that it is free-draining and non-frost susceptible, as well as being virtually non-corrosive as far as the reinforcing elements are concerned. It also is relatively stable, eliminating post-construction movements. Fine-grained materials can be used as fill but a slower construction schedule is necessary.

11.7 Railroads

Railroads have played and continue to play an important role in national transportation systems, although

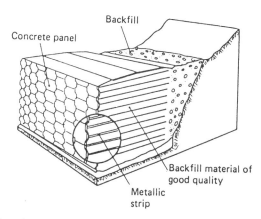

Concrete panel

Backfill

Backfill material of good quality

Metallic strip

Fig. 11.16   Reinforced earth system.

the construction of new railroads on a large scale is something that belongs to the past. Nonetheless, railroads continue to be built such as those associated with high speed networks. A vital part of the work concerned with a high speed railroad, with trains travelling at speeds of up to 300 km h^{-1}, is the trackbed support. In other words, the dynamic behaviour of foundations and earthworks involves a detailed understanding of the soil–structure interaction. This distinguishes a modern high speed railway from other railways or highways. Obviously, the grades and curvature of railroads impose stricter limits than do those associated with highways. Furthermore, underground systems are being, and will continue to be, constructed beneath many large cities in order to convey large numbers of people from one place to another quickly and efficiently so providing some relief to congested surface traffic.

Topography and geology are as important in railroad construction as in highway construction. As noted, a very stable trackbed and earthworks are necessary for a high speed railroad. Accordingly, drainage is an important aspect of trackbed design so that surface water is efficiently removed and groundwater level is maintained below the subgrade. Another important factor is trackbed stiffness. Consequently, where sudden changes of stiffness are likely to occur, notably between embankments and bridges, this can involve the inclusion of service blocks, and layers of cement stabilized and well graded granular material to provide a gradual transition in stiffness. Deep dry soil mixing can be used where embankments are constructed over soft clays and peats.

Railroads obviously can be affected by geohazards. In rugged terrain in particular, trains may be interrupted for a time by rock falls, landslides or mudflows. Areas that are prone to such hazards need to be identified and, where possible, stabilization or protective measures carried out (see Chapter 3). In some areas, because of the nature of the terrain, it may be impossible to stabilize entire slopes. In such cases warning devices can be installed that are triggered by mass movements so trains can be halted. Lengths of track that are subjected to mass movements should be inspected regularly. Remedial works in landslipped areas can include some combination of subsurface drainage, redesign and/or reinforcement of slopes (see Section 11.1). Railway services also may be interrupted seriously by flooding.

Railway track formations normally consist of a layer of coarse aggregate, the ballast, in which the sleepers are embedded. The ballast may rest directly on the subgrade or, depending on the bearing capacity, on a layer of blanketing sand (Fig. 11.12). The function of the ballast is to provide a free-draining base that is stable enough to maintain the track alignment with minimum of maintenance. The blanketing sand provides a filter that prevents the migration of fines from the subgrade into the ballast due to pumping. The ballast must be thick enough to retain the track in position and to prevent intermittent loading from deforming the subgrade whilst the aggregate beneath the sleepers must be able to resist abrasion and attrition. Hence, strong, good wearing, angular aggregates are required such as provided by many dense siliceous rocks. The thickness of the ballast can vary from as low as 150 mm for lightly trafficked railroads up to 500 mm on railroads that carry high speed trains or heavy traffic. The blanketing layer of sand normally has a minimum thickness of 150 mm.

Under repeated loading, differential permanent strains develop in the ballast of a rail track that bring about a change in the rail line and level. If the voids in the ballast are allowed to fill with fine-grained material, then a failure condition can develop. The fines may be derived by pumping from the sub-ballast or subgrade if these become saturated. Accordingly, railway track ballast requires regular attention to maintain line and level.

11.8 Bridges

Like tunnels, the location of bridges may be predetermined by the location of the routeways of which they form part. Consequently, this means that the ground conditions beneath bridge locations must be adequately investigated. This is especially the case when a bridge has to cross a river. The geology beneath a river should be correlated to the geology on both banks and drilling beneath the river should go deep enough to determine in place solid rock. The geological conditions may be complicated by the presence of a buried channel beneath a river. Buried channels generally originated during the Pleistocene epoch when valleys were deepened by glacial action and sea levels were at lower positions. Subsequently, much of these valleys were occupied by various types of sediment, which may include peat. The data obtained from the site investigation should enable the bridge, piers and abutments to be designed satisfactorily.

The ground beneath bridge piers has to support not only the dead load of the bridge but also the live load of the traffic that the bridge will carry, as well as accommodating the horizontal thrust of the river water when bridges cross rivers. The choice of foundations usually is influenced by a number of factors. For example, the existence of sound rock near the surface allows spread foundations to be used without the need for widespread piling whereas piled foundations are adopted for flood plains and rivers where alluvial deposits overlie bedrock. Because of the possibility of strong currents and high tidal range in an estuary, precast concrete open-bottom caissons, which are floated out and put in place by specially adapted barges, may be used for piers. Such caissons provide permanent formwork shells for the concrete infill. Alternatively, piers can be designed as cellular structures supported by cylindrical caissons.

The anchorages for suspension bridges have to resist very high pull-out loads. For example, the Hessle anchorage, on the north side of the river for the Humber Bridge, England, has to resist a horizontal pull of 38 000 tonnes. Resistance is derived from friction at the soil or rock/concrete interface at the base of the anchorage, from the passive resistance at the front and from wedge action at the sides.

When a bridge is constructed across a river, the effective cross-sectional area of the river is reduced by the piers. This leads to an increase in the flow velocity of the river that together with the occurrence of eddies around the piers enhances scouring action. Less scouring generally takes place where a river bed is formed of cobbles and gravel than where it is formed of sand or finer-grained material. Scouring of river bed materials around bridge piers has caused some bridges to fail. During floods damage is caused by very high peak flows, by build-up of debris at the bridge and by excessive scour around supporting caissons/piers (Fig. 11.17). The problem of scouring is accentuated in estuaries, especially where the flow patterns of the ebb and flood tides are different.

Bridges obviously are affected by ground movements such as subsidence. Subsidence movements can cause relative displacements in all directions and so subject a bridge to tensile and compressive stresses. Although a bridge can have a rigid design to resist such ground movements, it usually is more economic to articulate it thereby reducing the effects of subsidence. In the case of multi-span bridges, the piers should be hinged at the top and bottom to allow for tilting or change in length. Jacking points can be used to maintain the level of the deck. As far as shallow abandoned room and pillar workings are concerned, it usually is necessary to fill voids beneath a bridge with grout.

Seismic forces in earthquake prone regions can cause damage to bridges and so must be considered in bridge design (Fig. 11.18). Most seismic damage to low bridges has been caused by failures of substructures resulting from large ground deformations or liquefaction. Indeed, it appears that the worst

Fig. 11.17   The collapsed John Ross Bridge over the Tugela River, Natal, South Africa.

Fig. 11.18   Luanhe Bridge on the Beijing-Yuguan Highway, China, damaged by the Tangshan earthquake of 1976.

damage is sustained by bridges located on soft ground, especially that capable of liquefaction. Failure or subsidence of backfill in a bridge approach, leading to an abrupt change in profile, can prevent traffic from using the approach even if the bridge is undamaged. Such failure frequently exerts large enough forces on abutments to cause damage to substructures. On the other hand, seismic damage to superstructures due purely to the effects of vibrations is rare. Nonetheless, as a result of substructure failure, damage can occur within bearing supports and hinges, which combined with excessive movement of substructures can bring about collapse of a superstructure. By contrast, the effects of vibrations can be responsible for catastrophic failures of high bridges that possess relatively little overall stiffness. Arch-type bridges are the strongest whereas simple or cantilever-beam bridges are the most vulnerable to seismic effects. Furthermore, the greater the height of substructures and the larger the number of spans, the more likely it is for a bridge to collapse.

11.9   Foundations for buildings

The design of foundations embodies three essential operations, namely, calculating the loads to be transmitted by the foundation structure to the soils or rocks supporting it, determining the engineering performance of these soils and rocks, and then designing a suitable foundation.

11.9.1   Foundation structures

Footings represent the commonest type of foundation structure and distribute the load to the ground over an area sufficient to suit the pressures to the properties of the soil or rock. Their size therefore is governed by the strength of the foundation materials. If the footing supports a single column it is known as a spread or pad footing whereas a footing beneath a wall is referred to as a strip or continuous footing. The allowable bearing capacity of the ground beneath a footing must be chosen to provide an adequate factor of safety against shear failure in the soil and ensure that settlements are not excessive. Settlement of a footing due to a given load per unit area of its base depends upon the compressibility of the foundation materials. Settlements of structures on sands and gravels are largely completed by the end of the construction period in contrast to some clay soils where settlements can continue to take place for a significant time. If sand or gravel is dense or moderately dense and if it is not underlain by more compressible soils, quite large pressures can be used with safety. The ultimate bearing capacity of clay soils can be estimated with reasonable accuracy from their shear strength and settlement from their consolidation characteristics. If footings, especially of lightweight buildings, are to be constructed on fine-grained soil, it is necessary to determine whether or not the soil is likely to swell or shrink according to any seasonal variations of moisture content. Significant variations below a depth of about 2 to 3 m are rather rare.

A raft permits the construction of a satisfactory foundation in materials whose strength is too low for the use of footings. The chief function of a raft is to spread the load of a structure over as large an area of ground as possible and thus reduce the bearing pressure to a minimum. In addition, a raft provides a degree of rigidity that reduces differential movements in the superstructure. The settlement of a raft foundation does not depend on the load of the structure that it supports. It depends on the difference between this load and the mass of the soil that is removed prior to the construction of the raft, provided that any heave produced by the excavation is inconsequential. A raft can be built at a sufficient depth so that the mass of soil removed equals the mass of the building. Accordingly, such rafts sometimes are termed floating or buoyancy foundations. The success of this type of foundation structure in overcoming difficult soil conditions has led to the use of deep raft and rigid frame basements for a number of high buildings on clay soils.

When the soil immediately beneath a proposed structure is too weak or too compressible to provide adequate support, the loads can be transferred to more suitable material at greater depth by means of piles (Fig. 11.19). Such end-bearing piles must be capable of sustaining the load with an adequate factor of safety, without allowing settlement detrimental to the structure to occur. In addition, friction along the sides of piles also contributes towards their load carrying capacity. Indeed, friction is likely to be the

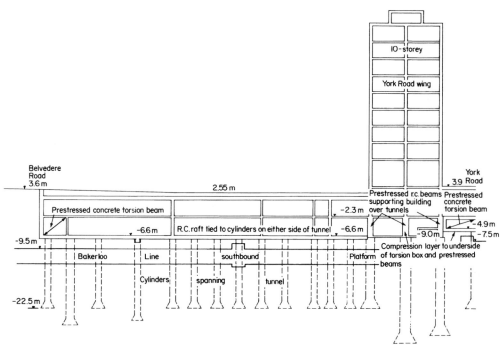

Fig. 11.19 Cross-section of the Shell Centre, London, showing the Bakerloo Line underground tunnels in relation to the building. (Courtesy of the Institution of Civil Engineers.)

predominant factor for piles in clay soils and silts while end-bearing provides the carrying capacity for piles terminating in or on gravel or rock. Piles may be divided into three main types according to the effects of their installation, namely, displacement piles, small-displacement piles and non-displacement piles. Displacement piles are installed by driving and so their volume has to be accommodated below ground by vertical and lateral displacements of soil that may give rise to heave or compaction, which may have detrimental effects upon neighbouring structures. Driving also may cause piles, which are already installed, to lift. Driving piles into clay soil may affect its consistency. In other words, the penetration of the pile combined with the vibrations set up by the falling hammer, destroy the structure of the clay soil and initiate a new process of consolidation that drags the piles in a downward direction. In fact, they may settle on account of their contact with the remoulded mass of clay soil even if they are not loaded. Sensitive clay soils are affected in this way while insensitive clay soils are not. Small-displacement piles include some piles that may be used in soft alluvial ground of considerable depth. They also may be used to withstand uplift forces. They are not suitable in stiff clays or gravels. Non-displacement piles are formed by boring. The hole so formed may be lined with casing that is or is not left in place. When working near existing structures that are founded on loose sands or silts, particularly if these are saturated, it is essential to avoid the use of methods that cause dangerous vibrations that may give rise to quick conditions.

11.9.2 Foundation design

Foundation design is primarily concerned with ensuring that movements of a foundation are kept within limits that can be tolerated by the proposed structure without adversely affecting its functional requirements. Hence, the design of a foundation structure requires an understanding of the local geological and groundwater conditions, and more particularly, an appreciation of the various types of ground movement that can occur. In order to avoid shear failure or substantial shear deformation of the ground, the foundation pressures used in design should have an adequate factor of safety when compared with the ultimate bearing capacity of the foundation. The ultimate bearing capacity is the value of the loading intensity that causes the ground to fail in shear. If this is to be avoided, then a factor of safety must be applied to the ultimate bearing capacity, the value obtained being the safe bearing capacity. However, this value may still mean that there is a risk of excessive or differential settlement. Thus, the allowable bearing capacity is the value that is used in design, this taking into account all possibilities of ground deformation and failure, and so its value is normally less than that of the safe bearing capacity. The value of ultimate bearing capacity depends on the type of foundation structure, as well as the soil properties, notably the shear strength. For example, the dimensions, shape and depth at which a footing is placed all influence the bearing capacity. With uniform soil conditions, the ultimate bearing capacity increases with depth of installation of the foundation structure. This increase is associated with the confining effects of the soil, the decreased overburden pressure at foundation level and with the shear forces that can be mobilized between the sides of the foundation structure and the ground. Generally, with a factor of safety of 3, bearing capacity failure almost invariably is ruled out in clay soils. However, due consideration must be given to the amount of settlement that is likely to occur. More particularly, it is important to have a reliable estimate of the amount of differential settlement that can take place. If estimated differential settlement is excessive, then it may be necessary to change the layout of a structure or the type of foundation structure. In some cases appreciable differential settlements are provided for by designing articulated structures capable of taking differential movements of individual sections without damaging the structure. The position of the water table in relation to the foundation structure also has an important influence on the ultimate bearing capacity. For instance, high groundwater levels lower the effective stresses in the ground so that the ultimate bearing capacity is reduced.

Settlements may be reduced by the correct design of the foundation structure. This may include larger or deeper foundations. Also, settlements can be reduced if the site is preloaded or surcharged prior to construction or if the soil is subjected to dynamic compaction or vibrocompaction.

If a rock mass contains few defects the allowable bearing pressure at the surface may be taken conservatively as the unconfined compressive strength of the intact rock. Most rock masses, however, are affected by joints or weathering that may significantly alter their strength and engineering behaviour. The great variation in the physical properties of weathered rock and the non-uniformity of the extent of weathering, even at a single site, permit few generalizations concerning the design and construction of foundation structures. The depth to bedrock and the degree of weathering must be determined. If the weathered residuum plays the major role in the regolith, rock fragments being of minor consequence, then design of footings or rafts should be according to the matrix material. Piles can provide support at depth. Settlement is rarely a limiting condition in foundations on most fresh rocks. It consequently does not entail special study except in the case of special structures where settlements must be small. The problem then generally resolves itself into one of reducing the unit bearing load by widening the base of structures or using spread footings. Severe settlements, however, may take place in low-grade compaction shales.

Subsidence can be regarded as the vertical component of ground movement caused by mining operations although there is a horizontal component (see Chapter 6). Tensile stresses, compressive stresses and tilt are other components of subsidence. Subsidence can and does have serious effects on buildings and therefore calls for special constructional design in site development or in existing buildings necessitates extensive remedial measures.

11.9.3 Treatment of foundations

Grouting has been used to improve foundation conditions. It refers to the process of injecting under pressure, setting fluids into pores, fissures or cavities in the ground. The process is widely used in foundation engineering in order to reduce seepage of water or to increase the mechanical performance of the soils or rocks concerned. If the sealing and strengthening actions are to be successful, then grouting must

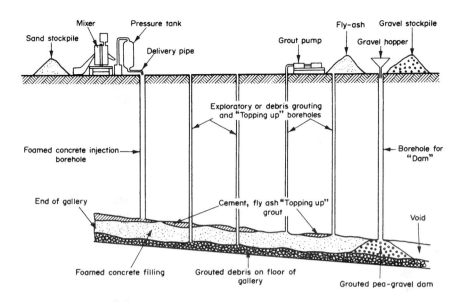

Fig. 11.20 Bulk grouting of old mine workings.

extend a considerable distance into the ground. This is achieved by injecting the grout into a special array of groutholes. The groutability of soil and therefore the choice of particulate or suspension grouts is influenced by the pore size of the soil or the width of fissures in the rock masses to be treated. Generally, cement grouts are limited to soils with pore dimensions greater than 0.2 mm and cannot enter a fissure smaller than about 0.1 mm. Because chemical grouts are non-particulate, their penetrability depends primarily on their viscosity. The shape of an opening also affects groutability. In order that the grout can achieve effective adhesion the sides of the voids or fissures must be clean. If they are coated with clay, then they need to be washed prior to grouting. Cavities in the ground may be filled with bulk grouts (i.e. gravel, sand, cement mixtures, to which pulverized fuel ash, PFA, may be added) or foam grouts (Fig. 11.20).

Vibroflotation is used to improve loose sands below foundation structures. The process may reduce settlement by more than 50% and the shear strength of treated sands is increased substantially. Vibrations of appropriate form eliminate the inter-granular friction of loosely packed sands thereby converting them into a dense state. A vibroflot, suspended from a crane, is used to penetrate the sand and can operate efficiently below the water table. The best results have been obtained in fairly coarse sands that contain little or no silt or clay, since these reduce the effectiveness of the vibroflot. However, vibrocompaction is used more frequently than vibroflotation since it can be applied to a much wider range of ground conditions. In this technique columns of coarse backfill are formed at individual compaction centres within the ground. In this way the ground is stiffened and undergoes less settlement on loading. The vibroflot is used to compact the columns (Fig. 11.21). Vibrocompaction is commonly used in normally consolidated compressible clay soils, saturated silts, and alluvial and estuarine soils. Stone columns have been formed successfully in clay soils with undrained cohesive strengths as low as 7 kPa.

Dynamic compaction brings about an improvement in the mechanical properties of a soil by the repeated application of very high intensity impacts to the surface. This is achieved by dropping a large weight, typically 10 to 20 tonnes, from a crawler crane, from heights of 15 to 40 m, at regular intervals across the surface (Fig. 11.22). Repeated passes are made over a site, although several tampings may be made at each imprint during a pass. Each imprint is backfilled after tamping. The first pass at widely spaced centres improves the bottom layer of the treatment zone and subsequent passes then consolidate the upper layers. In finer materials the increased pore water pressures must be allowed to dissipate between passes, which may take several weeks. Care must be taken in establishing the treatment pattern, tamping energies and the number of passes for a particular site, and this should be accompanied by in situ testing as the work proceeds. Coarse granular fill requires more energy to overcome the possibility of bridging action, for similar depths, than finer material. Before subjecting sites that previously were built over to dynamic compaction, underground services, cellars, etc. should be located.

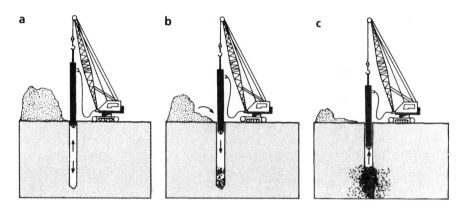

Fig. 11.21 Formation of a stone column by vibrocompaction. (a) Sinking the vibrator into the subsoil up to the depth where sufficient load bearing capacity is encountered. (b) Aggregates are placed into the hole made by the vibrator. (c) It is necessary to repeat this process as may times as may be required to achieve a degree of compaction of the surrounding soil and the aggregate as to ensure that no further penetration of the vibrator can be effected.

Fig. 11.22 Dynamic compaction in progress on a site in Dartford, Kent, England. This weight (tamper) weighs 10 tonnes (Courtesy Keller Ground Engineering).

Further Reading

General

Anon. 1976. *Manual of Applied Geology for Engineering*. Institution of Civil Engineers, Telford Press, London.

Aswathanarayana, U. 1995. *Geoenvironment: An Introduction*. A.A. Balkema, Rotterdam.

Attewell, P.B. and Farmer, I.W. 1976. *Principles of Engineering Geology*. Chapman and Hall.

Bell, F.G. 1980. *Engineering Geology and Geotechnics*. Butterworth, London.

Bell, F.G. 1983. *Fundamentals of Engineering Geology*. Butterworth, London.

Bell, F.G. 1998. *Environmental Geology: Principles and Practice*. Blackwell Science, Oxford.

Bell, F.G. 2004. *Engineering Geology and Construction*. E. & F.N. Spon, London.

Bennett, M.R. and Doyle, P. 1997. *Environmental Geology: Geology and the Human Environment*. Wiley, Chichester.

Blyth, F.G.H. and De Freitas, M.H. 1984. *A Geology for Engineers*. Seventh Edition, Arnold, London.

Coates, D.R. 1981. *Environmental Geology*. Wiley, New York.

Flawn, P.T. 1975. *Environmental Geology*. Harper and Row, New York.

Howard, A.D. and Remson, I. 1978. *Geology in Environmental Planning*. McGraw-Hill, New York.

Johnson, R.B. and De Graff, J.V. 1988. *Principles of Engineering Geology*, Wiley, New York.

Lahee, F.H. 1961. *Field Geology*. McGraw-Hill, New York.

Legget, R.F. and Karrow, P.F. 1983. *Handbook of Geology in Civil Engineering*. McGraw-Hill, New York.

Legget, R.F. and Hatheway, A.W. 1988. *Geology and Engineering*. Third Edition. McGraw-Hill, New York.

Keller, E.A. 1992. *Environmental Geology*. Macmillan Publishing Co., New York.

Krynine, P.D. and Judd, W.R. 1956. *Principles of Engineering Geology and Geotechnics*. McGraw-Hill, New York.

O'Halloran, D., Green, C., Harley. M., Stanley, M. and Knill, J.L., (Eds.). 1994. *Geological and Landscape Conservation*. Geological Society, London.

Paige, S. (Ed.). 1950. *Applications of Geology to Engineering Practice*. Berkey Volume. Geological Society of America, New York.

Rahn, P.R. 1996. *Engineering Geology: An Environmental Approach*. Second Edition, Prentice Hall, Upper Saddle River, New Jersey.

Trask, P.D. (Ed.). 1950. *Applied Sedimentation*. Wiley, New York.

Zaruba, Q. and Mencl, V. 1976. *Engineering Geology*. Academia, Prague.

Chapter 1

Duff, P. McL. D. (Ed.). 1993. *Holmes' Principles of Physical Geology*. Fourth Edition. Chapman and Hall, London.

Friedman, G.M., Sanders, J.E. and Kopaska-Merkel, D.C. 1992. *Principles of Sedimentary Deposits*. Macmillan Publishing Co., New York.

Fry, N. 1994. *The Field Description of Metamorphic Rocks*. Wiley, Chichester.

Hall, A. 1996. *Igneous Petrology*. Second Edition. Longman Group Ltd, Harlow, Essex.

Hobbs, B., Means, W.D. and Williams, P.F. 1976. *An Outline of Structural Geology*. Wiley, New York.

Krumbein, W.C. and Sloss, L.L. 1963. *Stratigraphy and Sedimentation*. Second Edition. W.H. Freeman and Co., San Francisco.

Ollier, C.D. 1983. *Weathering*. Longman, Harlow.

Pettijohn, F.J. 1975. *Sedimentary Rocks*. Harper and Row, New York.

Price, N.L. 1966. *Fault and Joint Development in Brittle and Semi-Brittle Rock*. Pergamon Press, Oxford.

Ramsey, J.G. 1967. *Folding and Fracturing of Rocks*. McGraw-Hill, London.

Reid, H.H. and Watson, J. 1961. *Introduction to Geology*. Volume 1, Macmillan, London.

Shrock, R.R. 1948. *Sequence in Layered Rocks*. McGraw-Hill, New York.

Thorpe, R.S. and Brown, G.C. 1995. *The Field Description of Igneous Rocks*. Wiley, Chichester.

Tucker, M.E. 2003. *The Field Description of Sedimentary Rocks*. Wiley, Chichester.

Chapter 2

Alexander, D. 1993. *Natural Disasters*. University College Press Limited, London.

Anon. 1985. *Geology for Urban Planning*. Economic Commission for Asia and the Pacific. United Nations, New York.

Bell, F.G. 1998. *Environmental Geology: Principles and Practice*. Blackwell, Oxford.

Bridge, D.M., Brown, M.J. and Hooker, P.G. 1996. *Wolverhampton Environmental Survey: an Integrated Geoscientific Study*. Technical Report WE/95/49, British Geological Survey, Keyworth, Nottinghamshire.

Burrough, P.A. 1986. *Principles of Geographical Information Systems for Land Resource Analysis*. Oxford University Press, Oxford.

Canter, L. 1977. *Environmental Impact Assessment*. McGraw-Hill, New York.

Carroll, B. and Turpin, T. 2002. *Environmental Impact Assessment Handbook*. Thomas Telford Press, London.

Coates, D.R. 1981. *Environmental Geology*. Wiley, New York.

Cooke, R.U. and Doornkamp, J.C. 1990. *Geomorphology in Environmental Management*. Clarendon Press, Oxford.

Culshaw, M.G., Bell, F.G., Cripps, J.C. and O'Hara, M. (Eds.). 1987. *Planning and Engineering Geology, Engineering Geology Special Publication No 4*. Geological Society, London.

Dearman, W.R. 1991. *Engineering Geological Maps*. Butterworth-Heinemann, Oxford.

Demek, J. 1972. *Manual of Detailed Geomorphological Mapping*. Academia, Prague.

Griggs, G.B. and Gilchrist, J.A. 1983. *Geological Hazards, Resources and Environmental Planning*. Wadsworth, Belmont, California.

Hays, W.W., (Ed.). 1981. *Facing Geologic and Hydrologic Hazards – Earth Science Considerations*. United States Geological Survey, Professional Paper 1240-B, B103-B108, Washington, D.C.

Howard, A.D. and Remson, I. 1978. *Geology in Environmental Planning*. McGraw-Hill, New York.

Legget, R.F. 1973. *Cities and Geology*. McGraw-Hill, New York.

Levensen, D. 1980. *Geology and the Urban Environmental*. Oxford University Press, New York.

Longley, P.A., Goodchild, M.F., Macguire, D.J. and Rhind, D.W. 2001. *Geographic Information Systems and Science*. Wiley, Chichester.

McCall, G.J.H., De Mulder, E.F.J. and Marker, B.R. (Eds.). 1996. *Urban Geoscience*. A.A. Balkema, Rotterdam.

McHarg, I.L. 1969. *Design with Nature*. National History Press, Garden City, New York.

Robinson, G.D. and Spieker, A.M. (Eds.). 1978. *Nature to be Commanded – Earth Science Maps Applied to Land and Water Management*. Professional Paper 966, United States Geological Survey, Washington, DC.

Star, J and Estes, J. 1990. *Geographical Information Systems*. Prentice-Hall, Englewood Cliffs, New Jersey.

Tan, B.K. and Ran, J.L. (Eds) 1986. *Landplan II: Role of Geology in Planning and Development of Urban Centres in Southeast Asia.* Association of Geoscientists for International Development, Report Series No 12, Bangkok.

Utgard, R.O., McKenzie, G.D. and Foley, D. (Eds) 1978. *Geology in the Urban Environment.* Burgess, Minneapolis.

Chapter 3

Anderson, M.G. and Richards, K.S. (Eds.). 1987. *Slope Stability,* Wiley, New York.

Anon. 1971. *The Surveillance and Prediction of Volcanic Activity.* UNESCO, Paris.

Anon. 1976. *Disaster Prevention and Mitigation, Volume 1, Volcanological Aspects.* United Nations, New York.

Anon. 1978. *The Assessment and Mitigation of Earthquake Risk.* UNESCO, Paris.

Bagnold, R.A. 1941. *Physics of Windblown Sand.* Methuen, London

Baker, V.R., Kochel, R.C. and Patton, P.C. (Eds.). 1988. *Flood Geomorphology,* Wiley, New York.

Bascom, W. 1964. *Waves and Beaches.* Doubleday and Co., New York.

Bell, F.G. 1999. *Geological Hazards: Their Assessment, Avoidance and Mitigation.* E. & F.N. Spon, London.

Beven, K. and Carling, F. (Eds.). 1989. *Floods: Hydrological, Sedimentological and Geomorphological Implications.* Wiley, Chichester

Bolt, B.A. 1993. *Earthquakes.* W.H. Freeman, New York.

Bolt, B.A., Horn, W.L., McDonald, G.A. and Scott, R.F. 1975. *Geological Hazards.* Springer-Verlag, New York.

Brenner, D.J. 1989. *Radon, Risk and Remedy.* W.H. Freeman, New York.

Bromhead, E.N. 1992. *The Stability of Slopes.* Second Edition. Blackie, Glasgow.

Brookes, A. 1988. *Channelized Rivers; Perspectives for Environmental Management.* Wiley, Chichester.

Brunsden, D. and Prior, D.B. (Eds.). 1984. *Slope Instability.* Wiley-Interscience, Chichester.

Coates, D.R. (Ed.). 1976. *Geomorphology and Engineering.* Allen and Unwin, London.

Collinge, V. and Kirby, C. (Eds.). 1987. *Weather, Radar and Flood Forecasting.* Wiley, Chichester.

Cooke, R.U. and Doornkamp, J.C. 1990. *Geomorphology in Environmental Management.* Second Edition, Clarendon Press, Oxford.

Cooke, R.U., Brunsden, D., Doornkamp, J.C. and Jones, D.K.C. 1982. *Urban Geomorphology in Drylands.* Oxford University Press, Oxford.

Cooke, R.U., Warren, A and Goudie, A. 1993. *Desert Geomorphology.* University College London Press, London.

Crandell, D.R., Booth, B., Kusumadinata, K., Shimozuru, D., Walker, G.P.L. and Westercamp, D. 1984. *Source Book for Volcanic Hazards Zonation,* UNESCO, Paris.

Cruden, D.M. and Fell, R. (Eds.). 1997. *Landslide Risk Assessment.* A.A. Balkema, Rotterdam.

Derbyshire, E., Gregory, K.J. and Hails, J.R. 1981. *Geomorphological Processes.* Butterworth, London.

El-Sabh, M.I. and Murty, T.S., (Eds.). 1988. *Natural and Man-Made Hazards,* D. Riedel Publishing Co, Dordrecht.

Embleton, C. and King, C.A.M. 1968. *Glacial and Periglacial Geomorphology.* Arnold, London.

Eyles, N., (Ed.). 1983. *Glacial Geology: An Introduction for Engineers and Earth Scientists.* Pergamon Press, Oxford.

Firth, C.R. and McGuire, W.J., (Eds.). 1999. *Volcanoes in the Quaternary.* Special Publication No. 161, Geological Society, London.

Flint, R.F. 1967. *Glacial and Pleistocene Geology.* Wiley, New York.

Fookes, P.G., Lee, E.M., and Milligan, G. (Eds.). 2005. *Geomorphology for Engineers.* Whittles Publishing Ltd., Caithness.

Ford, D.C. and Williams, P.W. 1989. *Karst Geomorphology and Hydrology.* Unwin Hyman, London.

Glennie, K.N. 1970. *Desert Sedimentary Environments.* Elsevier, Amsterdam.

Goudie, A.S. 1973. *Duricrusts in Tropical and Subtropical Latitudes.* Oxford University Press, Oxford.

Gregory, K.J. and Walling, D.E. 1973. *Drainage Basin Forms and Process, a Geomorphological Approach.* Edward Arnold, London.

Griggs, G.B. and Gilchrist, J.A. 1983. *Geologic Hazards, Resources and Environmental Planning.* Wadsworth, Belmont, California.

Hails, J.R. (Ed.). 1977. *Applied Geomorphology.* Elsevier, Amsterdam.

Hays, W.W. 1981. *Facing geologic and hydrologic hazards: earth science perspectives.* United States Geological Survey, Professional Paper 1240B, Washington, D.C.

Hoffman, G.T., Howell, T.A. and Solomon, K.H. 1990. *Management of Farm Irrigation Systems.* American Society Agricultural Engineers Monograph, St Joseph, MI.

Hooke, J. 1998. *Coastal Defence and Earth Science Conservation.* Geological Society, London.

Ingle, J.G. 1966. *Movement of Beach Sand.* Elsevier, Amsterdam.

Kay, R. and Alder, J. 1999. *Coastal Planning and Management.* E. & F.N. Spon, London.

King, C.A.M. 1976. *Beaches and Coasts.* Arnold, London.

Komar, P.O. 1976. *Beach Processes and Sedimentation.* Prentice-Hall, Eaglewood Cliffs, New Jersey.

Latter, J.H., (Ed.). 1989. *Volcanic Hazards: Assessment and Monitoring.* Springer-Verlag, Berlin.

Lee, E.M. and Jones, D.K.C. 2004. *Landslide Risk Assessment.* Thomas Telford Press, London.

Lee, E.M. and Clark, A.R. 2002. *Investigation and Management of Soft Rock Cliffs.* Thomas Telford Press, London.

Legget, R. F., (Ed.). 1975. *Glacial Till – An Interdisciplinary Study.* Special Publication No. 12, Royal Society of Canada, Ottawa.

Leopold, L.B., Wolman, M.G. and Miller, L.P. 1964. *Fluvial Processes in Geomorphology.* Freeman and Co., San Francisco.

Lomnitz, C. 1994. *Fundamentals of Earthquake Prediction.* Wiley, New York.

Lomnitz, C. and Rosenblueth, E. 1976. *Seismic Risk and Engineering Decisions.* Elsevier, Amsterdam.

MacDonald, G.A. 1972. *Volcanoes.* Prentice-Hall, Englewood Cliffs, New Jersey.

McCall, G.J.H., Laming, D.J.C. and Scott, S.C. (Eds.). 1992. *Geohazards: Natural and Man-Made.* Chapman and Hall, London.

Maund, J.G. and Eddleston, M. (Eds.). 1998. *Geohazards in Engineering Geology, Engineering Geology Special Publication No. 15.* Geological Society, London.

Medvedev, S.V. 1965. *Engineering Seismology.* Israel Program for Scientific Translations, Jerusalem.

Mogi, K. 1985. *Earthquake Prediction.* Academic Press, Orlando, Florida.

Morgan, C.S., (Ed.). 1985. *Landslides in South Wales Coalfield,* Polytechnic of Wales, Pontypridd.

Mortimore, R.N. and Dupperet, A. (Eds.). 2004. *Coastal Chalk Cliff Instability.* Engineering Geology Special Publication No. 20, Geology Society, London.

Muir Wood, A.M. and Flemming, C.A. 1981. *Coastal Hydraulics,* Second Edition, Macmillan, London.

Newson, M.D. 1994. *Hydrology and the River Environment.* Clarendon Press, Oxford.

Pilarczyk, K.W., (Ed.). 1990. *Coastal Protection,* A.A. Balkema, Rotterdam.

Richter, C.F. 1956. *Elementary Seismology,* W.H. Freeman, San Francisco.

Rittmann, A. 1962. *Volcanoes and their Activity.* Translated by Vincent, E.A., Interscience, London.

Schuster, R.L. and Krizek, R.J., (Eds.). 1978. *Landslides: Analysis and Control,* Transportation Research Board, Special Report 176, National Academy of Sciences, Washington, D.C.

Sharp, C.F.S. 1938. *Landslides and Related Phenomena.* Columbia University Press, New York.

Silvester, R. 1974. *Coastal Engineering.* Elsevier, Amsterdam.

Thorn, R.B. and Simmons, J.C.F. 1971. *Sea Defence Works.* Butterworth, London.

Tufnell, L. 1984. *Glacial Hazards.* Longman, Harlow, Essex.

Turner, A.K. and Schuster, R.L. (Eds.). 1996. *Landslide Investigation and Mitigation.* Transportation Research Board, Special Report 247, National Research Council, Washington, D.C.

Varnes, D.J. 1984. *Landslide Hazard Zonation: A Review of Principles and Practice, Natural Hazards 3.* UNESCO, Paris.

Veder, C. 1981. *Landslides and Their Stabilization.* Springer-Verlag, New York.

Verstappen, J. Th. 1983. *Applied Geomorphology: Geomorphological Surveys for Environmental Development.* Elsevier, Amsterdam.

Waltham, A.C., Bell, F.G. and Culshaw, M.G. 2004. *Sinkholes and Subsidence: Karst and Cavernous Rocks in Engineering and Construction.* Praxis Press/Springer, Chichester/New York.

Ward, R. 1978. *Floods: a Geographical Perspective.* Macmillan, London.

Weigel, R.L., (Ed.). 1970. *Earthquake Engineering.* Prentice Hall, Englewood Cliffs, New Jersey.

Wilson, E.M. 1983. *Engineering Hydrology.* Third Edition. Macmillan, London.

Wisler C.O. and Brater, E.F. 1967. *Hydrology.* Wiley, New York.

Zaruba, Q. and Mencl. V. 1982. *Landslides and Their Control.* Second Edition, Elsevier, Amsterdam.

Chapter 4

Anon. 1993. *Guidelines for Drinking Water Quality.* World Health Organization. Geneva.

Anon. 1980. *Directive Relating to the Quality of Water intended for Human Consumption (80/778/EEC).* Commission of the European Community, Brussels.

Brassington, R. 1988. *Field Hydrogeology.* Wiley, Chichester.

Brown, R.H., Konoplyantsev, A.A., Ineson, J and Kovalevsky, V.S. (eds.). 1972. *Groundwater Studies.* An International Guide for Research and Practice, Studies and Reports in Hydrology 7, UNESCO, Paris.

Campbell, M.D. and Lehr, J.H. 1973. *Water Well Technology.* McGraw-Hill, New York.

Chow, V.T. (Ed.). 1986. *Handbook of Applied Hydrology.* McGraw-Hill, New York.

Clark, L. 1988. *The Field Guide to Water Wells and Boreholes.* Wiley, Chichester.

Cripps, J.C., Bell, F.G. and Culshaw, M.G. (Eds.). 1986. *Groundwater in Engineering Geology,* Engineering Geology Special Publication No. 3. Geological Society, London.

Davis, S.D. and De Weist, R.J.H. 1966. *Hydrogeology.* Wiley, New York.

De Weist, R.J.H. 1967. *Geohydrology.* Wiley, New York.

Fetter, C.W. 1980. *Applied Hydrogeology.* Merrill, Columbus, Ohio.

Freeze, R.A. and Cherry, J.A. 1979. *Groundwater.* Prentice Hall, Englewood Cliffs, New Jersey.

Gower, A.M. (Ed.). 1980. *Water Quality in Catchment Ecosystems.* Wiley, Chichester.

Hamill, L. and Bell, F.G. 1986. *Groundwater Resource Development.* Butterworth, London.

Helweg, O.J. 1985. *Water Resources, Planning and Management.* Wiley, New York.

Hem, J.D. 1985. *Study and Interpretation of the Chemical Characteristics of Natural Water.* Water Supply Paper 2254, United States Geological Survey, Washington, D.C.

Hiscock, K.M. Rivett, M.O. and Davison, R.M. (Eds.). 2002. *Sustainable Groundwater Development.* Special Publication 193, Geological Society, London.

Kunst, S., Kruse, T. and Burmester, A. (Eds.) 2002. *Sustainable Water and Soil Management.* Springer, Berlin.

Linsley, R.K. and Franzini, J.B. 1972. *Water Resources Engineering.* McGraw-Hill, New York.

Price, M. 1985. *Introducing Groundwater.* Allen and Unwin, London.

Robins, N.S. (Ed.). 1998. *Groundwater Pollution, Aquifer Recharge and Vulnerability,* Special Publication No. 130. Geological Society, London.

Sawyer, C.N. and McCarty, P.L. 1967. *Chemistry for Sanitary Engineers.* Second Edition, McGraw-Hill, New York.

Sherard, J.L., Woodward, R.L. Gizienski, S.F. and Clevenger, W.A. 1967. *Earth and Earth Dams.* Wiley, New York.

Skeat, W.D. (Ed.). 1969. *Manual of British Water Engineering Practice.* Volume 2, Institution of Water Engineers, London.

Todd, D.K. 1980. *Groundwater Hydrology,* Second Edition. Wiley, New York.

Walters, R.C.S. 1962. *Dam Geology.* Butterworth, London.

Walton, W.C. 1970. *Groundwater Resource Evaluation.* McGraw-Hill, New York.

Wilson, E.M. 1983. *Engineering Hydrology,* Third Edition, Macmillan, London.

Younger, P.L. and Robins, N.S. (Eds.). 2002. *Mine Water Hydrogeology and Geochemistry*. Special Publication 198, Geological Society, London.

Chapter 5

Anon, 1971. *Guide for Interpreting Engineering Uses of Soils*. United States Department of Agriculture, Washington, D.C.

Anon, 1975. *Soil Taxonomy*. Agricultural Handbook No 436, United States Department of Agriculture, Washington, D.C.

Anon. 1977. *Desertification: Its Causes and Consequences*. UNESCO, Pergamon, Oxford.

Anon, 1980. *Soil Map of the World*. Vol. 1 Legend. FAO-UNESCO, Paris.

Bell, F.G. 2000. *Engineering Properties of Soils and Rocks*, Fourth Edition. Blackwell, Oxford.

Davidson, D.A. 1980. *Soils and Land Use Planning*. Longman, London.

Dent, D. and Young, A. 1981. *Soil Survey and Land Evaluation*. Allen and Unwin, London.

Dregne, H.E. 1983. *Desertification of Arid Lands*. Harwood, New York.

Fitzpatrick, E.A. 1992. *An Introduction to Soil Science*. Second Edition. Longman Scientific and Technical, Harlow Essex.

Foth, H.D. and Ellis, B.G. 1997. *Soil Fertility*, Second Edition. CRL Press, Boca Raton, Florida.

Greenland, D.J. and Lal, R., (Eds.). 1977. *Soil Conservation and Management in the Humid Tropics*. Wiley, Chichester.

Hansen, V.E., Israelsen, O.W. and Stringham, G.E. 1980. *Irrigation Principles and Practice*, Fourth Edition. Wiley, New York.

Hudson, N.W. 1981. *Soil Conservation*. Batsford, London.

Jacks, G.V. and Whyte, R.O. 1939. *The Rape of the Earth: A World Survey of Soil Erosion*. Faber and Faber, London.

Jensen, M.E. 1973. *Consumptive Use of Water and Irrigation Water Requirements*. American Society of Civil Engineers, New York.

Kirby, M.J. and Morgan, R.P.C. (Eds.). 1980. *Soil Erosion*. Wiley, Chichester.

Klingebeil, A.A. and Montgomery, P.H. 1966. *Land Capability Classification*. Soil Conservation Service Agricultural Handbook 210, United States Department Agriculture, Washington, D.C.

Kunst, S., Kruse, T. and Burmester, A. (Eds.) 2002. *Sustainable Water and Soil Management*. Springer, Berlin.

Lilienthal, D. 1944. *Tennessee Valley Authority*. Penguin, London.

Morgan, R.P.C., 2005. *Soil Erosion and Conservation*. Third Edition, Blackwell, Oxford.

Morgan, R.P.C. and Rickson, R.J. (Eds). 1995. *Slope Stabilization and Erosion Control: A Bioengineering Approach*. E. & F.N. Spon, London, 95–131.

Lal, R., (Ed.). 1986. *Soil Quality and Soil Erosion*. CRC Press, Boca Raton, Florida.

Rimwanich, E. (Ed) 1988. *Land Conservation for Future Generations*. Department of Land Development, Ministry of Agriculture, Bangkok.

Schwab, G.O., Fangmeier, D.D., Elliot, W.J. and Frevert, R.K. 1993. *Soil and Water Conservation Engineering*, Fourth Edition. Wiley, New York.

Sumner, M.E., (Ed.). 1991. *Handbook of Soil Science*. CRC Press, Boca Raton, Florida.

Thornes, J.B. (Ed.).1990. *Vegetation and Erosion*. Wiley Chichester.

Chapter 6

Anon. 1970. *The Prevention of Spontaneous Combustion*. Institution of Mining Engineers, London.

Anon, 1973. *Spoil Heaps and Lagoons*. Technical Handbook, National Coal Board, London.

Anon, 1975. *Subsidence Engineer's Handbook,* National Coal Board, London.

Anon, 1976. *Reclamation of Derelict Land: Procedure for Locating Abandoned Mine Shafts*. Department of the Environment, London.

Anon, 1977. *Ground Subsidence.* Institution of Civil Engineers, London.

Anon. 1979. *Current Geotechnical Practice in Mine Waste Disposal.* American Society of Civil Engineers, New York.

Anon, 1982. *Treatment of Disused Mine Shafts and Adits.* National Coal Board, London.

Anon. 1986. *The Effect of Underground Mining on Surface.* International Society for Rock Mechanics, South African National Group, Johannesburg.

Anon. 1992. *Construction over Mined Areas.* South African Institution of Civil Engineers, Yeoville.

Bell, F.G. 1975. *Site Investigation in Areas of Mining Subsidence.* Newnes-Butterworth, London.

Bell, F.G. and Donnelly, L.J. 2006. *Mining and its Impact on the Environment.* Spon Press, London.

Bell, F.G., Culshaw, M.G., Cripps, J.C. and Lovell, M.A. (Eds). 1988. *Engineering Geology of Underground Movements.* Engineering Geology Special Publication No 5. Geological Society, London.

Borchers, J.W. (Ed.). 1998. *Land Subsidence Case Studies and Current Research,* Special Publication No 8. Association of Engineering Geologists, Star Publishing Company, Belmont, California.

Brauner, G. 1973a. *Subsidence due to Underground Mining, Part I, Theory and Practice in Predicting Surface Deformation.* Bureau of Mines, Department of the Interior, US Government Printing Office, Washington, D.C.

Brauner, G. 1973b. *Subsidence due to Underground Mining: Part II, Ground Movements and Mining Damage.* Bureau of Mines, Department of the Interior, US Government Printing Office, Washington, D.C.

Brodie, M.J., Broughton, L.M. and Robertson, A. 1989. *British Columbia Acid Mine Drainage Task Force, Draft Technical Guide.* Steffan, Robertson and Kirsten, Vancouver.

Down, C.G. and Stocks, J. 1977. *Environmental Impact of Mining.* Applied Science Publishers Ltd, London

Healy, P.R. and Head, J.M. 1984. *Construction over Abandoned Mine Working.* Construction Industry Research and Information Association, Special Publication 32, London.

Holzer, T.L. (Ed.). 1984. *Man-induced Land Subsidence.* Reviews in Engineering Geology. American Geological Society, New York.

Kratzsch, H. 1983. *Mining Subsidence Engineering.* Springer-Verlag, Berlin.

MacFarland, M.C. and Brown, J.C. 1993. *Study of Stability Problems and Hazard Evaluation in the Missouri Portion of the Tri-State Mining Area.* United States Bureau of Mines, Washington, D.C.

Ladwig, J. 1985. *Hydrologic Aspects of Acid Mine Drainage Control.* Circular No. IC-9027, United States Bureau of Mines, Washington, D.C.

Rainbow, A.K.W. (Ed.), 1990. *Reclamation, Treatment and Utilization of Coal Mining Wastes.* Balkema, Rotterdam.

Saxena, S.K. (Ed).1979. *Evaluation and Prediction of Subsidence.* Proceedings Speciality Conference of the Society American of Civil Engineers, Gainsville. Florida.

Vartanyan, G.S., (Ed.). 1989. Mining and Geoenvironment. UNESCO-UNEP, Paris

Vick, S.G. 1983. *Planning, Design and Analysis of Tailings Dams.* Wiley, New York.

Walker, S. 1999. *Uncontrolled Fires in Coal and Coal Wastes.* International Energy Association Coal Research, London.

Waltham, A.C. 1989. *Ground Subsidence.* Blackie, Glasgow.

Walton, G. 1991. *A Handbook on the Design of Tips and Related Structures.* Her Majesty's Stationery Office, London.

Whittacker, B.N. and Reddish, D.J. 1989. *Subsidence: Occurrence, Prediction and Control.* Elsevier, Amsterdam.

Younger, P.L. and Robins, N.S. (Eds.) 2002. *Mine Water Hydrogeology and Geochemistry.* Special Publication No. 198. Geological Society, London.

Chapter 7

Anon. 1987. *Geotechnical Practice for Waste Disposal '87.* Geotechnical Special Publication No 13, American Society of Civil Engineers, New York.

Anon. 1988. *Draft for Development, DD175: 1988, Code of Practice for the Identification of Potentially Contaminated Land and its Investigation.* British Standards Institution, London.

Anon. 1995. *A Guide to Risk Assessment and Risk Management for Environmental Protection.* Department of the Environment, Her Majesty's Stationery Office, London.

Attewell, P.B. 1993. *Ground Pollution; Environmental Geology, Engineering and Law.* E. & F.N. Spon, London.

Barber, C. 1982. *Domestic Waste and Leachate.* Notes on Water Research No. 31, Water Research Centre, Medmenham.

Bentley, S.F. (Ed.).1996. *Engineering Geology of Waste Disposal,* Engineering Geology Special Publication No 11, Geological Society, London.

Bouazza, A., Kodikara, J. and Parker, R. (Eds.). 1997. *Environmental Geotechnics.* A.A. Balkema, Rotterdam.

Cairney, T. and Hobson, D. (Eds.). 1998. *Contaminated Land, Problems and Solutions.* Second Edition, Spon Press, London.

Cecille, L. (Ed.). 1991. *Radioactive Waste Management and Disposal.* Elsevier Applied Science, London.

Crawford, J.F. and Smith, P.G. 1985. *Landfill Technology.* Butterworth, London.

Daniel, D.E. (Ed.). 1993. *Geotechnical Practice for Waste Disposal.* Chapman and Hall, London.

Davies, M.C.R. (Ed.). 1991. *Land Reclamation: An End to Dereliction.* Elsevier Applied Science. London.

Dixon, N., Murray, E.J. and Jones, D.R.V., (Eds.). 1998. *Geotechnical Engineering of Landfills.* Thomas Telford Press, London.

Fell, R., Phillips, A. and Gerrard, C. (Eds.). 1993. *Geotechnical Management of Waste and Contamination.* A.A. Balkema, Rotterdam.

Fox, H.R., Moore, H.M. and McIntosh, A.D. (Eds.). 1998. *Land Reclamation: Achieving Sustainable Benefits.* A.A. Balkema, Rotterdam.

Genske, D.D. 2003. *Urban Land: Degradation, Investigation, Remediation.* Springer-Verlag, Berlin.

Hester, R.E. and Harrison, R.M., (Eds.). 2001. *Assessment and Reclamation of Contaminated Land.* Thomas Telford, Press.

ICRCL. 1987. *Guidelines on the Assessment and Redevelopment of Contaminated Land: Guidance Note 59/ 83.* Interdepartmental Committee on the Redevelopment of Contaminated Land, Second Edition, Department of the Environment, Her Majesty's Stationery Office, London.

Krauskopf, K.B. 1988. *Radioactive Waste Disposal and Geology.* Chapman and Hall, London.

Lerner, D.N. and Walton, N.R.G. (Eds.). 1998. *Contaminated Land and Groundwater: Future Directions,* Engineering Geology Special Publication No. 14. Geological Society, London.

Mather, J., Banks, D., Dumpleton, S. and Fermore, M. (Eds.). 1998. *Groundwater Contaminants and their Migration.* Special Publication No 128. Geological Society, London.

Oweis, I.S and Khera, R.P. 1990. *Geotechnology of Waste Management.* Butterworth, London.

Porteous, A., (Ed.). 1985. *Hazardous Waste Management Handbook.* Butterworth, London.

Sara, M.N. 1994. *Standard Handbook for Solid and Hazardous Waste Facilities Assessment.* Lewis Publishers, Boca Raton, Florida.

Sarsby, R.W. (Ed.). 1995. *Waste Disposal by Landfill, Green '93 - Geotechnics Related to the Environment.* A.A. Balkema, Rotterdam.

Sarsby, R.W. (Ed.). 1998. *Green 2, Contaminated and Derelict Land.* Thomas Telford Press, London.

Sarsby, R.W. and Meggyes, T., (Eds.). 2001. *Green 3, The Exploitation of Natural Resources and the Consequences.* Thomas Telford Press. London.

USEPA, 1988. *Guidance for Conducting Remedial Investigations and Feasibility Studies under CERCLA.* Office of Emergency and Remedial Response, U.S. Government Printing Office, Washington, D.C.

USEPA, 1989. *Methods for Evaluating the Attainment of Clean-up Standards, Volume 1: Soils and Solid Media.* EPA/540/2-90/011. Office of Policy, Planning and Evaluation, U.S. Government Printing Office, Washington, D.C.

Yong, R.N. and Thomas, H.R. (Eds.). 1999. *Geoenvironmental Engineering: Ground Contamination, Pollutant Management and Remediation.* Thomas Telford Press, London

Chapter 8

Allum, J.A.E. 1966. *Photogeology and Regional Mapping*, Pergamon, Oxford.

Anon. 1970. Logging of cores for engineering purposes. Engineering Group Working Party Report. *Quarterly Journal Engineering Geology*, 5, 1–24.

Anon 1978. *Terrain Evaluation for Highway Engineering and Transport Planning.* Transport Road Research Laboratory, Report SR448, DOE, Crowthorne.

Anon 1999. *Code of Practice for Site Investigations, BS 5930.* British Standards Institution, London.

Anon 1988. Engineering geophysics. Engineering Group Working Party Report. *Quarterly Journal Engineering Geology*, 21, 207–273.

Assaad, F.A., LaMoreaux, P.E., Hughes, T.H., Wangfang, Z and Jordan, H. 2004. *Field Methods for Geologists and Hydrogeologists*. Springer, Berlin.

Bell, F.G. 1975. *Site Investigations in Areas of Mining Subsidence.* Newnes-Butterworth, London.

Bell, F.G., Culshaw, M.G., Cripps, J.C. and Coffey, J.R. (Eds). 1990. *Field Testing in Engineering Geology.* Engineering Geology Special Publication No. 6. The Geological Society, London.

Bristow, C.S. and Jol, H.M. (Eds.). 2003. *Ground Probing Radar in Sediments.* Special Publication 211, Geological Society, London.

Carter, M. and Symons, M.W. 1989. *Site Investigations and Foundations Explained.* Pentech Press, London.

Clayton, C.R.I., Simons, N.E. and Matthews, M.C. 1996. *Site Investigation: A Handbook for Engineers.* Second Edition, Blackwell Scientific Publications, Oxford.

Craig, C., (Ed.). 1996. *Advances in Site Investigation Practice.* Thomas Telford Press, London.

Dent, D. and Young, A. 1981. *Soil Survey and Land Evaluation.* Allen and Unwin, London.

Griffiths, D. and King. R.E. 1984. *Applied Geophysics for Engineers and Geologists.* Pergamon Press, Oxford.

Griffiths, J.S. (Ed.). 2002. *Mapping in Engineering Geology.* Special Publication 211, Geological Society, London.

Griffiths, J.S. (Ed.). 2001. *Land Surface Evaluation for Engineering Practice.* Engineering Geology Special Publication No. 18. Geological Society, London.

Hatheway, A.W. and McClure, C.R., (Eds.). *Geology in the Siting of Nuclear Power Plants.* Reviews in Engineering Geology, Volume IV, Geological Society of America, Boulder, Colorado.

Hawkins, A.B. (Ed.) 1986. *Site Investigation Practice: Assessing BS 5930.* Engineering Geology Special Publication No 2. The Geological Society, London.

Hvorslev, M.J. 1949. *Subsurface Exploration and Sampling for Civil Engineering Purposes.* American Society Civil Engineers Report, Waterways Experimental Station, Vicksburg.

Kennie, T.J.M. and Matthews, M.C. (Eds.). 1985. *Remote Sensing in Civil Engineering.* Surrey University Press, London.

McCann, D.M., Eddleston, D., Fenning, P.J. and Reeves, G.M. (Eds.). 1997. *Modern Geophysics in Engineering Geology.* Engineering Geology Special Publication No 12. Geological Society, London.

McDowell, P.W. et al. 2002. *Geophysics in Engineering Investigations.* Engineering Geology Special Publication No. 19. Construction Industry Research and Information Association (CIRIA)/ Geological Society, London.

Milson, J. 1989. *Field Geophysics.* Wiley, Chichester.

Mitchell, C.W. 1991. *Terrain Evaluation.* Second Edition, Longman, London.

Reynolds, J.M. 2004. *An Introduction to Applied and Environmental Geophysics.* Second Edition, Wiley, Chichester.

Sabins, F.F. 1996. *Remote Sensing – Principles and Interpretation.* W. Freeman and Co., San Francisco.

Sharma, P.V. 1997. *Environmental and Engineering Geophysics.* Cambridge University Press, Cambridge.

Simons, N.E., Menzies, B.K. and Matthews, M.C. 2001. *A Short Course in Geotechnical Site Investigation*. Thomas Telford Press, London.

Chapter 9

Anon. 1986. *Guide to Rock and Soil Descriptions*. Geoguide 3. Geotechnical Control Office, Government Publication Centre, Hong Kong.

Bell, F.G. 2000. *Engineering Properties of Soils and Rocks*. Fourth Edition. Blackwell Science, Oxford.

Bieniawski, Z.T. 1989. *Engineering Rock Mass Classifications*. Wiley-Interscience, New York.

Blight, G.E., (Ed.). 1997. *Mechanics of Residual Soils*. A.A. Balkema, Rotterdam.

Brand, E.W. and Brenner, R.P. (Eds.) 1981. *Soft Clay Engineering*. Elsevier, Amsterdam.

Cripps, J.C., Culshaw, M.G., Coulthard, J.M. and Henshaw, S. (Eds.). 1992. *Engineering Geology of Weak Rock*. Engineering Geology Special Publication No 8, Balkema, Rotterdam.

Deere, D.M. and Miller, R.P. 1966. *Engineering Classification and Index Properties for Intact Rock*. Technical Report AFWL-TR-116, Air Force Weapons Laboratory, Kirtland Air Force Base, New Mexico.

Farmer, I.W. 1983. *Engineering Behaviour of Rocks*. Chapman and Hall, London.

Foster, A., Culshaw, M.G., Cripps, J.C., Moon, C.F. and Little, J. (Eds.). 1991. *Quaternary Engineering Geology*. Engineering Geology Special Publication No 7, The Geological Society, London.

Fookes, P.G. (Ed.).1997. *Tropical Residual Soils*. Geological Society, London.

Fookes, P.G. and Parry, R.H.G., (Eds.). 1994. *Engineering Characteristics of Arid Soils*. A.A. Balkema, Rotterdam.

Gillott, J.E. 1968. *Clay in Engineering Geology*. Elsevier, Amsterdam.

Goodman, R.E. 1993. *Engineering Geology: Rock in Engineering Construction*. Wiley, New York.

Jefferson, I., Murray, E.J., Faragher, E. and Fleming, P.R., (Eds.). 2001. *Problematic Soils*. Thomas Telford Press, London.

Mitchell, J.K. 1988. *Fundamentals of Soil Behaviour*. Second Edition. Wiley, New York.

Santi, P.M. and Shakoor, A., (Eds.). 1997. *Characterization of Weak Rock and Weathered Rock Masses*. Special Publication No. 9. Association of Engineering Geologists, College Station, Texas.

Washburn, A.L. 1973. *Periglacial Processes and Environments*. Arnold, London.

Wilson, R.C.L., (Ed.). 1983. *Residual Deposits: Surface Related Weathering Processes and Materials*. Special Publication No. 11, Geological Society, London, Blackwell Scientific Publications, Oxford.

Chapter 10.

Keeling, P.S. 1969. *The Geology and Mineralogy of Brick Clays*. Brick Development Association, Stoke-on-Trent.

Latham, J.-P., (Ed.). 1998. *Advances in Aggregates and Armourstone Evaluation*. Engineering Geology Special Publication No. 13. Geological Society, London.

McNally, G. 1998. *Soil and Rock Construction Materials*. E. & F.N. Spon, London.

Prentice, J.E. 1990. *Geology of Construction Materials*. Chapman and Hall, London.

Reeves, G.M., Cripps, J.C. and Sims, I. (Eds.). 2006. *Clay Materials in Construction*. Engineering Geology Special Publication No. 21, Geological Society, London.

Scott, P.W. and Bristow, C.M. (Eds.). 2002. *Industrial Minerals and Extractive Industry Geology*. Special Publication 211, Geological Society, London.

Simpson, J.L. and Horrobin, P.J. 1970. *The Weathering and Performance of Building Materials*. Medical and Technical Publishing Co., Aylesbury, Bucks.

Smith, M.R. and Collis, L., (Eds.). 2001. *Aggregates: Sand, Gravel and Crushed Rock Aggregates*. Third Edition. Engineering Geology Special Publication No 9. Geological Society, London.

Smith, M.R., (Ed.). 1999. *Stone: Building Stone, Rock Fill and Armourstone in Construction*. Engineering Geology Special Publication No. 16, Geological Society, London.

Weiss, T. and Vollbrecht, A. (Eds.). 2003. *Natural Stone, Weathering Phenomena, Conservation Strategies and Case Studies.* Special Publication 205, Geological Society, London.

Winkler, E.M. 1973. *Stone: Properties, Durability in Man's Environment.* Springer-Verlag, New York.

Chapter 11

Anon. 1952. *Soil Mechanics for Road Engineers.* Transport and Road Research Laboratory, Her Majesty's Stationery Office, London.

Anon. 1965. *Soils and Geology – Pavement Design for Frost Conditions.* Technical Manual TM 5-818-2, United States Corps of Engineers, Department of Army, Washington, D.C.

Anon. 1975. *Settlement of Structures.* British Geotechnical Society, Pentech Press, London.

Anon. 1986. *Code of Practice for Earthworks.* BS6031:1986, British Standards Institution, London.

Anon. 1996. *Building on Soft Soils.* A.A. Balkema, Rotterdam.

Attewell, P.B and Taylor, R.K. (Eds.). 1984. *Ground Movements and their Effects on Structures.* Surrey University Press, London.

Barton, N., Lien, R. and Lunde, J. 1975. *Engineering Classification of Rock Masses for the Design of Tunnel Support.* Publication 106, Norwegian Geotechnical Institute, Oslo.

Bell, A.L., (Ed.). 1994. *Grouting in the Ground.* Thomas Telford Press, London.

Bell, F.G. 1978. *Foundation Engineering in Difficult Ground.* Butterworth, London.

Bell, F.G. (Ed.). 1987. *Ground Engineer's Reference Book.* Butterworth, London.

Bell, F.G. (Ed.). 1992. *Engineering in Rock Masses.* Butterworth-Heinemann, Oxford.

Bell, F.G. 1993. *Engineering Treatment of Soils.* E. & F.N. Spon, London.

Bell, F.G. 2004. *Engineering Geology and Construction.* E. & F.N. Spon, London.

Bell, F.G., Cripps, J.C., Culshaw, M.G. and Lovell, M.A. (Eds.). 1988. *Engineering Geology of Ground Movements.* Engineering Geology Special Publication No. 5. Geological Society, London.

Bieniawski, Z.T. 1989. *Engineering Rock Mass Classifications.* Wiley-Interscience, New York.

Bickel, J.O. and Kuesel, T.R. 1982. *Tunnel Engineering Handbook.* Van Nostrand Reinhold, New York.

Chandler, R.J., (Ed.). 1991. *Slope Stability Engineering – Developments and Applications.* Thomas Telford Press, London.

Church, H.K. 1981. *Excavation Handbook.* McGraw-Hill, New York.

Davies, M.C.R., (Ed.). 1997. *Ground Improvement Geosystems.* Thomas Telford Press, London.

Eddleston, M, Walthall, S. Cripps, J.C. and Culshaw, M.G. (Eds.). 1995. *Engineering Geology of Construction.* Engineering Geology Special Publication No. 10, Geological Society, London.

Fang, H-Y., (Ed.). 1991. *Foundation Engineering Handbook.* Chapman and Hall, New York.

Harris, C.S., Hart, M.B., Varley, P.M. and Warren, C.D., (Eds.). 1996. *Engineering Geology of the Channel Tunnel.* Thomas Telford Press, London.

Harris, J.S. 1995. *Ground Freezing in Practice.* Thomas Telford Press, London.

Henn, R.W. 1996. *Practical Guide to Grouting of Underground Structures.* Thomas Telford, Press, London.

Hoek, E. and Bray, J. 1981. *Rock Slope Engineering.* Third Edition. Institution of Mining and Metallurgy, London.

Hoek, E. and Brown, E.T. 1980. *Underground Excavations in Rock.* Institution of Mining and Metallurgy, London.

Hoek, E., Kaiser, P.K. and Bawden, W.F. 1998. *Support of Underground Excavations in Hard Rock.* A.A. Balkema, Rotterdam.

Houlsby, A.C. 1990. *Construction and Design of Cement Grouting.* Wiley-Interscience, New York.

Hudson, J.A. and Harrison, J.P. 1997. *Engineering Rock Mechanics: An Introduction to Principles.* Pergamon, Oxford.

Johnson, G.H. (Ed.). 1981. *Permafrost, Engineering Design and Construction.* Wiley, Toronto.

Kezdi, A. and Rethati, L. 1988. *Handbook of Soil Mechanics, Volume 3: Soil Mechanics of Earthworks, Foundations and Highway Engineering.* Elsevier, Amsterdam.

Langeford, U. and Kilstrom, K. 1973. *The Modern Technique of Rock Blasting.* Wiley, New York.

McGregor, K. 1967. *The Drilling of Rock.* C R Books Ltd, A Maclaren Co., London.

Megaw. T.M. and Bartlett, J.V. 1982. *Tunnels: Planning and Design.* Volume 2, Methuen, London.

Muir Wood, A.H. 2000. *Tunnelling.* E. & F.N. Spon, London.

Ortigao, J.A.R. and Sayao, A. (Eds.). 2004. *Handbook of Slope Stabilisation Engineering.* Springer, Berlin.

Powrie, W. 1997. *Soil Mechanics: Concepts and Applications.* E. & F.N. Spon, London.

Proctor, R. and White, T., (Eds.). 1946. *Rock Tunneling with Steel Supports.* Commercial Stamping and Shearing Co, Youngstown, Ohio.

Puller, M. 1996. *Deep Excavations: A Practical Manual.* Thomas Telford Press, London.

Raison, C. (Ed.). 2004. *Ground and Soil Improvement.* Thomas Telford Press, London.

Simons, N.E. and Menzies, B.K. 2000. *A Short Course in Foundation Engineering.* Thomas Telford Press, London.

Simons, N.E. and Menzies, B.K. 2001. *A Short Course in Soil and Rock Slope Engineering.* Thomas Telford Press, London.

Stacey, T.R. and Page, C.H. 1986. *Practical Handbook for Underground Rock Mechanics.* Trans Tech Publications, Clausthal-Zelterfeld.

Szechy, K. 1966. *The Art of Tunnelling.* Akademiai Kiado, Budapest.

Tomlinson, M.J. 1996. *Foundation Design and Construction.* Sixth Edition. Longmans, London.

Tomlinson, M.J. 1993. *Pile Design and Construction Practice.* E. & F.N. Spon, London.

Wahlstrom, E.E. 1980. *Tunneling in Rock.* Elsevier, Amsterdam

Waltham, A.C., Bell, F.G. and Culshaw, M.G. 2005. *Sinkholes and Subsidence: Karst and Cavernous Rocks in Engineering and Construction.* Praxis Press/Springer, Chichester/New York.

Whittacker, B.N. and Frith, R.C. 1990. *Tunnelling: Design, Stability and Construction.* Institution of Mining and Metallurgy, London.

Wyllie, D.C. 1999. *Foundations on Rock.* Second Edition. E. & F.N. Spon, London.

Wyllie, D.C. and Mah, C.W. 2004. *Rock Slope Engineering: Civil and Mining.* Spon Press, London.

Some journals of interest

Advances in Agronomy
Bulletin of Engineering Geology and the Environment
Canadian Geotechnical Journal
Catena
Coastal Engineering
Earth Surface Processes and Landforms
Engineering Geology
Environmental and Engineering Geoscience
Environmental Geochemistry and Health
Environmental Geology
Environmental Management
Environmental Science and Technology
Geophysics
Geotechnical and Geological Engineering
Geotechnique
Geotextiles and Geomembranes
Ground Engineering
Ground Water
Ground Water Monitoring Review

International Journal of Rock Mechanics and Mining Science
International Journal Surface Mining, Reclamation and Environment
Journal American Water Resources Association
Journal Coastal Research
Journal Contaminant Hydrology
Journal Geophysical Research
Journal of Hydrology
Journal Institution Highway Engineers
Journal Soil Water Conservation
Journal Institution Water Engineers, Scientists and Managers
Monumentum
Natural Hazards
Photogrammetric Record
Proceedings American Society Civil Engineers, Journal Geotechnical and Geoenvironmetal Engineering
Proceedings American Society Civil Engineer, Journal Hydraulics Division
Proceedings American Society Civil Engineers, Journal Irrigation and Drainage Engineering
Proceedings American Society Civil Engineers, Journal Transportation Engineering
Proceedings American Society Civil Engineers, Journal Waterway, Port, Coastal and Ocean Engineering
Proceedings Institution Civil Engineers, Geotechnical Engineering
Proceedings Soil Science Society America
Quarterly Journal Engineering Geology and Hydrogeology
Rock Mechanics and Rock Engineering
Sedimentology
Soil Science
Transactions American Geophysical Union
Transactions America Society Agricultural Engineers
Transactions Institution Mining and Metallurgy, Section A, Mining Industry
Transactions Institution Mining and Metallurgy, Section B, Applied Earth Sciences
Tunnels and Tunnelling
Underground Space
Waste Management
Waste Management and Research
Water Pollution Control
Water Resources Management
Water Resources Research

Glossary

Aggradation: the build-up of sediment as a result of deposition.

Allophane: an amorphous clay mineral.

Alluvium: unconsolidated sediments deposited by a stream or river e.g. gravels, sands, silts and clays.

Alpha particles: particles emitted from the nucleus of an atom during radioactive decay.

Amphibolite: a metamorphic rock consisting mainly of the amphibole and plagioclase minerals, with little or no quartz.

Andesitic: referring to andesite, a dark coloured, fine grained extrusive (volcanic) rock.

Artesian conditions: exist where the piezometric level is above the ground surface and so groundwater will overflow from a well.

Asperities: outward protruding irregularities on the surface of a discontinuity.

Attrition: reduction in size of particles carried by air or water when they collide with each other.

Auger: a hand-held or mechanical tool for boring holes.

Backwash: movement of sediment down a beach as a wave recedes.

Band drain: consists of a flat plastic core containing drainage channels surrounded by a thin filter layer; they are installed vertically in clayey soils to increase the rate of consolidation and frequently are used beneath embankments.

Barrage: a dam constructed across an estuary.

Basaltic: pertaining to basalt, the commonest basic volcanic rock, fine grained and dark in colour.

Batholith: a very large mass of intrusive igneous rock such as granite.

Bayesian approach: a statistical method involving the concept of probability distributions of random variables.

Beidellite: a clay mineral belonging to the montmorillonite group. It is an aluminium montmorillonite.

Bentonite: a clay material composed primarily of montmorillonite, which is an expansive clay mineral.

BOD: (biochemical oxygen demand) the amount of oxygen removed from aquatic environments rich in organic material by the metabolic requirements of aerobic micro-organisms.

Bombs: the largest type of pyroclastic material.

Borrow material: soil that is dug from a local pit for construction purposes.

Buffer capacity: the capacity of an ionic compound, which when added to a solution allows it to resist changes in acidity or alkalinity and therefore stabilizes its pH value.

Bulking factor: intact density per unit volume divided by disturbed density per unit volume.

Buried channel: an old river channel that has been filled subsequently with deposits.

Caldera: a large circular shaped volcanic depression that may have formed by volcanic explosion and/or collapse.

Capillary fringe: the lower subdivision of the zone of aeration immediately above the water table in which the water is held by capillary pressure i.e. suction pressure.

Capillary rise: the height above the water table to which water will rise under capillary pressure.

Carcinogenic: pertaining to a substance that has the potential to cause cancer.

Catchment area: the area that collects and drains rainwater i.e. the drainage basin.

Cation exchange capacity (CEC): the reversible replacement of certain cations by others on the surface of a clay mineral.

Chelating agents: agents that cause chelation i.e. the retention of a metallic ion by two atoms of a single organic molecule.

Cleavage: closely spaced planes that impart fissility to a rock mass, which are characteristically developed in slate.

COD: (chemical oxygen demand) the amount of oxygen required for oxidation of all oxidizable compounds in a body of water.

Cohesive soil: soil that derives its strength from cohesive (molecular) forces i.e. silt and clay.

Creep deformation: deformation that occurs under a constant, non-changing load such as overburden pressure.

Crown hole: a depression, often conical in shape, formed at the surface due to subsidence e.g by void migation or shaft collapse.

Curie point: the temperature above which thermal agitation prevents spontaneousmagnetic ordering (i.e. the temperature at which there is a transition from ferromagnetism to paramagnetism).

Dacite: a fine grained extrusive igneous rock with a similar composition to andesite but with less calcic feldspar.

Daughter products: atoms produced by radioactive decay.

Deflocculation: the process whereby tightly held clayey particles of soil are dispersed.

Deuteric: refers to a process or an effect in an igneous rock that occurs in the later stages of its formation.

Diagenesis: chemical, physical and biological changes undergone by a sediment after its initial deposition.

Diaphragm wall: a wall constructed in a trench filled with bentonite slurry that is replaced by concrete poured from a tremie pipe, the upper part of the concrete wall may be reinforced by a steel cage.

Discontinuities: a structural break in a rock mass representing a plane of weakness (unless cemented).

Dolerite: a fine to medium grained basic igneous rock, commonly found in dykes and sills, which primarily consists of plagioclase and pyoxene minerals.

Dolostone: a carbonate rock in which 50% or more of the material is dolomite [(MgCa)$_2$CO$_3$].

Drainage density: ratio of the total stream length of all the streams in a drainage basin to the area of the basin.

Drumlin: a rounded oval mound formed of glacial debris that was shaped by an ice sheet.

Duricrust: a hard crust occurring at the surface formed by the precipitation of salts found in arid and semi-arid regions.

Dyke: a steeply inclined minor igneous intrusion that cuts through the rocks in which it occurs.

EDM: electronic distance measurement.

Effective pressure: total pressure (i.e. load of overburden) minus pore water pressure.

Effluorescence: an encrustation of salts, usually whitish in colour, precipitated on the surfaces of rocks or soils (and bricks and concrete) as water is evaporated.

Elastic modulus (Young's modulus): the ratio of stress to strain during elastic deformation of material.

Eluvial: said of a soil horizon from which material has been removed by leaching.

Esker: a long narrow sinuous ridge, normally composed of sand and gravel, which was deposited by a subglacial or englacial stream flowing in an ice tunnel as a glacier retreated.

Fault: a surface or zone of fracture in rock masses along which displacement has occurred.

Fault breccia: material occupying a fault zone and which consists of angular fragments set in a fine-grained matrix.

Fault step: an interruption in the surface due to subsidence occurring at a fault.

Flood routing: progressive determination of the timing and size of a flood wave at successive points along a river.

Foot wall: the underside of a fault or orebody.

Frictional soil: soil that derives its strength from frictional forces e.g. sands and gravels.

Fumarole: a small vent on a volcano from which gases are emitted.

Gabbro: a dark coloured, coarse-grained basic plutonic igneous rock composed principally of the plagioclase and pyroxene minerals.

Gamma-rays: electromagnetic waves that may accompany the emission of either alpha or beta particles during radioactive decay.

Gas chromatography: a process whereby gases are separated by passing them over a solid or liquid phase.

Geocomposite: a geosynthetic material consisting of an inner more porous material sandwiched between two fabrics, which is used for drainage purposes.

Geodimeter: an electronic distance measuring instrument.

Geographical information systems (GIS): a form of technology capable of capturing, storing, retrieving, editing, analysing, comparing and displaying spatial environmental information.

Geohazard: a geological hazard, either natural or man-made, which threatens either humans or the environment in which they live.

Geo grid: a geosynthetic material formed by either punching holes in a thick extruded sheet followed by unidirectional or bidirectional drawing to form an orthogonal grid, or by joining thick strands of extruded filaments at required cross-over points.

Geomat: a geosynthetic material with high tensile strength and excellent drape qualities.

Geomembrane: a non-porous sheet that has a very low permeability (in engineering terms impermeable) usually formed of polyethylene.

Geotextiles: man-made fabrics, generally made from plastics but also may be made from natural materials, used in contruction.

Goethite: a brownish coloured earthy iron bearing mineral [FeO(OH)] that forms as a result of weathering.

Geoweb: a geosynthetic material formed by weaving thick strands together.

Gneiss: a foliated metamorphic rock with a mineral composition similar to granite.

Granitic: pertaining to granite, a light coloured, coarse grained igneous rock of plutonic origin i.e. formed deep with the Earth's crust.

Groyne: a wall constructed of timber, concrete or steel, running approximately normal to the shoreline to impede the movement of longshore drift and thereby inhibit coastal erosion.

Groundwater divide: a ridge in the water table from which the groundwater moves in different directions.

Half-life: the time taken for of an element to decay during radioactive breakdown.

Halloysite: a clay mineral.

Halogens: the chemical elements bromine, chlorine, fluorine, iodine.

Hanging wall: the overlying side of a fault or ore body.

Hornfels: a fine grained metamorphic rock that forms at the contact of an igneous intrusion.

Hydraulic conductivity: the coefficient of permeability.

Hydraulic gradient: the rate of change of pressure head per unit of distance of flow at a given point and in a given direction.

Hydrograph: a graph showing stream flow with respect to time.

Hydrothermal activity: the activity of heated water within the ground.

Hypabyssal: mode of occurrence of minor intrusions such as sills and dykes.

Ignimbrite: a rock that forms from hot ash flows and nuees ardentes, the material having been fused or welded together.

Illite: a clay mineral [$K_{2-3}Al8(Al_{2-3}Si_{13-14})O_{40}(OH)_8$].

Illuvial: pertaining to illviation, i.e. the accumulation of precipitated or leached material in a lower soil horizon.

Interfluve: the ridge between two river valleys.

Jarosite: a hydrous iron sulphate that commonly forms from the breakdown of pyrite [$KFe_3(SO_4)_2(OH)_6$]

Jet grouting: a method of grouting whereby soil is either mixed with or replaced by grout by water being jetted form the grout pipe.

Joints: a fracture within a rock mass, one of the commonest types of discontinuities.

Kame: a steep-sided mound of sandy-gravelly material deposited by a subglacial stream at the snout of a glacier.

Kame terrace: a terrace-like ridge of sandy-gravelly material deposited by a meltwater stream flowing between the side of a glacier and the valley wall or lateral moraine.

Kaolinite: a clay mineral [$Al_4Si_4O_{10}(OH)_8$].

Karst landscapes: the topography associated with soluble rocks, notably limestones.

Lahar: a mudflow associated with volcanic activity.

Lamination: the state of being laminated i.e. consisting of thin layers or laminae as in shales and micaceous sandstones.

Lapilli: pyroelastic material commonly the size of walnuts.

Levees: deposits formed at the side of a river that may be reinforced for river control.

Longshore drift (littoral drift): lateral movement of sediment by waves along a shore.

Magma: molten rock material formed with the Earth's crust or mantle from which igneous rocks are derived on cooling and solidification.

Mantle: the zone of the Earth below the crust and above the core. Marl: a calcareous mudrock.

Meander: a sinuous curve with a stream or river.

Montmorillonite: a clay mineral, which on wetting can expand significantly [$Al_4Si_8O_2o(OH)_4.nH_2O$].

Moraine: a glacial deposit of unsorted, unstratified material consisting of till. A lateral moraine forms at the side of a glacier, a medial moraine is formed from two lateral moraines when two glaciers merge, a terminal moraine forms at the snout of a glacier, an englacial moraine forms within a glacier, when englacial material falls to the base of a glacier it forms a ground moraine, a subglacial moraine is formed from material worn from the valley floor over which a glacier travels.

Mutagenic: pertaining to a substance that has the potential to cause mutation.

Normally consolidated clay: a clay deposit from which none of the overburden has been removed by erosion.

Nuees ardentes: rapidly moving cloud of gas and pyroclastics erupted from a volcano.

Orogenic belt: fold mountain belt.

Overconsolidated clay: a clay on which the overburden pressure in the past was greater than at present, this normally being brought about by removal of material by erosion.

Oxbow lake: an abandoned cut-off meander.

Perched water table: a relatively small water table that occurs above the main water table, which may be impounded by a lens of clay.

pF value: value on a logarithmic scale of soil suction.

Phreatic zone: zone of saturation occurring beneath the water table or phreatic level.

Phreatic explosion: a volcanic explosion of steam and muddy material caused when groundwater is heated by an underlying igneous source.

Phyllite: a fissile rock formed during regional metamorphism from argillaceous sedimentary rocks, it is of low to intermediate grade falling between slate and mica schist.

Piezometric level: the level to which water rises under piezometric pressure.

Piezometric pressure: the pressure that causes water to rise in a standpipe sunk in the ground.

Plinian eruption: the most violent type of central vent volcanic eruption.

Plutonic: mode of occurrence of major intrusions, that is, batholiths.

Point bar: a low arcuate ridge of sand and/or gravel that is deposited on the inside of a meander.

Point load test: indirect tensile test whereby a core or an irregular shaped specimen is failed by loading between two conical shaped platens.

Pozzolan: a material that when used in cement allows it to harden underwater.

Precambrian: geological time prior to the Cambrian period i.e. over approximately 600 million years ago.

Pumice: a light coloured, highly glassy vesicular rock normally having the composition of rhyolite.

Pyrite: an iron sulphide mineral often referred to as fools gold.

Quartzite: a metamorphic rock composed primarily of quartz and formed from the metamorphism of sandstone.

Quick absorption test: weight of water absorbed by an oven dried rock sample when it is immersed for a limited period, divided by its dry weight and expressed as a percentage.

Reinforced earth: earth that is reinforced by strips of metal or geotextile and may be used in slope protection or embankment construction.

Residual stress: stress within rock or soil that has not been dissipated.

Rip-current: a strong narrow surface or near-surface current of short duration and high velocity that flows seaward from the shore through the breaker zone, more or less normal to the shoreline.

River terrace: a level surface in a river valley, which is often paired, and represents a stage in the dissection of a flood plain.

Rockhead: the upper surface of rock beneath the soil cover.

Regolith: the weathered or detrital layer of material overlying rock.

Sabkha: a salt-flat characteristic of arid regions.

Saltation: a mode of sediment transportation in which the sediment moves in a series of jumps.

Sandwicks: drains formed by filling stockings made of woven bonded polypropylene with graded sand that are installed in small diameter holes to enhance the rate of consolidation of clayey soils.

Schist: a metamorphic rock characterized by schistosity i.e. the preferred orientation of minerals of platey (e.g. micas) and prismatic (e.g. hornblende) habits.

Schmidt hammer: a non-destructive portable testing device for measuring rock hardness, which expends a given amount of stored energy from a loaded spring that indicates the amount of rebound of a hammer mass within the instrument.

Scoria: cindery, vesicular pyroclasts.

Secant piles: piles that are constructed so that they interlock and therefore form an impermeable barrier.

Seismicity: movements associated with an earthquake.

Sinkhole: a depression in a limestone terrain that is formed by dissolution of the limestone and into which a stream often runs.

Slurry trench: a trench that is excavated under bentonite slurry, the slurry preventing the walls of the trench from collapsing.

Solum: uppermost part of a weathered profile, that is, the residual soil in which the original textures and structures of the parent rock are lost.

Stope: a working in a metalliferous mine.

Strike: the direction normal to the true dip in dipping strata.

Superficial deposits: young unconsolidated sediments occurring at the surface.

Swash: movement of sediment up a beach by a wave.

Tephra: a general term for pyroclasts.

Teratogenic: pertaining to a substance that has the potential to cause malformation in humans.

Thixotropic: pertaining to thixotropy i.e. the property of certain substances (e.g. some clays) to revert from a gel to a sol when shaken.

Traction: movement of sediment along the floor of a river or the sea by water, or the ground surface by wind.

Tuff: indurated volcanic ash.

UNESCO: United Nations Educational, Scientific and Cultural Organization.

Vadose zone: zone of aeration above the water table.

Valley bulge: an anticlinal fold formed by the mass movement of argillaceous material in a valley bottom, the movement being caused by flowage of material from beneath the load of competent strata forming the valley sides towards the area where stress has been relieved, namely, the valley.

Vein: a mineral filling in a discontinuity.

Vesicles: small more or less spherical cavities that occur in some lavas and formed by entrapped gas.

Wave refraction: the process by which a wave as it approaches the shore and so moves into shallower water is turned from its original direction.

Yield point: the stress at which deformation of a material changes from the elastic to the plastic stage.

Further definitions can be obtained from:

Challinor, J.A. 1978. *A Dictionary of Geology.* University of Wales Press, Cardiff.

Hancock, P.L. and Skinner, B.J. 2000. *The Oxford Companion to the Earth.* Oxford University Press, Oxford.

Jackson, E.A. (Ed.). 1997. *Glossary of Geology.* American Geological Institute, Boulder. Colorado.

Kearney, P. 1996. *The New Penguin Dictionary of Geology.* Penguin, London.

Index